Springer Proceedings in Physics 23

Springer Proceedings in Physics

Managing Editor: H. K. V. Lotsch

Volume 1 *Fluctuations and Sensitivity in Nonequilibrium Systems*
Editors: W. Horsthemke and D. K. Kondepudi

Volume 2 *EXAFS and Near Edge Structure III*
Editors: K. O. Hodgson, B. Hedman, and J. E. Penner-Hahn

Volume 3 *Nonlinear Phenomena in Physics* Editor: F. Claro

Volume 4 *Time-Resolved Vibrational Spectroscopy*
Editors: A. Laubereau and M. Stockburger

Volume 5 *Physics of Finely Divided Matter* Editors: N. Boccara and M. Daoud

Volume 6 *Aerogels* Editor: J. Fricke

Volume 7 *Nonlinear Optics: Materials and Devices* Editors: C. Flytzanis and J. L. Oudar

Volume 8 *Optical Bistability III* Editors: H. M. Gibbs, P. Mandel, N. Peyghambarian, and S. D. Smith

Volume 9 *Ion Formation from Organic Solids (IFOS III)* Editor: A. Benninghoven

Volume 10 *Atomic Transport and Defects in Metals by Neutron Scattering*
Editors: C. Janot, W. Petry, D. Richter, and T. Springer

Volume 11 *Biophysical Effects of Steady Magnetic Fields*
Editors: G. Maret, J. Kiepenheuer, and N. Boccara

Volume 12 *Quantum Optics IV* Editors: J. D. Harvey and D. F. Walls

Volume 13 *The Physics and Fabrication of Microstructures and Microdevices*
Editors: M. J. Kelly and C. Weisbuch

Volume 14 *Magnetic Properties of Low-Dimensional Systems*
Editors: L. M. Falicov and J. L. Morán-López

Volume 15 *Gas Flow and Chemical Lasers* Editor: S. Rosenwaks

Volume 16 *Photons and Continuum States of Atoms and Molecules*
Editors: N. K. Rahman, C. Guidotti, and M. Allegrini

Volume 17 *Quantum Aspects of Molecular Motions in Solids*
Editors: A. Heidemann, A. Magerl, M. Prager, D. Richter, and T. Springer

Volume 18 *Electro-optic and Photorefractive Materials* Editor: P. Günter

Volume 19 *Lasers and Synergetics* Editors: R. Graham and A. Wunderlin

Volume 20 *Primary Processes in Photobiology* Editor: T. Kobayashi

Volume 21 *Physics of Amphiphilic Layers* Editors: J. Meunier, D. Langevin, and N. Boccara

Volume 22 *Semiconductor Interfaces: Formation and Properties*
Editors: G. Le Lay, J. Derrien, and N. Boccara

Volume 23 *Magnetic Excitations and Fluctuations II* Editors: U. Balucani, S. W. Lovesey, M. G. Rasetti, and V. Tognetti

Springer Proceedings in Physics is a new series dedicated to the publication of conference proceedings. Each volume is produced on the basis of camera-ready manuscripts prepared by conference contributors. In this way, publication can be achieved very soon after the conference and costs are kept low; the quality of visual presentation is, nevertheless, very high. We believe that such a series is preferable to the method of publishing conference proceedings in journals, where the typesetting requires time and considerable expense, and results in a longer publication period. Springer Proceedings in Physics can be considered as a journal in every other way: it should be cited in publications of research papers as *Springer Proc. Phys.*, followed by the respective volume number, page number and year.

Magnetic Excitations and Fluctuations II

Proceedings of an International Workshop,
Turin, Italy, May 25–30, 1987

Editors: U. Balucani, S. W. Lovesey,
M. G. Rasetti, and V. Tognetti

With 96 Figures

Springer-Verlag Berlin Heidelberg New York
London Paris Tokyo

Dr. Umberto Balucani
Istituto di Elettronica, Quantistica, CNR, via Panciatichi 56/30, I-50127 Firenze, Italy

Professor Stephen W. Lovesey
Rutherford Appleton Lab., Chilton, Oxfordshire, OX 11 0QX, U.K.

Professor Mario G. Rasetti
Dipartimento di Fisica Politecnico di Torino, Corso Duca degli Abruzzi, 24, I-10129 Torino, Italy

Professor Valerio Tognetti
Dipartimento di Fisica, Università di Firenze, Largo E. Fermi, I-50125 Firenze, Italy

ISBN-13: 978-3-642-73109-9 e-ISBN-13: 978-3-642-73107-5
DOI: 10.1007/978-3-642-73107-5

Library of Congress Cataloging-in-Publication Data. Magnetic excitations and fluctuations II: proceedings of an international workshop, Turin, Italy, May 25–30, 1987 / editors, U. Balucani . . . [et al.]. (Springer proceedings in physics; v. 23) Sponsored by the Institute for Scientific Interchange (ISI) Includes index. 1. Statistical mechanics – Congresses. 2. Mathematical physics – Congresses. 3. Spin excitations – Congresses. I. Balucani, U. (Umberto), 1942–. II. Institute for Scientific Interchange. III. Series. QC174.7M34 1987 539.7'2-dc19 87-28852

Softcover reprint of the hardcover 1st edition 1987

2153/3150-543210

Preface

An international workshop on Elementary Excitations and Fluctuations in Magnetic Systems was held in Turin for five days beginning 25 May, 1987. The workshop followed much the same format as the one with the same title held in San Miniato in 1984 (proceedings: Springer Series in Solid-State Sciences, Vol. 54), that most participants contributed talks and provided papers for the proceedings. While many of the participants had attended the first workshop, 15 of the 40 invited review papers were presented by scientists who had not.

The majority of the talks reported theoretical work concerned with the introduction of new techniques. However, experimental work was also well represented, not least because many of the reported theoretical studies were motivated by experimental findings, and a highlight of the workshop was an extremely stimulating session devoted to recent neutron scattering measurements, on various systems, that exploited polarization analysis.

The fine venue of the workshop, Villa Gualino, with its excellent facilities and spacious accommodation, helped to produce a delightful relaxed and friendly atmosphere. For the use of Villa Gualino and significant financial support we are indebted to our host organization, the Institute for Scientific Interchange (ISI). Additional financial support came from the Consiglio Nazionale delle Ricerche (CNR), Centro Interuniversitario di Struttura della Material del Ministero della Pubblica Istruzione (CISM-MPI) and Gruppo Nazionale di Struttura della Materia (GNSM-CNR).

The permanent staff of the ISI, Tiziana Bertoletti (Coordinator) and Carmen Novella (Secretary) took on the lion's share of the administration and organization of the workshop. Preparation of the proceedings for publication was competently and swiftly handled by Carla Pardini (Florence). The creature comforts of good food, a good bar service and well-prepared accommodation were provided, with much charm, by Lina De Nisco and her helpers.

Finally, we thank all the participants for their attendance and spirited individual contributions to the success of the workshop.

Turin, June 1987

U. Balucani S.W. Lovesey
M. Rasetti V.Tognetti

Contents

Part I Statistical Mechanics and Mathematical Physics

Part II Random Field Problems, Spin Glasses and Chaotic Behaviour

Part III **Theory of Spin Excitations**

*Lecturers' names are printed in italics.

Part I

Statistical Mechanics and Mathematical Physics

Statistical Mechanics of the Integrable Models

R.K. Bullough[1], *D.J. Pilling*[1], *and J. Timonen*[2]

[1]Department of Mathematics, U.M.I.S.T., Manchester M60 1QD, United Kingdom

[2]Department of Physics, University of Jyväskylä, SF-40100 Jyväskylä, Finland

1. THE INTEGRABLE MODELS

There is an infinity of classically integrable models. The only ones we can consider here, and these only briefly, are:
the sine-Gordon (s-G) model

$$\phi_{xx} - \phi_{tt} = m^2 \sin\phi \quad , \tag{1.1}$$

the sinh-Gordon (sinh-G) model

$$\phi_{xx} - \phi_{tt} = m^2 \sinh\phi \; , \tag{1.2}$$

and the repulsive and attractive non-linear Schrödinger (NLS) models

$$-i\phi_t = \phi_{xx} - 2c\phi\,|\phi|^2. \tag{1.3}$$

The "attractive" NLS has real coupling constant $c < 0$; the "repulsive" has $c > 0$; ϕ is complex. In (1.1) and (1.2) m is a mass ($\hbar = c = 1$) and ϕ is real. These 4 integrable models are in one space and one time (1+1) dimensions. There are integrable models in 2+1 dimensions we cannot discuss here [1].

All the classically integrable models have a number of features in common. For example, one can associate a linear n×n matrix differential operator L with each model which maps from the field(s) ϕ to "scattering data" S, they are all Hamiltonian and the maps are canonical; they all have (continuously infinite) sets of commuting constants of the motion, and are all embedded in infinite dimensional Lie algebras; they are all complete integrable in the sense of LIOUVILLE-ARNOLD [2].

The last property means action-angle type variables can be found (namely from the spectral data S). The Hamiltonian can be expressed in terms of the action variables. For repulsive NLS the two Hamiltonians are

$$H[\phi] = \int [\phi^*_x\phi_x + c\phi^{*2}\phi^2]\,dx;\; \{\phi, \phi^*\} = i\delta(x-x') \tag{1.4}$$

$$H[p] = \int \omega(k)P(k)dk;\; \omega(k) = k^2. \tag{1.5}$$

The sinh-G has a similar pair: (1.5) is actually unchanged except that $\omega(k) = (m^2+k^2)^{\frac{1}{2}}$. The $P(k)$ are action variables, and there are $Q(k)$ such that $\{P(k), Q(k')\} = \delta(k-k')$; $0 \le P(k) < \infty$, $0 \le Q(k) < 2\pi$.

It is noteworthy that linear Schrödinger $i\phi_t = \phi_{xx}$ (LS) has Hamiltonian (1.5) with the same action-angle variables; linear Klein-Gordon $\phi_{xx} - \phi_{tt} = m^2\phi$ (K-G) coincides similarly with sinh-G. One might conclude from this that the linear and nonlinear theories have the same free energies. We shall resolve this problem in this note. However, our main point is to draw attention to the functional integral methods we describe in the paper. These methods depart substantially from the conventional ones.

2. THE STATISTICAL MECHANICS

We are concerned only with the free energy and so with the partition function Z. For a consistent approach to both the quantum and classical statistical mechanics of the integrable models we work with the functional integral for Z in Hamiltonian, not Lagrangian, form. Then we express it in terms of action-angle variables. The quantum form of Z for <u>all</u> classical Hamiltonians (1.5) is

$$Z = \int D\mu \, \exp\int_0^\beta d\tau[i \int_{-\infty}^{\infty} P(k)Q(k)_{,\tau} \, dk - H[p]]; \tag{2.1}$$

β^{-1} = temperature. The measure $D\mu$ proves to be [3]

$$D\mu = \lim_{N\to\infty} (2\pi)^{-N-1} \prod_{n=-\frac{1}{2}N}^{+\frac{1}{2}N} dP_n dQ_n \, . \tag{2.2}$$

The expression (2.1) derives by canonical transformation from the similar form in ϕ: for sinh-G or s-G

$$Z = \int D\Pi D\phi \, \exp \int_0^\beta d\tau[i \int \Pi(x)\phi(x)_{,\tau} \, dx - H[\phi]], \tag{2.3}$$

with $\{\Pi,\phi\} = \delta(x-x')$, and $D\mu$ is the proper measure of the transformation in the case of sinh-G [3] (but not in the case of s-G [4]). The Z, (2.3), is evaluated over the space-"time" torus $-\frac{1}{2}L \le x < \frac{1}{2}L$; $0 \le \tau < \beta$, periodic in x and τ. Periodicity in x is introduced for a proper thermodynamic limit: this restricts allowed wave vectors k to a set $\{\tilde{k}_n\}$ determined by an integral equation like (2.5) below. The P_n in (2.2) then corresponds as $P_n \leftrightarrow P(\tilde{k}_n)2\pi L^{-1} \leftrightarrow P(k)dk$ as $L\to\infty$; $Q_n \leftrightarrow Q(\tilde{k}_n) \to Q(k)$. The Hamiltonian H[p] in (2.1) is then, for the same reason,

$$H\,[p] = \sum_{n=-\frac{1}{2}N}^{+\frac{1}{2}N} \omega(\tilde{k}_n) \, P_n \tag{2.4}$$

and L = (N+1)a, with a a lattice spacing. For the <u>classical</u> statistical mechanics we use Floquet theory on an appropriate lattice (e.g. a sinh-Gordon lattice or an NLS lattice [5])as one method of showing that the $\tilde{k}_n$ are restricted by

$$\tilde{k}_n = 2\pi L^{-1} - L^{-1} \sum_{m\neq n} \Delta_c \, (\tilde{k}_n,\tilde{k}_m) \, P_m \, . \tag{2.5}$$

The argument defines the phase shifts Δ_c [5] (e.g. $\Delta_c(k,k') = -2c/(k-k')$ for repulsive NLS). We then use the classical limit ($\hbar\to 0$) of (2.1) which is

$$Z = \int D\mu \, \exp -\beta \, H[p]. \tag{2.6}$$

The phase shifts Δ_c vanish for the linearised equations LS and K-G. With $\tilde{k}_n = 2\pi n L^{-1} \to k$ as $L\to\infty$, evaluation of (2.6) is trivial. Similarly, without the condition (2.5) with the $\Delta_c \neq 0$, (2.6) necessarily yields the same free field results for the NLS and sinh-G. But when the $\Delta_c \neq 0$ and (2.5) applies, we can evaluate (2.6) by iteration to a different result which can be identified [3] with the iteration of

$$\varepsilon(k) = \omega(k) + (2\pi\beta)^{-1} \int_{-\infty}^{\infty} \{d\Delta_c(k,k')/dk\} \ln(\beta\varepsilon(k'))dk' \tag{2.7a}$$

for

$$FL^{-1} = (2\pi\beta)^{-1} \int_{-\infty}^{\infty} \ln(\beta\varepsilon(k))dk; \tag{2.7b}$$

$F = -\beta^{-1}\ln Z$. Moreover we have shown that, by choosing $\omega(k)$ for sinh-G, iteration of (2.7) coincides with the asymptotic expansion in $(m\beta)^{-1}$ of the classical free energy found from (2.3) in classical limit by the transfer integral method [3].

3. GENERALISATION OF THE BETHE ANSATZ METHOD

The P(k), Q(k) are real canonical variables. The Z calculated from (2.1) is real (and positive) if

$$\int_0^\beta P_n dQ_{n,\tau} d\tau = \oint P_n dQ_n = 2\pi m_n \tag{3.1}$$

in which $m_n = 0, 1, \ldots$ for each label n. Since $h = 2\pi$, this is Bohr quantisation. Whilst (3.1) is sufficient for a real free energy F, it is also (more - or - less) necessary. With it it is possible to replace (2.5) by its quantum form: Δ_c is replaced by the proper 2-body S-matrix shifts but there are certainly now at least two cases: $\Delta_c \to \Delta_f$ the fermi form [6] when, by (3.1), $P_n = 0, 1$ only: in this case (2.5) is exactly the condition on the k's demanded by the Bethe Ansatz (BA) and quantum inverse methods (QIM) [6]. Alternatively, $\Delta_c \to \Delta_b$ the bose form and $P_n = 0, 1, 2, \ldots$. The two phase shifts relate by

$$\Delta_b(k) = \Delta_k(k) + 2\pi\theta(-k) \tag{3.2}$$

and Δ_f is the appropriate smooth branch, e.g. $\Delta_f(k,k') = -2\tan^{-1}\{c/(k-k')\} \to -2\pi$ as $k-k' \to -\infty$ for NLS; $\theta(k)$ is the step function, $\theta = 1$, $k > 0$, $= 0$, $k < 0$. Evidently it is Δ_b which has the classical limit Δ_c.

We believe the bose form of (2.5), with $\Delta_c \to \Delta_b$ and $P_m = 0, 1, \ldots$, has been new. From (2.1), with the bose condition (3.1), it readily leads to

$$\varepsilon(k) = \omega(k) + (2\pi\beta)^{-1} \int_{-\infty}^{\infty} dk' (d\Delta_b(k,k')/dk) \ln(1-e^{-\beta\varepsilon(k')}) \tag{3.3a}$$

$$FL^{-1} = (2\pi\beta)^{-1} \int_{-\infty}^{\infty} dk \ln(1-e^{-\beta\varepsilon(k)}) \tag{3.3b}$$

and these quantum expressions have (2.7) as classical limit. By using $\Sigma_{n=-\frac{1}{2}N}^{+\frac{1}{2}N} P_n = N$, a classical, and quantum, constant of the motion, one can introduce a chemical potential μ. By using (3.2), one finds (3.3) then transforms exactly to the expressions found by YANG and YANG [7] which are in fermion description. Alternatively, the same result is found directly by functional integration from (2.1) with the fermi condition (3.1) and the fermi form of (2.5) together with the condition on N for μ. All of these results, classical, bose, or fermi apply to both sinh-G, repulsive NLS, and other models with H[p] given by (1.5).

4. REMARK ON THE SINE-GORDON MODEL

The classical Hamiltonian H[p] for s-G is more complicated than (1.5) since it adds to this expression, kink, antikink and breather contributions [4]. By using this in (2.6) we have found the analogue of (2.7) to which it forms a natural continuation [4]: two terms contribute to FL^{-1} and (2.7a) is replaced by a pair of coupled integral equations. We have also shown [8] that the classical limit of the n-1 BA equations for the s-G model (n = integer > 1) and its quantum free energy reduces to the same system of coupled integral equations and classical free energy. We have further shown that iteration of this system yields an asymptotic expansion coincident with that found by the transfer integral method [4]. Our argument involves transform of n-2 of the n-1 fermions of the BA integral equations to a single bose field as $n \to \infty$: the one bose and one fermi field are then in semiclassical approximation (the $\Delta_f \to \Delta_c$). The classical limit of this semiclassical system is then exactly the classical system found from (2.6). A different form of this calculation of the classical limit for s-G is given by CHEN, JOHNSON and FOWLER [9]. We have not yet constructed the bose equivalent form of the quantum BA equations for the s-G model.

5. REMARK ON THE FUNCTIONAL INTEGRAL

This paper summarises some results obtained for the quantum and classical partition functions of integrable models. The method is to express Z in a form like (2.1) in terms of action-angle variables. Although the method seems general, we confine our remarks here to the case when H[p] takes the form (1.5). In this case Z is given by (2.1) and its classical limit is (2.6), while the quantum form (2.1) becomes (2.6) under the reality condition (3.1). In each case the periodicity requires the modes k satisfy the quantum or classical forms of (2.5) (while in the case of the s-G model, for example, both (2.5) and H[p] need extension to include the soliton contributions [4]). In the fermion description of the quantum case based on (1.5), the conditions (2.5) and (3.1) together are exactly the BA and QIM conditions for the repulsive NLS model [6]: this suggests (3.1) is necessary as well as sufficient.

The quantum conditions (3.1) are not trivial: semiclassical quantisation is aroided through (2.5) where Δ_c becomes the exact quantum S-matrix phase shift. This is calculated independently by appeal to the Yang-Baxter relations [6]. The resultant structure of the quantum functional integral, which is (2.6) with (2.5) and (3.1), is then very different from Feynman's original form.

The functional integral itself plays a rather trivial role. We have generalised the Yangs' method [7] directly by defining an entropy S and a free energy $F = E - \beta^{-1}S$): E is determined by H[p]. One then sees that the functional integral acts only to minimize Z. This method, which we call a generalised BA method [3,4], makes no ansatz and relies only on the quantum forms of (2.5) in the quantum cases, and on (2.5) as written in the classical cases. This, or the alternative functional integral methods sketched in this paper, seem to be a genuine extension and generalisation of the BA method for the integrable models. At $\beta^{-1} = 0$, in the quantum case of the models, the method yields the quantum mechanics.

The generality of the methods is witnessed by the fact that in the classical cases we have now carried through the statistical mechanics of the Toda lattice and the Landau-Lifshitz model [10] as well as s-G [4].

REFERENCES

1. Z. Jiang, R. K. Bullough, S. V. Manakov: Physica 18 D 305 (1986) and references
2. V. I. Arnold: Mathematical Methods of Classical Mechanics (Springer-Verlag, Heidelberg, 1978)
3. R. K. Bullough, D. J. Pilling, J. Timonen: J. Phys. A: Math. Gen. 19 L955 (1986)
4. J. Timonen, M. Stirland, D. J. Pilling, Yi Cheng, R. K. Bullough: Phys. Rev. Lett. 56, 2233 (1986)
5. Yi Cheng: Ph.D Thesis, University of Manchester (1987)
6. H. B. Thacker: Rev. Mod. Phys. 53, 253 (1981)
7. C. N. Yang, C. P. Yang: J. Math. Phys. 10, 1115 (1969)
8. J. Timonen, R. K. Bullough, D. J. Pilling: Phys. Rev. B 34, 6523 (1986)
9. Niu Niu Chen, M. D. Johnson, M. Fowler: Phys. Rev. Lett. 56, 907 (1986)
10. Yu-Zhong Chen, D. J. Pilling, R. K. Bullough and J. Timonen: To be published (1987)

Lattice Equations, Hierarchies and Hamiltonian Structures

H.W. Capel and G.L. Wiersma

Institute for Theoretical Physics, University of Amsterdam,
Valckenierstraat 65, NL-1018 XE Amsterdam, The Netherlands

1. INTRODUCTION

By now many integrable systems with soliton solutions are known in two dimensions, for fields $u(n_1,n_2)$ defined at the sites (n_1,n_2) of a two-dimensional (2D) lattice, for time-dependent fields $u(n,t)$ defined at the sites n of a one-dimensional chain, and for fields $u(x,t)$ depending on two continuous variables x and t. The relation between an integrable partial differential equation and its integrable discrete versions can be treated in the framework of the direct linearization method [1,2] which is based on a linear integral equation with arbitrary measure and contour [3,4]. In the treatment use is made of Bäcklund transformations (BT's) which are generated by scalar multiplications of the free-wave function and/or measure in the integral equation [5]. The integrable lattice equations are obtained in the form of Bianchi identities expressing the commutativity of BT's. The integrable equations with one or more continuous variables and their direct linearizations are obtained applying suitable continuum limits to the lattice equation and the integral equation at the same time.

In the last few years more insight has been obtained in the structure of integrable systems. Here we deal with the two following aspects:

i) the generalization of integrable systems from 2 to 3 dimensions

ii) the hierarchies and hamiltonian structures associated with integrable systems with one or more continuous variables.

The derivation of three-dimensional (3D) lattice equations is discussed in section 2 on the basis of a linear integral equation containing a surface and a measure depending on two complex variables, cf. refs. [2,6,7]. In section 3 we derive a hierarchy of integrable equations for time-dependent fields at the sites of a 2D lattice. The derivation is based on the identification of a translation operator T with a vertex operator exp ∂ in which ∂ is an infinite sum of differentiations with respect to time variables multiplied by powers of the lattice parameter that tends to zero in the continuum limit, see also refs. [8,9] in which vertex operators have been used in the context of Lie algebras. In section 4 we evaluate the conserved quantities and show that they generate the equations of the hierarchy via a Poisson bracket. The various steps are illustrated by the integral equation of the KADOMTSEV-PETVIASHVILI (KP) equation and its discrete analogues. The 2D counterpart referring to the KORTEWEG-de VRIES (KdV) equation has been treated in ref. [10].

2. 3D LATTICE EQUATION

Consider the integral equation

$$u_k + \rho_k \iint_D d\zeta(\ell,\ell') \frac{u_\ell}{k+\ell'} = \rho_k \quad , \quad u = \iint_D d\zeta(\ell,\ell') u_\ell \quad . \tag{1}$$

Here D is a surface in the space of two complex variables ℓ and ℓ', $d\zeta(\ell,\ell')$ is a measure depending on ℓ and ℓ', ρ_k is a free-wave function depending on the complex spectral parameter k and on the coordinates of the system. The wave function u_k which can be solved from the integral equation depends on k and the coordinates as well. The surface D and the measure are to be chosen such that the solution of (1) at given ρ_k is unique (uniqueness condition). From u_k one can obtain the potential u depending only on the coordinates by an integration over the same surface with the same measure.

We introduce a simultaneous transformation T of the free-wave function and the measure

$$\rho_k \to T\rho_k = \rho_k \frac{p-k}{p+k} \quad , \quad d\zeta(\ell,\ell') \to Td\zeta(\ell,\ell') = d\zeta(\ell,\ell')\frac{p+\ell}{p+\ell'} \quad . \tag{2}$$

From eq. (1) and its counterpart with $T\rho_k, Td\zeta(\ell,\ell'), Tu_k$ and Tu, instead of ρ_k, $d\zeta(\ell,\ell')$, u_k and u, one can derive the linear relation

$$(p-k)T(u_k/\rho_k) = \big(p+(Tu)\big)(u_k/\rho_k) - y_k/\rho_k \quad , \tag{3}$$

in which y_k is the solution of the integral equation (1) with $k\rho_k$ instead of ρ_k at the right-hand side. The linear relation is proved by showing that left-hand side minus right-hand side (l.h.s. - r.h.s) satisfies the homogeneous integral equation. The uniqueness condition implies that l.h.s. - r.h.s. = 0.

We now consider 3 different transformations T_1,T_2,T_3 defined by (2) with p_1,p_2,p_3 instead of p. Eliminating y_k gives the 3 linear relations

$$(p_\alpha-k)T_\alpha\Big(\frac{u_k}{\rho_k}\Big)-(p_\beta-k)T_\beta\Big(\frac{u_k}{\rho_k}\Big) = V_\gamma \frac{u_k}{\rho_k} \quad , \quad V_\gamma = p_\alpha-p_\beta+(T_\alpha-T_\beta)u \tag{4}$$

with (α,β,γ) = (1,2,3),(2,3,1) and (3,1,2). From (4) one can derive the compatibility condition

$$(T_\alpha V_\alpha)/V_\alpha = \text{independent of } \alpha \quad , \quad \alpha = 1,2,3 \quad , \tag{5}$$

which can be worked out to give an algebraic (Bianchi) identity in terms of the potentials u, expressing the commutativity of the 3 BT's T_1,T_2,T_3.

Equation (5) has been derived without any specification of the coordinates of the system. Therefore we are free to identify the transformations T_1,T_2,T_3 with the 3 primitive translations on a 3D lattice. Thus starting from a potential u at a lattice site P we have the potentials T_1u,T_2u,T_3u at the lattice sites that are obtained from P after applying the primitive translations T_1,T_2,T_3. Equation (5) is then a difference equation on the 3D lattice. It is integrable in the sense that solutions can be obtained from the linear integral equation (1) with a free-wave function ρ_k and a measure $d\zeta(\ell,\ell')$ satisfying (2) with p_1,p_2,p_3 instead of p under the translations T_1,T_2,T_3. Its Lax representation is given by (4). Equation (5) can be shown to be the lattice analogue of the KP equation [2].

The 2D lattice equation can be obtained from (1) choosing a surface D = C × C' with ℓ C, ℓ' C' and the contour C' surrounding the ℓ of the contour C, and a measure $d\zeta(\ell,\ell') = \big(2\pi i(\ell'-\ell)\big)^{-1}d\lambda(\ell)d\ell'$. One obtains [2] the integral equation [5] for the KdV, i.e. (1) with $C, d\lambda(\ell), k+\ell$ instead of $D, d\zeta(\ell,\ell')$ and $k+\ell'$. From (2) one finds $Td\lambda(\ell) = d\lambda(\ell)$, so that we can define the inverse transformation T^{-1} for ρ_k by (2) with $T \to T^{-1}$, $p \to -p$. From T^{-1} one obtains a second linear relation. The 2D lattice equation follows from the compatibility of the two linear relations for $T = T_1,T_2$.

3. CONTINUUM LIMIT

Starting from the lattice KP we can derive various integrable equations with one or more continuous variables. As an example we consider a first continuum limit $p_3 \to \infty$, $T_3 \to 1 - (1/p_3)\partial_\tau$. From (5) we obtain

$$\partial_\tau(\Delta_1 - \Delta_2)u = \big(p_1 - p_2 + (\Delta_1 - \Delta_2)u\big)\Delta_1\Delta_2 u \quad , \tag{6}$$

$\Delta_1 = T_1 - 1$, $\Delta_2 = T_2 - 1$ being the difference operators associated with the primitive translations T_1 and T_2. The solutions of (5) can be found from (1) with ρ_k and $d\zeta(\ell,\ell')$ given by (2) for $T = T_1, T_2$ and the limiting behaviour under T_3, i.e.

$$\partial_\tau \ln \rho_k = 2k, \ \partial_\tau \ln d\zeta(\ell,\ell') = \ell'-\ell \quad . \tag{7}$$

By a second continuum limit one obtains a two-dimensional version of the Toda equation and a third continuum limit yields the KP.

Associated with (6) we can also derive an infinite hierarchy of integrable equations. For this purpose we introduce a vertex operator $T_3 = \exp \partial_3$, cf. refs. [8,9], in which $-\partial_3$ is a sum of differentiations with respect to time variables τ_r multiplied by powers p_3^{-r} of the lattice parameter p_3^{-1}, divided by r. The hierarchy will consist of equations $(\partial\phi/\partial\tau_r)$ for the fields $\phi \equiv \ln V_3$ at the sites of the 2D lattice generated by T_1 and T_2. We use the boundary condition that $(T_1T_2^{-1})^J$ has limits u^+ for $J \to \infty$ and u^- for $J \to -\infty$ that are invariant under the transformations T_1, T_2 and T_3. Then the relation $\phi = \ln\big(p_1 - p_2 + (T_1 - T_2)u\big)$ can be inverted to give

$$u - \tfrac{1}{2}(u^+ + u^-) = (\Delta_1 - \Delta_2)^{-1} (e^\phi - p_1 + p_2) \tag{8}$$

with

$$(\Delta_1 - \Delta_2)^{-1} f = \tfrac{1}{2} \sum_{j=0}^{\infty} (T_2{}^j T_1^{-j-1} - T_1{}^j T_2^{-j-1}) f \quad , \tag{9}$$

for any function f with $(T_1T_2^{-1})^J f \to 0$ for $J \to \pm\infty$. From the lattice KP (5) with the use of the vertex operator we obtain

$$(e^{\partial_3} - 1)\phi = (\Delta_1 - \Delta_2)^{-1} \Delta_1\Delta_2 \ln(V_2/V_1) \quad , \quad \partial_3 = -\sum_r (p_3^{-r})\partial_{\tau_r} \ , \tag{10}$$

which is the generating equation for the hierarchy. The r.h.s. can be expressed in terms of ϕ, using (4) for V_1 and V_2 and (8) for u.

Inserting the expression for ∂_3 and expanding both sides of (10) in powers of p_3^{-1} one obtains in the order p_3^{-r} an equation for $(\partial\phi/\partial\tau_r)$ in which the r.h.s. contains only derivatives of ϕ with respect to variables $\tau_{r'}$ with $r' < r$. Eliminating these derivatives with the lower order equations already derived, one obtains an equation of the form $(\partial\phi/\partial\tau_r) = F_r(\{\phi\})$ for field ϕ at a site P of the 2D lattice, in which $F_r(\{\phi\})$ is an expression involving the difference operators Δ_1, Δ_2, as well as the inverse operator $(\Delta_1-\Delta_2)^{-1}$ acting on the fields $\{\phi\}$ at P and nearby lattice sites. The first equation of the hierarchy (r = 1) is $(\partial\phi/\partial\tau_1) = (\Delta_1-\Delta_2)^{-1}\Delta_1\Delta_2 e^\phi$ which is equivalent to (6), but also the higher order equations can be worked out. We do not give the results here, but in section 4 we show how these equations can be generated via a hamiltonian structure.

4. HAMILTONIAN STRUCTURE

To connect the hierarchy (10) with a hamiltonian structure we consider the scattering problem given by eq. (4) with $\alpha = 1$, which is the part of the Lax representation that is not affected by taking the continuum limit. We introduce the monodromy along a diagonal on the 2D lattice given by $T_1(u_k/\rho_k) = S_k T_2(u_k/\rho_k)$, i.e. S_k gives the change in u_k/ρ_k when one moves one place along the diagonal as expressed by $T_1T_2^{-1}$. The scattering coefficient a(k) along the diagonal is the product of the monodromy factors S_k. We also define the generating function H_k :

$$H_k \equiv \sum_{j_1,j_2} k(k-p_2)^{-1} T_1^{j_1} T_2^{j_2} \ln S_k = \sum_s k^{-s} H^{(s)}, \tag{11}$$

which is obtained summing $\ln S_k$ over the sites of the 2D lattice. The function H_k

is expanded in powers of $(1/k)$ to give the various conserved quantities of the hierarchy. In fact, assuming that $(T_1T_2^{-1})^J(u_k/\rho_k)$ has limits α_k^+ and α_k^- for $J \to \infty$ and $J \to -\infty$, we have $T_3\, H_k = H_k$ for the generating function. Using the vertex operator and the expansion in powers of $(1/k)$ it follows that all the $H^{(s)}$ are conserved under all time evoluations τ_r.

The $H^{(s)}$ can be evaluated explicitly in terms of the fields ϕ. From (4) with $\alpha = 1$ one can derive an equation for the monodromy factors S_k

$$(T_1e^{\phi})S_k(T_2S_k - \gamma_k) = (T_2e^{\phi})(T_2S_k)(T_1S_k - \gamma_k) \tag{12}$$

with $\gamma_k = (p_2-k)/(p_1-k)$. Introducing D_k via $S_k \equiv \gamma_k(1+D_k/k)$ and using logarithms, the boundary conditions and (9) we obtain

$$\ln\big((p_2k^{-1} - 1)D_k\big) - \phi = (\Delta_1 - \Delta_2)^{-1}\,\Delta_2\,\ln(1 + k^{-1}D_k) \quad , \tag{13}$$

from which D_k can be formally solved as an expansion in powers of k^{-1}. Evaluating S_k one obtains with (11) the expansion of H_k yielding the various $H^{(s)}$. In this way we obtain e.g.

$$H^{(1)} = -\sum_{j_1,j_2} T_1^{j_1}\,T_2^{j_2}\,(e^{\phi} - p_1 + p_2) \quad ,$$

$$H^{(2)} = -\sum_{j_1,j_2} T_1^{j_1}\,T_2^{j_2}\,\big(e^{\phi}(\Delta_1-\Delta_2)^{-1}\,\Delta_2e^{\phi} + \tfrac{1}{2}e^{2\phi} + 2p_2e^{\phi} + \text{const.}\big) \quad , \tag{14}$$

in which the constant is given by $-\tfrac{1}{2}(p_1-p_2)(p_1+3p_2)$. The higher order $H^{(s)}$ can be worked out systematically.

To relate the hierarchy to a hamiltonian structure we start from a relation between the generating eq. (10) and the monodromy factors S_k. From (4) with $\alpha = 2$ and $\alpha = 3$ and $k = p_3$, we find

$$V_2/V_1 = -\,S_{p_3}(p_1-p_3)/(p_2-p_3) = -\,(1 + D_{p_3}/p_3)\,. \tag{15}$$

Differentiating (10) with respect to p_3^{-1} it can be shown that

$$\partial_\eta\phi = (\Delta_1-\Delta_2)^{-1}\,\Delta_1\Delta_2(\partial_{p_3^{-1}} - \partial_\eta)\ln(1 + p_3^{-1}\,D_{p_3}) \quad , \quad \partial_\eta = \sum_r p_3^{-(r-1)}\partial_{\tau_r} \quad , \tag{16}$$

in which ∂_η is obtained differentiating ∂_3 with respect to p_3^{-1}.

Finally, starting from (11) and (13) one can prove that [11]

$$(\partial_{p_3^{-1}} - \partial_\eta)\ln(1 + p_3^{-1}\,D_{p_3}) = p_3\,\frac{\partial H_{p_3}}{\partial\phi} \quad , \tag{17}$$

where the functions ϕ and D are to be taken at the same lattice site. From (17) we obtain the generating equation

$$\partial_\eta\phi = p_3(\Delta_1-\Delta_2)^{-1}\,\Delta_1\Delta_2\,\frac{\partial H_{p_3}}{\partial\phi} \quad . \tag{18}$$

Equation (18) establishes the hamiltonian structure of the hierarchy. In fact, using the expansions of ∂_η and H_k for $k = p_3$ we find that the time derivative of any field $F(\{\phi\})$ depending on the fields ϕ at the sites (n_1,n_2) of the 2D lattice under the time evolution τ_r is given by

$$\partial_{\tau_r} F = -\sum_{n_1,n_2} \frac{\partial F}{\partial\phi(n_1,n_2)} (\Delta_1-\Delta_2)^{-1} \Delta_1\Delta_2 \frac{\partial H^{(r)}}{\partial\phi(n_1,n_2)} = \{H^{(r)},F\} \quad . \tag{19}$$

The bracket in (19) can be shown to satisfy the usual requirements of a Poisson bracket such as the antisymmetry and the Jacobian identity.

Starting from the integral equation (1) we have derived an integrable 3D lattice version (5) of the KP. Taking a first continuum limit with a vertex operator we have obtained an infinite hierarchy of integrable equations for time-dependent fields at the sites of a 2D lattice. We have also established the hamiltonian structure, i.e. the equations of the hierarchy have been identified with the ones following from the conserved quantities (11) via a Poisson bracket (19). Other continuum limits can be treated as well [11], cf. also ref. [12] for the (bi)hamiltonian structure of the KP. Similar considerations can be anticipated to apply to other integrable systems and their 3D analogues for which the direct linearization method has been established, such as the sine-Gordon equation, the Nonlinear Schrödinger equation and the Isotropic Heisenberg Spin Model [1,2,4].

Acknowledgement: This investigation is part of the research program of the "Stichting voor Fundamenteel Onderzoek der Materie (FOM)" which is financially supported by the "Nederlandse Organisatie voor Zuiver-Wetenschappelijk Onderzoek (ZWO)".

REFERENCES

1. G.R.W. Quispel, F.W. Nijhoff, H.W. Capel, J. van der Linden: Physica 125A, 344 (1984)
2. F.W. Nijhoff, H.W. Capel, G.L. Wiersma: In Geometric Aspects of the Einstein Equations and Integrable Systems, ed. by R. Martini, Lecture Notes in Physics 239, 263 (Springer, Berlin, Heidelberg 1985)
3. A.S. Fokas, M.J. Ablowitz: Phys.Rev.Lett. 47, 1096 (1981)
4. F.W. Nijhoff, G.R.W. Quispel, J. van der Linden, H.W. Capel: Physica 119A, 101 (1983)
5. G.R.W. Quispel, F.W. Nijhoff, H.W. Capel, J. van der Linden: Physica 123A, 319 (1984)
6. S.V. Manakov: Physica 3D, 420 (1981)
7. A.S. Fokas, M.J. Ablowitz: Phys.Lett. 94A, 67 (1983)
8. J. Lepowski, R.J. Wilson: Comm.Math.Phys. 62, 43 (1978)
9. M. Jimbo, T. Miwa: Publ. RIMS, Kyoto, 19, 943 (1983)
10. G.L. Wiersma, H.W. Capel: Physica 142A, 199 (1987)
11. G.L. Wiersma, H.W. Capel: to be published.
12. A.S. Fokas, P.M. Santini: Stud.Appl.Math. 75, 179 (1986)

Modular Invariance and Two-Dimensional Critical Systems

C. Itzykson

Service de Physique Théorique, CEN-Saclay,
F-91191 Gif-sur-Yvette Cedex, France

1. Critical systems or massless field theories are highly sensitive to geometric effects. The idea that one can implement a local (as opposed to global) scale invariance follows from the work of Belavin Polyakov and Zamolodchikov coming after numerous studies on finite size scaling. These authors applied the techniques developed in the framework of string theories to the study of two dimensional statistical systems, with a prominent role played by the energy momentum tensor, the generator of coordinate transformations. A natural role is played by the infinitesimal conformal transformations generating an infinite Lie algebra, the Virasoro algebra with a "quantum" anomaly -the so called central charge- for which a rich representation theory exists.

An important step was made by J. Cardy, who noticed that rather than study the consistency of the short distance operator expansions, one could as well investigate the critical models in appropriate compact geometries in particular on a torus, characterized by a modular ratio of two basic periods τ ($\mathrm{Im}\,\tau > 0$). Changing the basis transforms τ by a rational unimodular transformation, which induces a definitely non trivial constraint of modular invariance on the partition expressed as

$$Z = \mathrm{tr}\left(q^{(L_0 - c/24)}\bar{q}^{(\bar{L}_0 - c/24)}\right), \tag{1}$$

Here $q = \exp 2i\pi\tau$, $2\pi(L_0+\bar{L}_0)$ plays the role of quantum Hamiltonian, and $2\pi(L_0-\bar{L}_0)/i$ of momentum operator. Finally the factor $(q\bar{q})^{-c/24}$ arises from a Casimir effect in a finite geometry. The trace in (1) can be split into contributions from the various (highest weight) irreducible representations of the Virasoro algebra, with "ground state" vector labelled by a pair of conformal weights $(h,\bar{h})$, eigenvalues of the pair L_0, $\bar{L}_0$. Thanks to Feigin and Fuchs and Rocha Caridi the corresponding sum over states, or characters, are known in factorized form $\chi_h(q)\ \bar{\chi}_{\bar{h}}(q)$. Conformal invariance then restricts the possible choice of representations appearing in (1) and corresponding to the central charge c. In particular if $c < 1$ and rational of the form $c = 1 - 6\,\frac{(p-p')^2}{pp'}$ (p and p' are coprime integers) it is possible to define minimal models with h and $\bar{h}$ running through the Kac table

$$h_{rs} = \frac{(rp-sp')^2-(p-p')^2}{4pp'} = h_{p'-r,p-s}\ , \tag{2}$$

$$1 \le r \le p'-1 \qquad 1 \le s \le p-1$$

in such a way that (1), written now as

$$Z = \sum \mathcal{N}_{h,\bar{h}}\ \chi_h(q)\ \bar{\chi}_{\bar{h}}(q), \tag{3}$$

involves only finitely many terms with the coefficients $\mathcal{N}$ integral and non-negative. Friedan and Shenker have shown that for p, p' successive integers, the representations are unitary. Using the known transformation properties of the conformal characters it was a challenge to classify all possible candidates of the form (3).

2. In collaboration with J.B. Zuber and A. Capelli we investigated this point of rather arithmetical nature. As was observed by Gepner, the problem is easily solved in terms of a related one, where the Virasoro algebra is replaced by the Kac-Moody SU_2 affine algebra (affine for short) and its integrable characters. We were surprised to discover, and then prove, that the corresponding invariants are classified by simply laced Lie algebras (A-D-E-classification). To illustrate the correspondence, let me just show one example. The affine characters are labeled by a level k (the analog of the central charge) and a lowest angular momentum ℓ (integer or half integer) such that $0 \le \ell \le \frac{k+1}{2}$. Use rather $\lambda = 2\ell+1$ as an index. Then for k =28 we found (inter alia) the solution

$$Z = |\chi_1+\chi_{11}+\chi_{19}+\chi_{29}|^2 + |\chi_7+\chi_{13}+\chi_{17}+\chi_{23}|^2 . \tag{4}$$

The 8 integers 1, 7, 11, 13, 17, 19, 23 and 29 (the relative primes to 30) are the Coxeter Dynkin exponents of the Lie algebra E_8! (add 1 and these are the degrees of the basic Casimir operators of this rank eight algebra). Furthermore any pair of invariants such that p = k+2, p' = k'+2 corresponds a Virasoro minimal invariant. This results in an exhaustive classification of all possible universality classes of models with $c < 1$ and finitely many primary (conformally covariant) observables including the familiar cases of the Ising or 3-states Potts models. Not only is it of interest to pursue these identifications from a microscopic point of view, as initiated by Huse, Belavin, Polyakov and Zamolodchikov, Dotsenko and Fateev, Friedan and Shenker, Cardy and in our laboratory by P. di Francesco, H. Saleur and J.B. Zuber, compute the correlation functions in various geometries... but it is also quite intriguing to unravel the connection between Lie algebras and integrable models, a work undertaken by V. Pasquier, which seems to have close connection with recent mathematical developments.

3. The techniques of two-dimensional conformal field theories can profitably be used to study perturbatively the vicinity of the critical point i.e. the critical massive region. With H. Saleur we were able to rederive a celebrated result of Fisher and Ferdinand, giving the partition function of the Ising model on a torus for T close to T_c. For instance, if m is the inverse correlation length, proportional to $T-T_c$ (the exponent ν is equal to 1 for the Ising model), if the critical Ising model partition function (of type (3)) is written ($c = \frac{1}{2}$)

$$Z_{\frac{1}{2}} = \sum_{ij} D_{ij}$$

$$D_{\frac{1}{2},\frac{1}{2}} = \left| q^{-1/48} \prod_1^{\infty} (1 + q^{n-\frac{1}{2}}) \right|^2$$

$$D_{0,\frac{1}{2}} = \left| q^{-1/48} \prod_1^{\infty} (1 - q^{n-\frac{1}{2}}) \right|^2 \qquad (5a)$$

$$D_{\frac{1}{2},0} = \left| q^{2/48} \prod_1^{\infty} (1 + q^{n}) \right|^2$$

and if the free bosonic field (c=1) partition function is written as

$$Z_1 = \frac{1}{\sqrt{\mathrm{Im}}\ |\eta(q)|^2} \quad , \quad \eta(q) = q^{1/24} \prod_1^{\infty} (1-q^n) \qquad (5b)$$

then the free energy of the Ising model expanded in m (A is the area of the Torus, γ Euler's constant), reads

$$\ell n\, Z_{Ising}(m) =$$

$$\ell n\, Z_{\frac{1}{2}} + \frac{mA^{\frac{1}{2}}}{Z_1 Z_{\frac{1}{2}}} + m^2 A \left[\frac{1}{4\pi}\, \ell n \left(\frac{Z_1 \sqrt{A}\, e^{\gamma}}{\pi} \right) - \frac{1}{2\pi} \sum \frac{D_{ij}\, \ell n D_{ij}}{Z_{\frac{1}{2}}} - \frac{1}{2} \frac{1}{(Z_1 Z_{\frac{1}{2}})^2} \right] + O(m^3 A^{3/2}) \qquad (5c)$$

exhibiting in the specific heat an additive renormalization, as well as an entropic contribution of the various "spin structures".

It is nice to see that perturbation theory around the critical point by relevant operators (dimension < 2) is both ultraviolet finite (except for possibly finitely many subtractions), infrared finite (if carried in a finite geometry such as a torus or a strip) and very likely convergent (since done in a finite region). It leads to an interesting deformation of the conformal structure with a flow in the central charge as well as the critical exponents. Unfortunately the expressions become very cumbersome beyond the lowest order.

As an example we were able to check these expressions for the ratio

$$R = -\frac{1}{3} \lim_{L\to\infty} \frac{m_4(L)}{[L m_2(L)]^2} \qquad (6a)$$

computed for an Ising model in the strip limit of a torus of size LT with $\langle M^2 \rangle$ and $\langle M^4 \rangle$ the first moments of the magnetization, and

$$m_2(L) = \lim_{T\to\infty} \frac{\langle m^2 \rangle}{LT}$$

$$m_4(L) = \lim_{T\to\infty} \frac{\langle M^4 \rangle - 3\langle M^2 \rangle^2}{LT} \qquad (6b)$$

Our "analytical" series (sums over eleven integers, fortunately extremely rapidly convergent!) yield

$$R = 2.46048 \pm 0.00005, \tag{6c}$$

while numerical measurements by Derrida, Burkhardt and Saleur give a value

$$R = 2.46044 \pm 0.00002 \tag{6d}$$

showing good agreement.

Of course one can study other models and various quantities.

4. This short summary only describes my personal involvement in a subject, which is by now a well developed "industry", and has obvious connections with string field theory. Numerous aspects would require a much more detailed description, both for the physical applications as well as the fascinating connections with several fields of mathematics. The most interesting part of the story lies perhaps ahead. Could the idea of local scale invariance be pursued in higher dimension, leading to a deeper understanding of critical phenomena?

References can be found in our publication, Nuclear Physics B280 FS 18) 445 (1987) with A. Cappelli and J.B. Zuber, and two preprints, one with the above authors (PhT 87/59) and one with H. Saleur (PhT 87/01).

Variational Principle for Quantum Effects on Nearly Sine-Gordon Chains

R. Giachetti[1] *and V. Tognetti*[2]

[1]Dipartimento di Matematica, Università di Cagliari
[2]Dipartimento di Fisica, Università di Firenze

1. Introduction.

Theoretical methods trying to reduce quantum statistical mechanics calculations to classical ones have recently raised large interest and found many applications [1-4]. Old approaches were revisited and improved, while new schemes have been proposed. A new approximate approach based on the path integral [5] was recently proposed both for a single particle in a potential as well as for interacting scalar fields [6,7]. It allows one to construct an effective potential to be inserted in the configurational integral so that quantum fluctuations can be taken into account for low coupling at all temperatures without resorting to the dilute soliton gas approximation. This method was successfully applied to evaluate quantum corrections to the specific heat of a sine-Gordon (SG) chain [6-8]. The soliton gas approximation is not valid in the temperature region where the specific heat presents a Schotty-like peak. When the results of these quantum corrections for the sine-Gordon chain were compared with the experimental data of real magnetic chains like $CsNiF_3$ and CHAB, this model appeared to be insufficient because the values of the easy-plane anisotropy are not strong enough to prevent out-of-plane fluctuations [9]. On the other hand, the quantum character of these systems has been fully realized [10]. All previous approaches considering the quantum contribution of the out-of-plane fluctuations were always confined to the dilute soliton gas approximation [9,11,12], while the soliton-soliton interactions must be taken into account for a realistic comparison.

In this paper we present detailed calculations of thermodynamical quantities for a one-dimensional chain described by the following Hamiltonian:

$$H(p,z) = \frac{1}{2} \sum_{a,b=-N}^{N} p_a K_{ab} p_b + \frac{1}{2} \sum_{a,b=-N}^{N} z_a L_{ab} z_b + g \sum_{a=-N}^{N} \mathcal{U}(z_a) . \tag{1}$$

The symmetric matrices K, L are assumed to be commuting and to satisfy periodic boundary conditions together with the requirement of translational invariance. The function $\mathcal{U}(z_a)$ describes a local nonlinear interaction with corresponding coupling constant g.

The non-diagonal form of the kinetic term is able to reproduce the correct linear dispersion relation of some real systems, like spin waves in the magnetic chain $CsNiF_3$. We shall therefore carry on the computations for this general case and, in order to give a feeling of the results in simpler circumstances, we shall illustrate the special cases of single degree of freedom double-well oscillator and of a Sine-Gordon field with a local kinetic energy, which corresponds to K being a multiple of the identity matrix.

2. Variational principle.

As $[K,L]=0$, there exists an orthogonal transformation which simultaneously diagonalizes the matrices K and L, $MKM^T = E$, $MLM^T = F$, with $E_{ab} = E_a\delta_{ab}$, $F_{ab} = F_a\delta_{ab}$. Performing then the canonical transformation $y_a = \sum_b M_{ab} z_b$, $q_a = \sum_b M_{ab} p_b$, and casting the functional integral in a Lagrangian form, we get

$$Z = e^{-\beta F} = \int_{x(0)=x(\beta\hbar)} \mathcal{D}[x(u)]\, e^{-S[x(u)]/\hbar} , \tag{2}$$

where the explicit expression for the action is

$$S[x(u)] = \int_0^{\beta\hbar} du \left(\frac{m}{2} \sum_{a=-N}^{N} \dot{x}_a^2 + \frac{m}{2} \sum_{a=-N}^{N} E_a F_a x_a^2 + g \sum_{a=-N}^{N} \mathcal{U}\left((N^T x)_a\right) \right) \tag{3}$$

and where we have defined $x_a = (mE_a)^{-1/2} y_a$, $N_{ab} = (mE_a)^{1/2} M_{ab}$, the 'mass' m being chosen to be $m = (\prod_a E_a)^{-1/(2N+1)}$ in such a way to preserve the measure of the functional integral.

According to the general procedure described in [6,8], we consider the convexity property for the free energy F, namely

$$F \leq F_0 + \frac{1}{\beta\hbar} < S - S_0 >_0 \ , \tag{4}$$

where F_0 is calculated by inserting in equation (2) an approximate action S_0, while the average $< S - S_0 >_0$ is performed with $e^{-S_0/\hbar}$ as a weight. Neglecting linear terms, uneffective by elementary symmetry arguments, the approximate action is chosen to be

$$S_0[x(u)] = \int_0^{\beta\hbar} du \left[\frac{m}{2} \sum_{a=-N}^{N} \dot{x}_a^2 + w(\overline{x}) + \frac{1}{2} \sum_{a,b=-N}^{N} (x_a(u) - \overline{x}_a) w_{ab}(\overline{x}) (x_b(u) - \overline{x}_b) \right], \tag{5}$$

where $\overline{x}_a$ is the average point of the path. The unknown functions $w(\overline{x})$ and $w_{ab}(\overline{x})$ are to be determined by the variational principle, i.e. by the minimum condition of the functional (4). The corresponding functional integral can be explicitly calculated by summing first over all the closed paths with a prescribed average $\overline{x}_a = \xi_a$, $(a = -N, N)$ and then integrating over all the possible values of $\xi \equiv (\xi_a)_{a=-N,N}$. The final result reads

$$e^{-\beta F_0} = \left(\frac{m}{2\pi\hbar^2\beta} \right)^{(2N+1)/2} \int d\xi \ e^{-\beta w(\xi)} \prod_{k=-N}^{N} \frac{f_k(\xi)}{\sinh f_k(\xi)} \quad , \tag{6}$$

where, according to [4], we have set $f_k(\xi) = (1/2)\beta\hbar\omega_k(\xi)$, while the 'frequency' $\omega_k(\xi)$ is defined together with the orthogonal matrix $U_{ka}(\xi)$ by the diagonalizing relation

$$\sum_{a,b=-N}^{N} U_{ka} w_{ab} U_{jb} = m\omega_k^2 \ \delta_{kj} . \tag{7}$$

Performing the minimization of the right hand side of (4) with respect to w and w_{ab} or, equivalently, with respect to w, f_k and U_{ka}, after a change of variables $\varsigma = N^T \xi$, we find that the average $< S - S_0 >_0$ turns out to be vanishing, so that is possible to define an effective potential to be inserted in the configurational integral in order to approximate the free energy F by means of F_0. Equation (6) can then be written in the following way:

$$e^{-\beta F_0} = \left(\frac{m}{2\pi\hbar^2\beta} \right)^{(2N+1)/2} \int d\varsigma \ \exp\{-\beta V_{\text{eff}}(\varsigma)\} , \tag{8}$$

where the effective potential reads

$$V_{\text{eff}}(\varsigma) = V(\varsigma) - g \sum_{n=1}^{\infty} \sum_{a=-N}^{N} (n-1) \left(\frac{D_a}{2} \right)^n \frac{\mathcal{U}^{(2n)}(\varsigma_a)}{n!} - \frac{1}{\beta} \sum_{k=-N}^{N} \ln \frac{f_k}{\sinh f_k} \ , \tag{9}$$

superscripts denoting derivations. The parameters

$$\frac{\alpha_k}{2} = \frac{\hbar^2\beta}{4mf_k^2} \left(f_k \ \coth f_k - 1 \right) \tag{10}$$

give a measure of the influence of the quantum fluctuations, being the differences between total and classical mean squares of the field components.

$$D_a = \sum_{k=-N}^{N} \left(\frac{\alpha_k}{2}\right) \left[\sum_{b=-N}^{N} U_{kb} N_{ba}\right]^2 \tag{11}$$

represents the temperature-dependent quantum renormalization factor. The frequencies are determined by the self-consistent equations

$$\begin{cases} \sum\limits_{a,b=-N}^{N} U_{ia} \left[F_a E_a \, \delta_{ab} + \frac{g}{m} \, P_{ab}\right] U_{kb} = \omega_k^2 \, \delta_{ab} \\ P_{ab} = \sum\limits_{c=-N}^{N} N_{ac} \left[\sum\limits_{n=1}^{\infty} \left(\frac{D_c}{2}\right)^{n-1} \frac{\mathcal{U}^{(2n)}(\varsigma_c)}{(n-1)!}\right] N_{bc} \quad . \end{cases} \tag{12}$$

For lowest temperatures when only the vacuum solution contributes to the free energy, D_a reduces to the mass renormalization factor of the field theory. For highest temperatures the potential V_{eff} agrees with the first quantum correction of the Wigner expansion [1].

3. The anharmonic oscillator.

The simplest application of our theory concerns a single particle in a potential. Since this system has only one degree of freedom, we need a single parameter $\alpha(\varsigma)$ corresponding to the frequency $f(\varsigma)$. The effective potential reads

$$V_{\text{eff}}(\varsigma) = \int d\eta \; V(\varsigma + \eta) \frac{e^{-\eta^2/\alpha}}{(\pi\alpha)^{1/2}} - \frac{m\omega^2\alpha}{4} - \frac{1}{\beta} \ln \frac{f}{\sinh f} \tag{13}$$

while the frequency turns out to be determined by the equation

$$\omega^2(\varsigma) = \frac{1}{m} \sum_{n=0}^{\infty} \left(\frac{\alpha}{4}\right)^n \frac{V^{(2n+2)}(\varsigma)}{n!} = \frac{1}{m} \int d\eta \; V^{(2)}(\varsigma + \eta) \frac{e^{-\eta^2/\alpha}}{(\pi\alpha)^{1/2}} \quad . \tag{14}$$

Equations equivalent to (13) and (14) have also been recovered later [13]. The effective potential for a double-well potential

$$V(\varsigma) = (\lambda/4)[\varsigma^2 - (\mu/\lambda)]^2, \quad (\lambda > 0) \tag{15}$$

is derived in [7] and turns out to be

$$V_{\text{eff}}(\varsigma) = V(\varsigma) - 3\lambda \left(\frac{\alpha}{4}\right)^2 - \frac{1}{\beta} \ln \frac{f}{\sinh f} \tag{16}$$

and the self-consistent equation for the frequency is

$$3(f\coth f - 1)/f^2 = (4/Q)^4 t^3 f^2 - 2(4/Q)^2 t(3y^2 + 1) \, , \tag{17}$$

which gives $\omega^2(\varsigma)$ in such a way to avoid singularities or unphysical results in the partition function at all temperatures. The parameter α remains always positive. Here $y = (\lambda/\mu)^{1/2}\varsigma$, $t = 4\lambda/\mu^2\beta$ and $Q = 4(2\hbar^2\lambda^2/m\mu^3)^{1/2}$ is the coupling parameter, being the ratio between the ground state energy of the harmonic oscillator and the height of the barrier.

Explicit calculations of the effective potential are shown in figure 1. In the low coupling case the effective potential becomes softer and softer for decreasing temperatures favouring temperature-activated jumps across the barrier. For higher coupling a change of symmetry of the potential is observed, although this result has to be taken with some care and considered as an extrapolation.

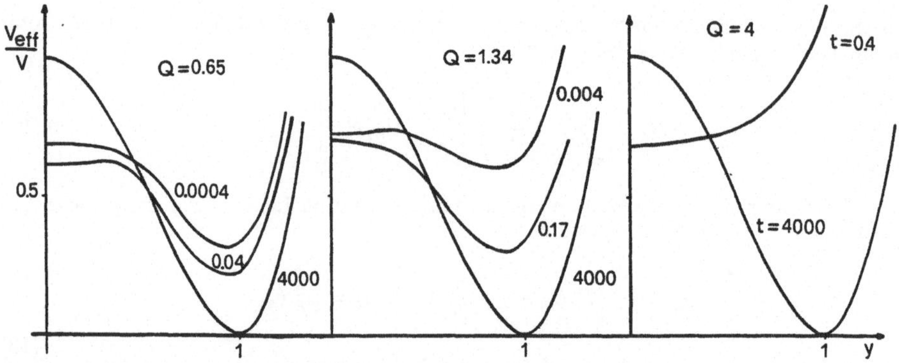

Fig. 1 - Effective potential for different couplings and temperatures.

4. Magnetic chains.

The one-dimensional ferromagnet in an applied magnetic field $CsNiF_3$ can be described by the following spin Hamiltonian ($S = 1$):

$$H = -2J\sum_i (\mathbf{S}_i \cdot \mathbf{S}_{i+1}) + A\sum_i (S_i^z)^2 - h\sum_i S_i^y \,, \tag{18}$$

where $h = g\mu_B H$ is the Zeeman field ($h \sim 1°$ K for the usual applied magnetic field). $A = 9°$ K is the easy-plane anisotropy parameter and $J = 11.8$ K is the exchange integral. Assuming an unitary lattice spacing, the dispersion relation for the spin waves reads

$$\omega^2(k) = 4\Omega_0^2[1/(4R^2) + \sin^2(k/2)][1 + b/(4R^2) + b\sin^2(k/2)] \,, \tag{19}$$

where $\Omega_0^2 = 4S^2J\widetilde{A}$, $R^2 = 2JS/h$ and $b = 4J/\widetilde{A}$, with $\widetilde{A} = A(1 - 1/2S)$. Performing an approximate Villain transformation and neglecting higher order terms which mix momenta and coordinates, we obtain a Hamiltonian of the form given in (1)

$$H = \sum_i \left\{ \widetilde{A}\left[\left(1 + \frac{b}{4R^2}\right) p_i^2 + \frac{b}{4}(p_i - p_{i+1})^2\right] + \frac{1}{4}\frac{\Omega_0^2}{\widetilde{A}}(\varphi_i - \varphi_{i+1})^2 + \frac{\Omega_0^2}{2\widetilde{A}R^2}(1 - \cos\varphi_i)\right\} \,. \tag{20}$$

The parameter b takes into account the out-of-plane fluctuations; for $b = 0$ we recover the SG model. The SG kink solitons are the nonlinear solutions also of the model (20) with energy $E_s = 8S\sqrt{2hJS}$. R is the length of the kink. Typical values of $CsNiF_3$ for an applied field of 5kG are $Q = 0.11$, $R = 5$ and $b = 10$. In this situation an useful expansion can be done [7,14], getting

$$e^{-\beta F_0} = \left(\frac{m}{2\pi\hbar^2\beta}\right)^{(2N+1)/2} \exp\{(2N+1)\beta(\Omega_0/R)^2 D/(4\widetilde{A})\} \left[\prod_k \frac{f_k}{\sinh f_k}\right] \int \prod_a (d\varphi_a) \exp\left\{-\frac{\beta}{4\widetilde{A}}\sum_a \left[\Omega_0^2(\varphi_a - \varphi_{a+1})^2 - \frac{2\Omega_0^2}{R^2} e^{-D/2}\cos\varphi_a\right]\right\} \,, \tag{21}$$

where D is calculated using equation (19) for f_k, i.e. the spectrum of the small oscillations around the vacuum. In this way the nonlinear contribution to the free energy can be calculated using a configurational integral, by a simple substitution $R \to R\, e^{D/4}$. Classical numerical approaches [15] and expansions [10] can be used.

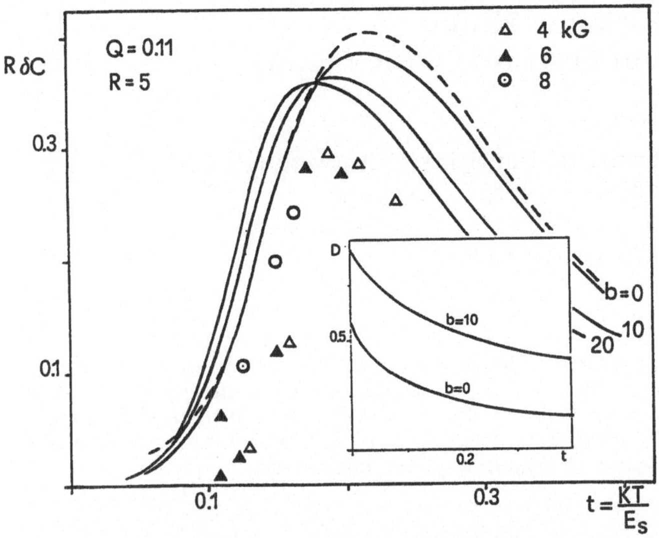

Fig. 2 - Nonlinear contribution to the specific heat vs. reduced temperature. Dashed line refers to complete classical SG result [15,16]. Full lines give quantum results. The experimental points are from [17]. The insert represents the quantum renormalization factor.

From figure 2 we observe that the quantum character of the system increases for finite values of b. For $b = 0$ the results of quantum SG obtained by Bethe Ansatz [18] and quantum Monte-Carlo [19] are perfectly recovered. For nonvanishing b, the position of the peak is shifted at lower temperatures with decreasing height of the maximum. The overall agreement with experimental data [17] improves. To conclude we stress that, although at this stage we have neglected higher order terms of the nonlinear part, however an important aspect of the quantum out-of-plane fluctuations has been considered and its relevance is proved.

References.

1. M. Hillery, R.F. O'Connell, M.O. Scully and E.P. Wigner, *Phys. Rep.* **106**, 122 (1984), R. Dickmann and R.F. O'Connell, *Phys. Rev. Lett.* **55**, 1703 (1985).
2. T. Tsuzuki, *Progr. Theor. Phys.* **70**, 975 (1983) and **72**, 956 (1984).
3. H. De Raedt and A. Lagendijk, *Phys. Rep.* **127**, 235 (1985).
4. M. Moraldi, M.G. Pini and A. Rettori, *Phys. Rev.* **A31**, 1971 (1984).
5. R.P. Feynman, *"Statistical Mechanics"*, (Benjamin, Reading, Mass., 1972).
6. R. Giachetti and V. Tognetti, *Phys. Rev. Lett.* **55**, 912 (1985).
7. R. Giachetti and V. Tognetti, *Phys. Rev.* **B33**, 7647 (1986).
8. R. Giachetti and V. Tognetti, *Int. J. Magn. Magn. Mat.* **54-57**, 861 (1986).
9. H.J. Mikeska and H. Frahm, *J. Phys.* **C19**, 3203 (1986).
10. M.G. Pini and A. Rettori, *Phys. Rev.* **B29**, 5246 (1984).
11. H.C. Fogedby, K. Osano and H.J. Jensen, *Phys. Rev.* **B34**, 3462 (1986).
12. H.J. Mikeska, *Phys. Rev.* **B26**, 5213 (1982).
13. R.P. Feynman and H. Kleinert, *Phys. Rev.* **A34**, 5080 (1986).
14. R. Giachetti and V. Tognetti, *Phys. Rev.* **B** (1987), in press.
15. T. Schneider and E. Stoll, *Phys. Rev.* **B22**, 5317 (1980).
16. K. Sasaki, Progr. Theor. Phys. **68**, 411 (1982) and **71**, 1169 (1984).
17. A.P. Ramirez and W.P. Wolf, *Phys. Rev.* **B32**, 1639 (1985).
18. M.D. Johnson and N.F. Wright, *Phys. Rev.* **B32**, 5798 (1985).
19. S. Wouters and H. De Raedt, these proceedings.

Quantum Theory of Planar Kinks in a Finite Easy Plane Ferromagnetic Chain

J.F. Lemmens[1] *and W.J.M. de Jonge*[2]

[1]Institute for Applied Mathematics, University of Antwerp (RUCA), Groenenborgerlaan 171, B-2020 Antwerpen, Belgium

[2]Department of Physics, Eindhoven University of Technology, NL-5600 MB Eindhoven, The Netherlands

An outline of a quantum approach to multi–kink profiles as coherent states in a ferromagnetic chain with an easy plane is given. Relying on the continuum and harmonic approximation the influence of the out–of–plane fluctuations on the stability and dynamics of multi–kink structures in a finite chain is studied. The multi–kink structure does not exhibit a "zero–frequency" Goldstone mode, this is a consequence of the out–of–plane fluctuations.

1 Introduction

Kink-like profiles, that can be described by the sine-Gordon equation, have received much attention during the last decade, especially in their relation to the non-linear excitations of one-dimensional ferromagnets with an easy plane. The properties of the kinks rely on the mapping [1] of the magnetic chain on the sine-Gordon system and the thermodynamical and response functions of the sine-Gordon system. These functions can be calculated classically [2] using transfer matrices or simulated for the classical model [3]. The heat capacity data of $CsNiF_3$ [4] and $[C_6H_{11}NH_3]CuBr_3$ shortly CHAB [5],[6] have been interpreted in terms of the sine-Gordon model, at least in a certain temperature and external field range. The inverse spin-lattice relaxation time, which is proportional to the soliton density [7], [8], [9] and neutron-scattering measurements of the central peak [10] [11] have also been interpreted in terms of this model. This interpretation yields a renormalisation of the so-called soliton rest energy. However, there has always been some controversy about this interpretation as stated by several authors [12], [13], [14], [15].

The mapping of the easy plane ferromagnet on the sine-Gordon model assumes an extreme anisotropy which is, as such, not present in the real system. Investigations of the influence of the lack of extreme anisotropy have led to the prediction of instabilities at a critical magnetic field [16], [17]. The measurement of the inverse spin-lattice relaxation time [8] shows a characteristic Arrhenius law with the renormalised soliton rest energy as an activation energy and did not show any marked deviation from this behaviour at the predicted critical field.

It has been conjectured that a considerable part of the deviation between the observed value of the soliton activation energy and the calculated soliton energy in a classical model is of quantum nature. Calculations by Maki [18] on the basis of the quantum sine-Gordon model and by Mikeska [19] on the basis of a more general quantum model predicted a reduction of the soliton rest energy. In these approaches the most important corrections of the soliton energy come from the zero-point quantum fluctuations and from normal ordering. Both calculations rely on methods used for the quantisation of classical field theories [20]. In relation to some experimental observations, it is found that the quantum sine–Gordon model is not superior to the "classical" sine–Gordon model [21]. Furthermore, for an antiferromagnetic chain it is shown [22] that the renormalisation crucial to obtain the reduction of the soliton energy leads also to a reduction of the linear excitations of the system.[23]. Both arguments suggest that it is important to consider the out–of–plane fluctuations on the same level as the in–plane fluctuations.We therefore studied the quantum fluctuations around a coherent state for an easy plane ferromagnet relying on well–established quantum methods. A detailed account on some aspects of this approach will be published soon [24].

2 Multi–kink profiles and the harmonic fluctuations.

We assume that the ferromagnetic chain can be described by the following Hamiltonian:

$$H = -J\Sigma_n S^z_n S^z_{n+1} + A\Sigma_n (S^z_n)^2$$
$$-J\Sigma_n \tfrac{1}{2}(S^+_n S^-_{n+1} + S^+_{n+1}S^-_n) - g\mu_B H \Sigma_n \tfrac{1}{2}(S^+_n + S^-_n) . \qquad (1)$$

The strong non-linearity in the equations of motion generated by (1) are introduced by the spin algebra. In the representation of Villain [25], [26] the Hamiltonian transforms to the following expression:

$$H = -J(S(S+1))\Sigma_n[\cos(\varnothing_n-\varnothing_{n+1})+g\mu_B H/(J\sqrt{S(S+1)})\cos(\varnothing_n)]$$
$$- J\Sigma_n S^z_n S^z_{n+1} - J\Sigma_n 1/2\, S^z_n [\cos(\varnothing_n-\varnothing_{n+1})+\cos(\varnothing_{n-1}-\varnothing_n)$$
$$+ g\mu_B H/(J\sqrt{S(S+1)})\cos(\varnothing_n) -2A/J]S^z_n + \text{higher order terms} . \qquad (2)$$

For a derivation of the higher order terms in equation (2) we refer to [22]. Because there is a close relationship between coherent states and the canonical transformations of the displaced oscillator type, we will use such transformations to introduce a site-dependent function ϕ_n [27] into the Hamiltonian (2) by displacing the operator $\varnothing_n$ via a canonical transformation . The transformed Hamiltonian can be decomposed as a sum of terms containing a definite number of operators as indicated by the superscript of H:

$$\exp(-U)H\exp(U) = H^0 + H^1 + H^2 + H^3 + H^4 + \ \ldots \ . \qquad (3)$$

It is seen that H^1 becomes identically zero if ϕ_n satisfies the following difference equation:

$$\sin(\phi_n-\phi_{n+1}) - \sin(\phi_{n-1}-\phi_n)+g\mu_B H/(J\sqrt{S(S+1)})\sin(\phi_n) = 0 . \qquad (4)$$

In the continuum approximation, the equation equivalent to (4) leads to the static part of the sine-Gordon equation:

$$\phi''(x)=m^2\sin(\phi(x)) , \qquad (5a)$$
$$m^2 = g\mu_B H/(Ja^2\sqrt{S(S+1)}) , \qquad (5b)$$

where " denotes the second derivative with respect to x. Introducing dimensionless units y=mx, H^2 reads in the continuum approximation as follows:

$$H^2 = JS(S+1)ma\int dy[\tfrac{1}{2}(\varnothing')^2+ \tfrac{1}{2}\varnothing^2\cos(\phi)]$$
$$Jma\int dy[\tfrac{1}{2}(S')^2+\tfrac{1}{2}S^2(\cos(\phi)- (\phi')^2 + \partial)] . \qquad (6)$$

The anisotropy term in (6) is given by $\partial=2A/(Jm^2a^2)$. We have denoted the continuum counterpart of the operator S^z_n by S.

A way to classify the solutions of (5) is to consider the first integral:

$$(\phi')^2 = 4\sin^2(\tfrac{1}{2}\phi)+c , \qquad (7)$$

where c is an integration constant that can take all real values greater than -4. The solutions with c negative will not be considered because it is known that they are unstable against small fluctuations [28].For positive c (7) can be integrated giving:

$$\pm(y-y_0)=k[F(\tfrac{1}{2}\phi_0+\tfrac{1}{2}\pi,k) - F(\tfrac{1}{2}\phi+\tfrac{1}{2}\pi,k)] , \qquad (8)$$

where F is the elliptic integral of the first kind and k is the modulus, y_0 is an integration constant, ϕ_0 is the value of ϕ in y_0. The periodicity of F is 2K, where K is the complete elliptic integral of the first kind [29] and k is related to c for c >0 by

$$k = 1/(1+c/4)^{1/2} . \qquad (9)$$

The solution (8) is periodic in the variable y with a period of 2kK.If a chain of length L=Na contains n kinks the number of kinks is equal to the chain length divided by the kink–length (i.e. the period of the solution in the variable x=y/m)

$$n = mL/(2kK) . \qquad (10)$$

The kink density (being the number of kinks divided by the number of magnetic ions N) is easily derived from (10) and is inversely proportional to the kink period 2kK. For a well-defined k we obtain the following relation:

$$\sin^2(\tfrac{1}{2}\phi) = cn^2(z) = 1-sn^2(z) , \qquad (11)$$

where sn and cn are respectively the sine and cosine amplitude function, together with dn they are the three basic Jacobian elliptic functions.Their argument z is related to y by

$$z = (y-y_0)/k - F(\tfrac{1}{2}\phi_0+ \tfrac{1}{2}\pi,k) . \qquad (12)$$

The propagator

$$\mathbb{D}(y\, t,y_0 t_0) =i\, \partial(t-t_0)<[\phi(y,t),\phi(y_0,t_0)]> \qquad (13)$$

describes the in–plane fluctuations of the system in the presence of a coherent state characterised by ϕ. The harmonic approximation for $\mathbb{D}$ follows then from the equations of motion for ϕ and S:

$$\ddot{\mathbb{D}}(y\, t,y_0 t_0) = 2JmaH_S\partial(y-y_0)\partial(t-t_0)$$
$$-(2Jma)^2S(S+1)\, H_S H_\phi\, \mathbb{D}(y\, t,y_0 t_0) . \qquad (14)$$

The operator H_ϕ can be written in terms of the sine amplitude function of modulus k:

$$H_\phi = - 1/(2k^2)\, [d^2/dz^2 + k^2(1-2sn^2(z))]. \qquad (15)$$

This is the linear operator of a generalised Lamé equation (with index 1) [30].Imposing periodic boundary conditions on the solutions of

$$H_\phi\Psi = e_\phi\Psi \qquad (16)$$

the spectrum e_ϕ has been calculated [31]. It has two branches, separated by a gap at the first Brillouin zone. It should be noted that this zone is induced by the periodic array of kinks. The operator H_S can also be written in terms of the sine amplitude function:

$$H_S =- 1/(2k^2)\, [d^2/dz^2 + k^2+4-\partial k^2-6k^2sn^2(z)]; \qquad (17)$$

this is again the linear operator of a generalised Lamé equation (with index 2). The spectrum e_S is generated by the eigenvalue problem:

$$H_S\, \Psi= e_S\, \Psi . \qquad (18)$$

Knowing that the eigenfunctions of H_S are combinations of elliptic functions it is possible to construct a nodeless wavefunction Ψ_0 for H_S:

$$H_S\Psi_0=e_{S0}\Psi_0 \qquad (19)$$

with

$$e_{S0}=\tfrac{1}{2}(\partial - 1 - 2/k^2(\sqrt{(4k^4-13k^2+13)}-1)), \qquad (20a)$$
$$\Psi_0 \simeq[dn^2(z)+\tfrac{1}{3}\,(\sqrt{(4k^4-13k^2+13)}- 4+ 2k^2)] . \qquad (20b)$$

The onset of the spectrum of the product operator $H_S\, H_\phi$ is related to e_{S0} as follows:

$$<\Psi_0H_S\, H_\phi\Psi_0>=e_{S0}<\Psi_0H_\phi\Psi_0> \qquad (21)$$

Because $<\Psi_0H_\phi\Psi_0>$ is positive for k less than 1, the spectrum of the operator H_SH_ϕ will have a negative part if e_{S0} is negative. Therefore the lowest eigenvalue of H_S will determine the stability of the coherent state. It is stable if e_{S0} is positive, the stability region in the ∂-k plane however is smaller than the stability region for a single kink. Eq (21) also determines the lowest value of the spectrum unless Ψ_0 turns out to be a product of two eigenfunctions of $H_\emptyset$ the frequency spectrum will have a gap.

3. Discussion and conclusions

An important point in our approach is the use of the continuum approximation. This approximation is legitimate if one can show that the solutions of the difference equation (4) are analytic functions of the site number.This is not always the case: **1)** The variable $\emptyset_n$ can take the value zero or π at the site n. This solution is of the so-called Heisenberg-Ising-type. According to [32] it can lead to stable solutions of the classical counterpart of our problem. **2)** If $\emptyset_n$ does not differ much from site to the neighboring site (4) can be written as a map, which has been studied [33]. This map has, depending on the value of m, solutions with the required properties. A full analysis and exploitation of the solutions of (4) along these lines is out the the scope of the paper, but indicates that in these matters the continuum approximation has to be handled with care.

The main results of the present study are: a complete quantum mechanical treatment of the spin chain allows that the excited sates of the chain can be classified using the "classical" solutions of the static sine–Gordon equation. These states are described by coherent state wave functions, which are considered as new vacuum states for the transformed Hamiltonian. The solutions of the static sine–Gordon equation, which are important, are multi–kink profiles described by elliptic functions with a modulus k, which is related with the kink density. The stability of these profiles is determined by the spectrum the quantum fluctuations which contain in–plane and out–of–plane fluctuations. Some states decouple from all others. For the low density limit the state corresponding with the translation mode has this property. For a kink density different from zero the states, which decouple, are oscillations.

The solutions of the static sine–Gordon equation,used to classify the excited states of the chain, are periodic arrays of either kinks or anti–kinks. As far as the periodicity of the array is concerned it is easy to show that a random array of kinks can lower its energy by adjusting the distance between the kink–centers, also the interaction between the kink and the linear excitations is minimal if the kinks form a periodic array.

In conclusion, we have given a method to obtain quantum states, which correspond with the periodic solutions of the static sine–Gordon equation. The spectrum of the second order fluctuations determine the stability of these multi–kink structures and the out–of–plane fluctuations are proven to be important for the qualitative behaviour of the system.

References

1. H.J.Mikeska: J. Phys. **C11**, L29, (1978).
2. J. F. Currie, J.A. Krumhansl, A.R.Bishop, S.E.Trullinger: Phys. Rev. B**22**, 477, (1980).
3. T.Schneider, E.Stoll: Phys. Rev. B**22**, 5317, (1980).
4. A.P.Ramirez, W.P.Wolf: Phys. Rev. Lett. **49**, 227, (1982).
5. K.Kopinga, A.M.C.Tinus, W.J.M.de Jonge: Phys.Rev. B**29**, 2868, (1984).
6. A.M.C.Tinus, W.J.M.de Jonge,K Kopinga: Phys.Rev. B**32**, 3154, (1985).
7. T.Goto: Phys. Rev. B**28**, 6347, (1983).
8. H.Benner, H.Seitz, J.Wiese, J.P.Boucher: J. Magn. Magn. Mater. **45**, 354, (1984).
9. K. Kopinga, W.J.M. de Jonge, C.H.W. Swüste, A.C. Phaff, R. Hoogerbeets, H Van Duyneveldt: *Solid State Sciences* Vol **54** (Springer, Berlin 1984), p 27.
10. J.K.Kjems, M.Steiner: Phys. Rev. Lett. **41**, 1137, (1978).
11. K. Kakurai, R. Pynn, B. Dorner, M. Steiner:J. Phys C. Solid State Phys.**17** L123, (1984).
12. J.M.Loveluck, T.Schneider, E.Stoll, J.Jauslin: Phys. Rev. Lett. **45**, 1505, (1981).
13. G.J.Reiter: Phys. Rev. Lett. **46**, 202, (1981).
14. S.T.Chui, K.B.Ma: Phys. Rev. B**27**, 4515, (1983).
15. M.G.Pini, E.Rettori: Phys. Rev. B**29**, 5246, (1984).
16. E.Magyari, H.Thomas: Phys. Rev. B**25**, 531, (1982).

17. P.,Kumar: Phys. Rev. B**25**, 483,(1982).
18. K.Maki: Phys. Rev. B**24,** 3991, (1981).
19. H.J.Mikeska: Phys. Rev. B**26**, 5213, (1982).
20. R.F.Dashen, B.Hasslacher, A.Neveu: Phys. Rev. D**11**, 3424, (1975).
21. M. Fowler, N.F. Wright, M.D.Johnson: *Solid State Sciences*, Vol **54** (Springer, Berlin 1984), p 99.
22. N.F.Wright, M.D.Johnson, M.Fowler: Phys.Rev.B**32**, 3169; (1985).
23. U.Heilmann, J.K.Kjems, Y.Endoh, G.F.Reiter,G.Shirane,R.J.Birgeneau: Phys. Rev. B**24**, 3939, (1981).
24. L.F.Lemmens, W.J.M.de Jonge: J. Phys A Math Gen **20**, ... (1987).
25. J.Villain: J. Phys. (Paris) **35**, 27,(1974).
26. F.M.D.Haldane: Phys. Lett. **93**A, 404,(1983).
27. F. Moussa, J.Villain: J.Phys.C. Solid State Phys **9** , 4433, (1976).
28. R.Giachetti,E.Sorace,V.Tognetti: Phys. Rev. B**30** , 3795, (1984).
29. M.Abramowitz,I.A.Stegun (editors): *"Handbook of mathematical functions"* chapter 16 and 17, (1964).
30. E.T.Whittaker, G.N.Watson : *"A Course of Modern Analysis"* (Cambridge U.P.,London,1935), chapters 20 and 23.
31. B.Sutherland: Phys.Rev.A**8** , 2514, (1973).
32. C.Etrich, H.C.Mikeska, E.Magyari, H.Thomas, R.Weber: Z.Phys B-Condensed Matter **62**, 97, (1985).
33. S. Aubry Phys. Rep. **103** , 127, (1984).

RAF Model and Solid Oxygen: Infinite Degeneracy, Soft Lines, No LRO in Classical Approximation and Order by Quantum Disorder

E. Rastelli and A. Tassi

Dipartimento di Fisica dell'Università, I-43100 Parma, Italy

1. Introduction

Exchange coupling competition in a Heisenberg Hamiltonian is the basic origin of non collinear helical configurations; furthermore, on suitable phase boundaries between different helical phases in the parameter space, infinite degeneracy of the ground state appears (for $S \to \infty$) corresponding to infinite unequivalent isoenergetic helices of wave vector $\vec{Q}$ belonging to lines $\mathcal{L}_{\vec{Q}}$ in the reciprocal space we call "degeneration lines". In the isotropic model the magnon dispersion curve shows "soft lines" vanishing for all wave vectors falling on $\mathcal{L}_{\vec{Q}}$ /1/. This causes a low frequency catastrophe that destroys long-range order (LRO), even in 3D, at any finite temperature leading to an unorthodox phase with possible algebraic decay of the correlation function /2/. Planar anisotropy leaves the infinite degeneracy of the ground state but it lifts the soft lines so restoring LRO. For anisotropy small enough we expect a conventional low-temperature ordered phase followed by a "degenerate helix" (DH)-like phase at intermediate temperature before the paramagnetic high-temperature phase is reached. Further exchange interactions remove the infinite degeneracy of the ground state /2/ and lift the soft lines, but a phase reminiscent of the classical scenario is still expected at intermediate temperature.

Only recently has zero-point motion been taken into account /3/: the degenerate helix fades on the classical degeneration line in the parameter space but in its neighborhood we have found a small but finite region of infinite degeneracy.

The very peculiar behaviour described above related to magic exchange competitions in loose-packed lattices, is also present in the classical rhombohedral Heisenberg antiferromagnet (RAF) with nearest neighbours (n.n.) in plane J and n.n. interplane J' interactions /4/. In spite of the customary expectation of an in-plane 120° three sublattice spin configuration we have found that the true ground state configuration is DH-like, which provides degeneration lines $\mathcal{L}_{\vec{Q}}$, soft lines of the magnon dispersion curve, and absence of LRO if no anisotropy is present.

We have worked out a comprehensive analysis in classical approximation for the RAF model: we have studied the ground state configuration, the magnon dispersion curve, the elastic and diffuse neutron scattering cross section by monocrystals and the elastic cross section by polycrystals.

We think that our results contribute to understand the magnetic behaviour of solid oxygen. Solid oxygen indeed is a molecular monoclinic antiferromagnet with LRO in the temperature range 0-24 K (α-phase) and a rhombohedral antiferromagnet (RAF) with short-range order (SRO) in the range 24-44 K (β-phase) /5/. We have found that the DH configuration could be easily recognized by elastic neutron scattering that claims LRO probably absent in β-oxygen /6/. On the contrary the diffuse scattering, even if monocrystals were available, could distinguish between a DH configuration and a stacking of uncorrelated layers with a 120° three-sublattice SRO, only via a careful examination of the peak profile that seems to be very hard to test by experiment. We find indeed that a DH configuration retains interplane correlation, whereas for a stacking of 2D layers any interplane correlation is obviously absent and this affects the peak profile.

2. Classical Rhombohedral Heisenberg Antiferromagnet: Ground State and Elementary Excitations

The Hamiltonian we consider is

$$H = -J \sum_{i\delta} \vec{S}_i \cdot \vec{S}_{i+\delta} - J' \sum_{i\delta'} \vec{S}_i \cdot \vec{S}_{i+\delta'} + D_A \sum_i (S_i^z)^2 , \tag{1}$$

where $\vec{\delta}$ and $\vec{\delta}'$ are vectors joining a site of a rhombohedral lattice with its in-plane and its out-of-plane n.n.; $J < 0$ and $J' \gtrless 0$ are the n.n. in plane and out-of-plane exchange couplings, respectively; D_A is the easy plane single ion anisotropy strength. We notice that the correct properties of the classical RAF model have been obtained only recently /4/. For the physically significant range $|j'| < 3$ where $j'=J'/J$, we find that the ground state shows infinite degeneration: one of the six degeneration lines $\mathcal{L}_{\vec{Q}}$ corresponding to infinite unequivalent helices is shown in Fig. 1 for selected values of j'. The remaining five lines are obtained by rotations of $n\pi/3$ (n=1,...,5) around the c-axis.

For small j' as is the case in β-oxygen, $\mathcal{L}_{\vec{Q}}$ is given by

$$\begin{aligned} Q_x &= -(2j'/a\sqrt{3}) \sin(cQ_z/3) , \\ Q_y &= 4\pi/3a + (2j'/a\sqrt{3}) \cos(cQ_z/3) , \\ Q_z & \text{ arbitrary.} \end{aligned} \tag{2}$$

Notice that the ground state configuration of the RAF model even for vanishing J' (Eq. (2)) differs deeply from its 2D correspondent triangular antiferromagnet ground state. The in-plane configuration tends to its 2D counterpart but interplane correlation exists. Furthermore, for $J' \neq 0$ even the in-plane spin pattern differs from the 120° three sublattice phase.

Any mechanism supporting LRO, such as planar anisotropy, should give peaks in elastic neutron scattering experiment on RAF, whereas no elastic scattering at all is expected in 2D systems.At intermediate temperature elastic scattering should give ridges along the degeneration line, which is a very novel picture.

In contrast with previous approaches based upon a 2D-like ground state /7/, we obtain well-defined spin wave excitations starting from our DH configuration.

The spin wave excitation energy is

$$E_k = (S_k D_k)^{\frac{1}{2}} , \tag{3}$$

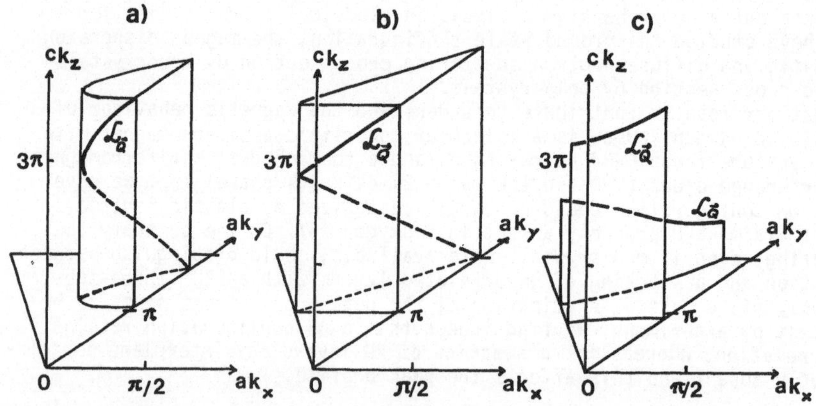

Fig. 1. Degeneration lines for j'=0.8 (a), 1 (b), 1.2 (c)

$$S_k = 2JS\sum_{\delta}(\cos\vec{Q}\cdot\vec{\delta} - \cos\vec{k}\cdot\vec{\delta}) + 2J'S\sum_{\delta'}(\cos\vec{Q}\cdot\vec{\delta}' - \cos\vec{k}\cdot\vec{\delta}') + D_A(2S-1), \tag{4}$$

$$D_k = 2JS\sum_{\delta}\cos\vec{Q}\cdot\vec{\delta}\,(1-\cos\vec{k}\cdot\vec{\delta}) + 2J'S\sum_{\delta'}\cos\vec{Q}\cdot\vec{\delta}'\,(1-\cos\vec{k}\cdot\vec{\delta}'). \tag{5}$$

It is clear to see that, in absence of anisotropy ($D_A=0$), $E_k=0$ for any $\vec{k}\in\mathcal{L}_{\vec{Q}}$. Soft lines appear and LRO is destroyed owing to a catastrophic population number. Planar anisotropy lifts the soft lines; nevertheless, the infinite degeneracy of the ground state is still present.

3. Neutron Scattering by RAF Mono- and Polycrystals

DH and normal helix (NH) configurations in presence of LRO provide dramatically different responses in elastic neutron scattering experiment. In scattering experiment by monocrystals one obtains indeed "Bragg lines" along the degeneration lines in the former case, customary Bragg peaks in the latter one. Even if only polycrystals are available, the elastic neutron scattering cross section gives sawtooth peaks for DH configuration and the well-known δ-like peaks for NH configuration /8/.

For a DH phase the elastic cross section reads

$$(d\sigma/d\Omega)_e = C_0\int_{Q_1}^{\infty} dQ\, \exp(-\alpha Q^2)(2-Q_1^2/Q^2)[Q(Q^2-Q_1^2)^{\frac{1}{2}}]^{-1}(\Gamma/\pi)[(k-Q)^2+\Gamma^2]^{-1}, \tag{6}$$

where $Q_1=4\pi/3a$. An exponential factor $C(Q)=C_0\exp(-\alpha Q^2)$ has been introduced in order to simulate the magnetic form factor and Debye-Waller factor, the instrumental resolution limitation is taken into account by replacing δ-function by a lorentzian.

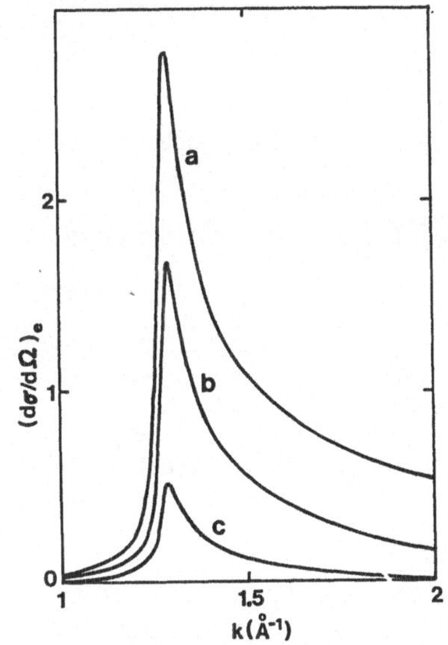

Fig. 2. $(d\sigma/d\Omega)_e$ in arbitrary units versus the scattering wave vector for α=0 (a), 0.3 (b), 1 (c)

Figure 2 shows $(d\sigma/d\Omega)_e$ for selected values of α. The lattice constant a of β-oxygen has been chosen so that $Q_1=1.27$ Å^{-1} and a typical half-maximum half-width $\Gamma=0.02$ Å^{-1} is assumed.

We notice that Eq.(6) compares favourably with the peak profile apparently present at zero energy transfer /6/. Unfortunately in β-oxygen LRO seems to be absent, even if the very large intensity observed at low (zero?) energy transfer seems to be more consistent with an elastic rather than with an inelastic origin.

Anyway we have also considered diffuse scattering because the more recent interpretation of the experimental data claims for absence of LRO in β-oxygen.

We have found that diffuse scattering by NH and DH monocrystals provide different peak profiles: in particular the central peak for small interplane coupling J' results to be lorentzian in the DH case, whereas tends to a square root behaviour in the NH case for scattering wave vector in the c-plane. Moreover no dispersion in the DH configuration of the central peak for scattering wave vector along the c-axis is present, at variance with respect to the NH-like behaviour /9/. Unfortunately this should be a not very promising perspective even if monocrystals were available which is not the case of solid oxygen.

For convenience of the reader we quote the diffuse neutron scattering cross section by a RAF monocrystal /9/:

$$S(\vec{k})=\pi(1+k_z^2/k^2)\left[(\kappa^2+|\vec{k}_\perp+\vec{Q}_\perp|^2)^{-\frac{1}{2}}+(\kappa^2+|\vec{k}_\perp-\vec{Q}_\perp|^2)^{-\frac{1}{2}}\right]+(1-k_z^2/k^2)4\pi/(\kappa^2+k^2)\,, \quad (7)$$

where $S(\vec{k})$ is the Fourier transform of the instantaneous spin-spin correlation function which is supposed decaying as $\exp(-\kappa r)/r$ where r is the spin-spin distance and κ is the inverse correlation length. The z-axis is along the c-axis and $\vec{k}_\perp$ and $\vec{Q}_\perp$ are the components of the scattering wave vector $\vec{k}$ and of the helix wave vector $\vec{Q}$ in the c-plane.

4. Order by Quantum Disorder in the RAF model

An interesting question concerns the effect of the zero-point motion /10/. The first quantum correction to the classical ground state energy is $\Delta(\vec{Q})=\sum_k E_k(\vec{Q})$ where E_k is given by Eq.(3). The weak dependence of the spin-wave energy on $\vec{Q}\in\mathcal{L}_{\vec{Q}}$ enters a weak modulation even in the ground state energy. We limit ourselves to an antiferromagnetic interplane coupling as it seems to be the case for solid oxygen /6/.

Quantum fluctuations for $D_A=0$ pin the helix corresponding to

$$Q_x=0,\ Q_y=(2/a)\cos^{-1}(-\tfrac{1}{2}+j'/2),\ Q_z=0,\ \text{for } j'<1\,, \quad (8)$$

$$Q_x=-2\pi/a_o\,,\ Q_y=2\pi/a_o\,,\ Q_z=0,\ \text{for } j'=1\,, \quad (9)$$

where a_o is the lattice constant of the f.c.c. lattice to which the rhombohedral lattice reduces in this case;

$$Q_x=(2/a\sqrt{3})\cos^{-1}\left[(j'^2-1)/8\right]^{\frac{1}{2}}\,,\ Q_y=\sqrt{3}Q_x\,,\ Q_z=(3/c)\cos^{-1}\left\{-\left[2\cos(aQ_x/2\sqrt{3})\cos(aQ_y/2)+\cos(aQ_x/\sqrt{3})\right]/j'\right\}\,,\ \text{for } 1<j'<3\,, \quad (10)$$

where c is three times the distance between n.n. planes.

We expect that quantum fluctuations also lift the soft lines in the magnon dispersion curve so that LRO is restored by quantum disorder.

Anyway this is still a zero-point motion effect so that we think that at intermediate temperature the whole scenario of the classical model should appear again, which leaves the hope of observing the rich phenomenology related to the DH configuration.

1. E.Rastelli, L.Reatto, A.Tassi: J.Phys.C16, L331 (1983)
2. E.Rastelli, L.Reatto, A.Tassi: in Magnetic Excitations and Fluctuations, ed.by S.W.Lovesey, U.Balucani, F.Borsa, V.Tognetti, Springer Series in Solid-State Sciences, vol.54 (Springer, Berlin,Heidelberg 1984) p.195
3. E.Rastelli, L.Reatto, A.Tassi: J.Phys.C19, 6623 (1986)
4. E.Rastelli and A.Tassi: J.Phys.C19, L423 (1986)
5. G.C.De Fotis: Phys.Rev.B23, 4714 (1981)
6. P.W.Stephens and C.F.Majkrzak: Phys.Rev.B33, 1 (1986)
7. R.J.Meier: Phys.Lett.112A, 341 (1985)
8. see for instance S.W.Lovesey: Theory of Neutron Scattering from Condensed Matter vol.2 (Clarendon, Oxford 1984)
9. E.Rastelli and A.Tassi: to be published
10. E.Rastelli and A.Tassi: J.Phys.C , Letters (to be published)

Phase Transitions in Ising Superlattices

D.W. Hone

Physics Department, University of California,
Santa Barbara, CA 93106, USA

The development of experimental techniques which allow for the controlled deposition of materials at the atomic monolayer level has led to extensive interest and activity in the creation and understanding of artificial layered structures. To date, work has focused largely on metals and semiconductors and the obvious importance of controlled electronic properties. But there is much to be learned from insulating magnetic superlattices, as well, and here studies are only just beginning. MILLS and collaborators [1], in particular, have developed the theory of the low-temperature elementary excitations to be expected in a variety of potential structures of this type.

Exchange coupled magnetic insulators are of all materials among the best understood theoretically and some of the best characterized experimentally. Moreover, one has an array of controllable fundamental parameters, including sign, magnitude and symmetry of the exchange interactions, spin magnitude, etc., within classes of materials (e.g., the rutile structure or the cubic perovskite structure antiferromagnetic fluorides) which might well be expected to grow comparably in superlattices.

Here we will explore the expected magnetic ordering transitions of binary magnetic superlattices: n_A layers or sheets of type A spins followed by n_B layers of type B spins, all then repeated periodically to form a crystal. We will focus on the behavior of T_c/J, the critical temperature in units of a suitably selected exchange parameter, as functions of the number of layers n_A and n_B in a unit "slab" and of the various exchange parameters. Of later interest will be the anisotropic growth of correlation lengths, dimensional crossover, etc. We restrict ourselves to ferromagnetic Ising models, with all spins $S_A = S_B = 1/2$, and with nearest neighbor exchange only. Because we are dealing with critical phenomena, this is not as restrictive as it might at first appear. The exchange symmetry of many materials which might reasonably be studied experimentally is uniaxial, and the sign of the exchange, typically antiferromagnetic, is not a vital issue for the questions being asked. Though the dependence on spin magnitudes, e.g., will ultimately be of interest, there remains ample richness of behavior in this model, and we point to the obvious benefits of a limited dimensionality of the parameter space to be studied. We will further restrict ourselves here largely to two, rather than three, spatial dimensions, because we have available a number of exact results at $d = 2$, as well as because of the obvious additional simplicity of calculations within the lower dimensionality. This will provide a reliable guide to later calculations and approximations for $d = 3$. The $d = 2$ system consists of n_A parallel adjacent chains of A spins followed by n_B such chains of B spins, all repeated periodically.

We will begin with standard mean field theory (MFT), the simplest possible calculation of T_c. It is also among the least accurate such calculation, and perhaps most seriously for the present study it falsely predicts a finite T_c even for a single chain of spins. It is possible to invoke more sophisticated effective field theories without this defect, and we will do so [2] in a future publication. But here we turn next to a real space renormalization group technique, the most reliable in predicting T_c and critical exponents, though incapable of examining magnetization distribution within layers or behavior away from T_c (which effective field theories *can* do, albeit within their limited accuracy). Some numerical (Monte Carlo) studies are also under way [3].

We study the Hamiltonian

$$\mathcal{H} = -\sum J_{pq}(i,j)\sigma_{ip}\sigma_{jq}, \qquad (1)$$

where the Ising spin variables $\sigma_{ip} = \pm 1$, p and q label the chain (for $d = 2$) or layer (for $d = 3$) of identical spins, and i and j label positions within the chain or layer. The exchange J_{pq} can be J_{AA} for nearest neighbor A spins, $J_{BB} \equiv jJ_{AA}$ for nearest neighbor B spins, or J_{AB} for nearest neighbors of opposite type spins. We take

$$J_{AB}^2 = J_{AA}J_{BB} = jJ_{AA}^2, \tag{2}$$

again to limit the number of free parameters; one would expect this to be a good approximation in any case. We will always take $j \geq 1$ (the B spins more strongly exchange coupled than the A spins). The critical temperature T_c is defined as the highest temperature at which the standard mean field equation for the magnetizations, $m_p = \langle \sigma_{ip} \rangle$, have non-trivial solutions (by symmetry there are at most $(n_A + n_B)/2$ inequivalent values of m_p). We define

$$K_{pq} \equiv J_{pq}/T_c. \tag{3}$$

Then the equation for T_c can be written as

$$\frac{\cosh \kappa(n_A - 1)}{\cosh \kappa(n_A + 1)} = \frac{\cos q(n_B + 1)}{\cos q(n_B - 1)}, \tag{4}$$

where

$$4K_{AA} = \mathrm{sech}^2 \kappa, \qquad 4K_{BB} = \sec^2 q. \tag{5}$$

The magnetization of a B (strongly coupled) spin, in a chain $x/2$ lattice constants from the center of the slab of B chains, is proportional $\cos qx$. Similarly, the magnetization of an A spin is proportional to $\cosh \kappa x$.

The fundamental problem with MFT becomes immediately evident. Consider single chains of B spins ($n_B = 1$) separated by increasingly larger numbers n_A of A spins. A little algebra gives the critical temperature from Eq. (4) as $K_{BB} = 0.5 - (4j)^{-1}$ (where we recall from above that $j \equiv J_{BB}/J_{AA}$). That is, T_c is controlled entirely by the isolated B chains and their near neighbors; even for non-magnetic A spins ($j \to \infty$) those B chains order at $K_{BB} = 1/2$, whereas physically it is clear that we should find the T_c appropriate to a 2-d pure A spin lattice in this limit.

For completeness and for later reference let us also consider within MFT the interesting sequence of superlattices with equal thicknesses of A and B slabs: $n_A = n_B = n$, as we change both n and the relative exchange $j = J_{BB}/J_{AA}$. Numerical solution of Eqs. (4,5) are readily obtained. The limit of large n is simple: q must go to zero ($q \lesssim \pi/2n$ to keep m_p of uniform sign across a slab of n B-spins), so $K_{BB}^* \approx 1/4$, and $K_{AA}^* \approx (4j)^{-1}$. We show the results for K_{BB}^* as a function of n for various exchange ratios: $j = 1.2$, 2, and 6, in Fig. 1; the

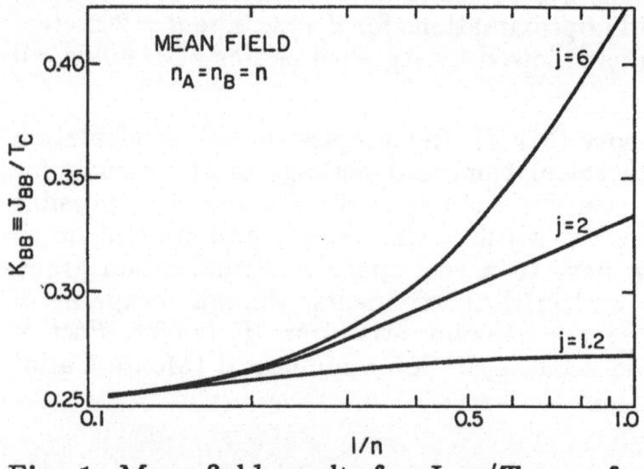

Fig. 1. Mean field results for J_{BB}/T_c as a function of number of layers n in both A and B slabs. Relative exchange $j = J_{BB}/J_{AA}$

asymptotic approach to a j-independent value of 1/4 at large n is clear. Again the strongly coupled B spins control T_c; the value of 1/4 is the MFT value for a pure 2-d Ising (square) lattice (though different, of course, from the exact value $K^* = 0.441$ for such a lattice).

We turn next to a real space renormalization group approach. We extend the MIGDAL-KADANOFF [4-6] decimation technique, using the recursion relations so derived to determine the (unstable) fixed point which defines T_c. It will prove useful to consider first the general 2-d rectangular $S = 1/2$ Ising model: spins on a square lattice with dimensionless exchange constants J/T given by K_x for nearest neighbor spin pairs along the x-axis and K_y for such pairs along the y-axis. Decimation proceeds first by moving K_y bonds so that chains in the y-direction separated by n_x units are now coupled with exchange strength $n_x K_y$ and those spins in between these chains are no longer coupled at all in the y-direction. Then these latter spins, coupled only along x, can be traced over, or "decimated", leaving an effective Hamiltonian involving only the spins coupled along y by $K'_y = n_x K_y$ and with an effective coupling along x of

$$K'_x = \tanh^{-1}(\tanh K_x)^{n_x}. \tag{6}$$

Now the K'_x bonds are moved similarly along y to leave coupled horizontal chains separated by n_y units. Again traces are taken to decimate the intervening spins, leaving a new effective rectangular Ising Hamiltonian, with

$$K''_x = n_y \tanh^{-1}(\tanh K_x)^{n_x} \qquad K''_y = \tanh^{-1}(\tanh n_x K_y)^{n_y}. \tag{7}$$

These recursion relations depend on the choice of scaling factors n_x and n_y. But if we extend the scaling to non-integer values, and in fact scale each time by an amount infinitesimally greater than the identity transformation:

$$n_x \to \lambda_x = 1 + \epsilon_x; \qquad n_y \to \lambda_y = 1 + \epsilon_y, \tag{8}$$

then the fixed point recursion relation $K''_i = K_i \equiv K^*_i$ at T_c (where $i = x, y$) can be written to first order in ϵ_i as

$$\begin{aligned} \epsilon_x \sinh(2K^*_x)\ln(\tanh K^*_x) + \epsilon_y(2K^*_x) &= 0 \\ \epsilon_x(2K^*_y) + \epsilon_y \sinh(2K^*_y)\ln(\tanh K^*_y) &= 0. \end{aligned} \tag{9}$$

By definition, $K^*_y = \mu K^*_x$ (the ratio of exchanges is the same at all temperatures; we defined that ratio as μ above), and Eq. (9) determines both the unique proper scaling ratio ϵ_x/ϵ_y to reach a fixed point (which also determines the relative growth of correlation lengths in the x and y direction as T_c is approached) and the value T_c through the secular equation for the pair (9):

$$F(K^*_x)F(K^*_y) = 1, \qquad \text{where} \qquad F(t) \equiv -\frac{\sinh 2t}{2t}\ln(\tanh t), \tag{10}$$

which is equivalent to the *exact* answer [7],

$$\sinh(2K^*_x)\sinh(2K^*_y) = 1 \tag{11a}$$

$$\text{or} \qquad -\ln\tanh K^*_y = 2K^*_x. \tag{11b}$$

(Note: the latter relation is symmetric under interchange of K_x and K_y, though the form doesn't make that apparent.)

The reason that this approach does provide the exact result for T_c for the rectangular Ising model appears not to be understood – and it does *not* give the exact critical exponents. Nevertheless, we now extend the approach to the superlattice with some confidence in the reliability of the results for T_c. The chains of like spins (A spins, e.g.) are taken aligned along the y-direction. We first carry out a finite scale bond moving – by $(n_A + n_B)$ units – and decimation, so that the *superlattice* is renormalized to a rectangular Ising *lattice*. After a subsequent moving of x-bonds by a scale factor $\gamma(n_A + n_B)$ (with γ to be determined) the coupling constants of the effective lattice become

$$K_x'' = \gamma(n_A + n_B)\tanh^{-1}\left[\tanh^2 K_{AB}\tanh^{n_A-1}K_{AA}\tanh^{n_B-1}K_{BB}\right]$$
$$K_y'' = \tanh^{-1}\left[\tanh(n_A K_{AA} + n_B K_{BB})\right]^{\gamma(n_A+n_B)} . \quad (12)$$

We now carry out an infinite succession of infinitesimal transformations, as on the general Ising rectangular lattice above, to go to arbitrarily large length scales. The fixed point determining T_c is most conveniently defined now in the asymmetric form (11b), which from Eq. (12) (and using the identity $\tanh(-\ln\sqrt{\tanh x}) = \exp(-2x)$) gives immediately

$$e^{-2(n_A+jn_B)K^*} = \tanh^2(\sqrt{j}K^*)\tanh^{n_A-1}K^*\tanh^{n_B-1}(jK^*). \quad (13)$$

Here we have used Eq. (2) to relate the three exchange values and have written $K^*_{AA} \equiv K^*$ for simplicity of notation. This form demonstrates explicitly that the scale factor γ disappears from the critical point condition, as it must; only the succession of transformations up to arbitrarily large length scales, not the single initial finite scale change, can determine T_c.

As before, we consider the limiting sequences of superlattices (i) $n_B = 1$ with increasing n_A, and (ii) $n_A = n_B = n$, in each case for various exchange ratios $J_{BB}/J_{AA} = j$. In the first instance, with single B chains separated by n_A chains of A spins, we expect to approach the pure A lattice behavior at large n_A. Both the numerical solutions of Eq. (13), shown in Fig. 2, and the form of that equation for $n_B = 1$, $n_A \to \infty$: $\exp(-2K^*) = \tanh K^*$ (the pure lattice form; see Eq. (11b)), exhibit this behavior – in marked contrast to the MFT result above: $K^*(MFT) = (2j-1)/4$.

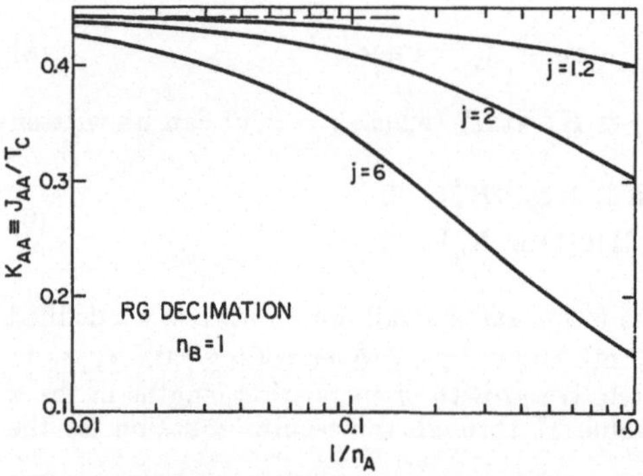

Fig. 2. Renormalization group results for J_{AA}/T_c as a function of n_A, the number of A chains separating single isolated B chains. The dashed line is the pure A result, $K = 0.441$.

In the case of equal thicknesses $n_A = n_B = n$ there are opposing tendencies. The thicker slabs of B spins tend to order at increasingly higher temperatures as n increases, but they are at the same time being separated by thicker layers of the more weakly coupled A spins, which tend to lower T_c. Ultimately, the latter effect dominates, but the quantitative change is small: T_c differs by only a few percent as n goes from 1 to ∞, where the condition is

$$\sinh(2K^*)\sinh(2jK^*) = 1, \quad (14)$$

precisely the same as Eq. (11a) for the rectangular lattice with $K_x = K_{AA}$ and $K_y = K_{BB}$. With the K_{AB} bonds becoming a negligible fraction of the total as $n \to \infty$, apparently the central feature is that half the bonds have strength J_{AA} and half J_{BB}; their relative location seems to be of less importance in setting T_c. The result *is* the same [8] for a system where the exchange is *randomly* J_{AA} or J_{BB} with equal probability. Having said this, we must introduce a general caveat. The critical point determined by this – or any – renormalization group method is the point of true singularities for a system of infinite size. Not only is no physical system infinite; in the present case each slab is becoming arbitrarily large, and the alert reader may well be questioning why we are not seeing a transition at the pure B spin

2-d Ising lattice critical temperature. The response functions (susceptibility, etc.) will indeed be large and highly peaked in temperature there; they are not, however, truly singular, so this does not show up as a fixed point in this particular analysis. We will explore this behavior further in future publications.

In extending these calculations to $d = 3$ we expect the MFT (and effective field) theory to become somewhat more accurate, as usual. However, the decimation approach in its original form becomes *less* useful as the spatial dimensionality increases [5,6]. We will need to explore systematic extensions of the Migdal recursion relations such as those proposed by MARTINELLI and PARISI [5]. We expect to compare all the approximations with the numerical Monte Carlo studies.

I gratefully acknowledge helpful conversations with Dr. Robert Sugar and with Dr. Antonio Siqueira, who will be collaborating on the extension of the work reported here.

References

1. R. E. Camley, T. S. Rahman, and D. L. Mills, *Phys. Rev. B* **27**, 261, (1983); L. L. Hinchey and D. L. Mills, *Phys. Rev.* **B33**, 3329; **B34**, 1689 (1986).

2. D. Hone and A. Siqueira, to appear.

3. M. Jarrell and D. Hone, work in progress.

4. L. P. Kadanoff, *Ann. Phys.* **100**, 359, (1976); A. A. Migdal, *Zh. Eksp. Theor. Fiz.* **69**, 810; 1457 (1975).

5. G. Martinelli and G. Parisi, *Nucl. Phys.* **B180**, 201, (1981).

6. B. Hu, *Phys. Rep.* **91**, 233, (1982).

7. L. Onsager, *Phys. Rev.* **65**, 117, (1944). A very readable discussion is given by D. Mattis, **The Theory of Magnetism** (Harper, New York, 1965) p. 267.

8. M. McCoy and T. T. Wu, **The Two-Dimensional Ising Model** (Harvard, Cambridge, 1973) p. 353.

A Fermionic Treatment of the Frustrated Ising Model

J.A. Blackman and J. Poulter

Department of Physics, University of Reading, Whiteknights, P.O. Box 220, Reading, RG6 2AF, United Kingdom

1. Background

The role of "frustrated plaquettes" in determining the ground state properties of the ±J Ising spin glass was first pointed out by TOULOUSE [1] in 1977. Since then many numerical studies have appeared (for a review, see BINDER and YOUNG [2]), but an understanding of the ground state remains incomplete.

In two dimensions, the statistical mechanics of the Ising model can be cast in the form of a non-interacting fermionic field theory (SCHULTZ et al [3], GREEN and HURST [4], SAMUEL [5]). For a perfect system, of course, this makes it possible to obtain an analytic solution. In the case of the ±J model, frustration manifests itself in a particularly compelling way within the fermionic representation as was shown by BLACKMAN [6]. Numerical studies on small samples were described by POULTER and BLACKMAN [7]. The present contribution extends that work to much larger samples.

We can write the partition function for the assembly of N spins as

$$Z = 2^N \prod_{\langle ij\rangle} \cosh K_{ij} \int d\eta \, \exp\Big(\sum_{\alpha\beta} \eta_\alpha A_{\alpha\beta} \eta_\beta\Big). \tag{1}$$

$K_{ij} = J_{ij}/kT$ where J_{ij} is the bond strength for the pair of sites $\langle ij \rangle$ and the product is over all bonds of the lattice. The integral is over 4N Grassmann variables (4 per lattice site). The matrix $A_{\alpha\beta}$ has constant elements (±1) connecting variables associated with a single lattice site; variables at the ends of the bond $\langle ij \rangle$ are connected by an element whose value is $\tanh K_{ij}$. The integral over Grassmann variables in (1) can be written also in terms of the determinant of A[5].

In studying the ground state properties of the ±J model, the eigenvalues of A in the low-temperature limit have particular significance. For the perfect system (all + bonds), the eigenvalues lie in a band between $2-\sqrt{2}$ and $2+\sqrt{2}$. In the presence of frustration, however, "local modes" appear below the band. Their eigenvalues approach zero in the $T\to 0$ limit as

$$\varepsilon = \tfrac{1}{2} X \exp(2rJ/kT); \tag{2}$$

X is a constant and r is an integer (≥1). Both depend on the distribution of frustrated plaquettes over the lattice. The number of local modes is equal to the number of frustrated plaquettes and their eigenvectors have weight only on these plaquettes.

Much of the physics of frustration is expressed by the values taken by X and r. The contribution to the ground state free energy from the "wrong bonds" is given by

$$F = 2J \sum_d r_d \tag{3}$$

and the entropy of the ground state is determined by

$$S = k \sum_{d} \ln X_d \, . \tag{4}$$

The summation is over all defect eigenstates.

The eigenstates at T = 0 for an arbitrarily large, arbitrarily disordered system can be determined trivially by inspection. Near to zero, degenerate state perturbation theory can be used to obtain the eigenvalues in the form given in (2).

2.Calculations

In the ±J model, the ground state is ferromagnetic if the concentration of negative bonds is less than a certain value p_c. One line of argument (e.g. DOMANY [8]) suggests that p_c is the concentration below which non-frustrated plaquettes percolate. We consider here the effect of varying p, but prefer to emphasize the frustrated plaquettes and, in particular, the eigenstates associated with them. As p increases these eigenstates become less localized and eventually extend over the whole system. We argue that it is the onset of delocalization that determines p_c.

We consider first a small lattice (20x20 spins) from our earlier study [7]. Figure 1 shows a randomly generated configuration in which p is 5% (below p_c). Squares containing a cross or a number are frustrated. Only 2nd and 3rd (r=2,3) order states are shown explicitly. There are 4 pairs of r=2 states and 1 pair with r=3. The plaquettes on which they have some weight are indicated. At this value of p the states are localized to small regions of the lattice.

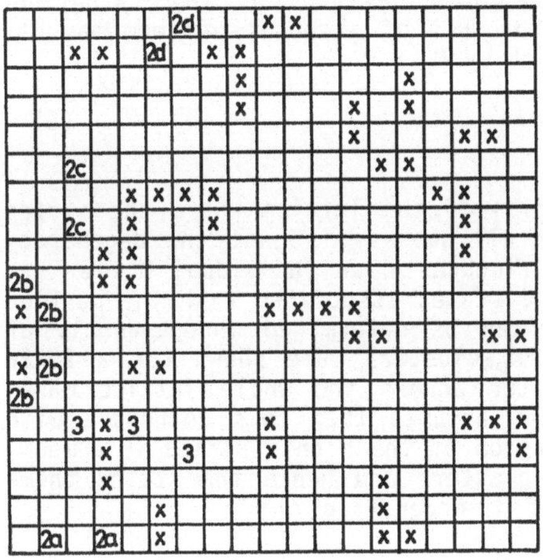

Fig.1. A 20x20 lattice with p=5%. Pairs of second and third order states have weights on squares labelled 2 and 3 respectively. The letters a,b,c,d distinguish different pairs of states.

A value for p of 15% is marginally above most estimates of p_c. In Fig.2 we display some results of a calculation on a 100x100 lattice with 15% negative bonds. Most of the details are omitted to avoid confusion. In this particular example there are 3774 eigenstates with values of r ranging from 1 to 8. Six of the eigenstates are displayed by indicating with different symbols the positions of the frustrated plaquettes on which they have weight. In particular, note the state (with r = 8) represented by crosses. It extends across the whole system, while the others, which are of lower order, are more localized. It is important to emphasize that there is considerable spatial overlap of the states (several having weight on the same frustrated plaquette). Six non-overlapping ones have been chosen for figure 2 purely for illustrative purposes.

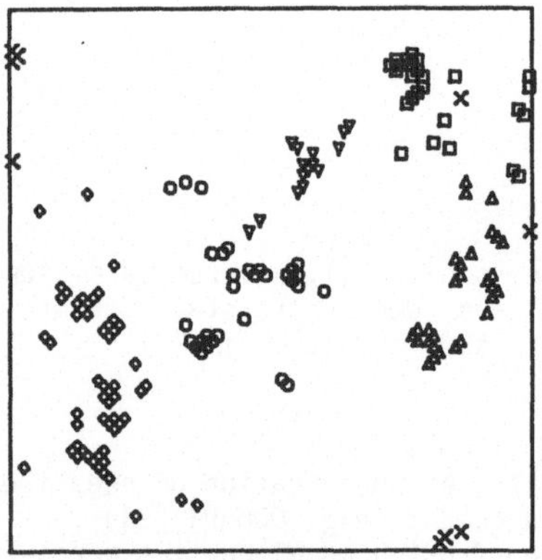

Fig.2. A 100x100 lattice with p=15%. Six eigenstates are indicated by different symbols.

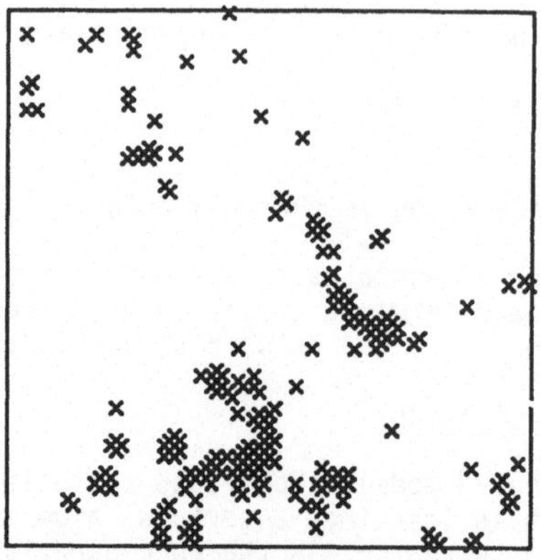

Fig.3. A 100x100 lattice with p=50%. The extension of one eigenstate is displayed.

At p=50%, where frustration is a maximum, there are fewer high order states: r=5 or 6 is typically the highest value, whereas at 12-15% states with r=8 or higher are not unusual in the 100x100 samples. At 50%, however, there is much more extension in the lower order states than at 15%. An example is given in figure 3. The crosses indicate the positions of the frustrated plaquettes over which just one of the r=5 states is extended.

For some purposes the eigenstates should be considered as a group. The X's defined in (2) are generally non-integer, although a product of a subset of them may be. Let us denote such a subset by R and the product by M_R :

$$M_R = \prod_{d \varepsilon R} X_d \ . \qquad (5)$$

A set R is associated with a particular region of the lattice where the states $d \varepsilon R$ have weight. The product of the M_R over all subsets is just the degeneracy of the ground state of the system. The significance of an individual M_R is that it represents the number of alternative ground states associated with the particular region of the lattice and this number can be determined independently of the rest of the lattice. Some examples of "M_R-clusters" are shown for a 20x20 lattice in [7]. The extension of an M_R-cluster across the whole lattice marks the onset of "non-localized" disorder and one assumes that this coincides with the disappearance of ferromagnetism. The extension of the high r states is a measure of the extension of their associated M_R-clusters and hence the states themselves are a convenient signal for the onset of disorder at p_c.

An alternative argument about the significance of extended states can be made as follows. Consider a change of scale in a renormalization group sense. Localized eigenstates will renormalize essentially to point defects and will be irrelevant in the destruction of long-range order. Extended states will remain extended, however, and these are the ones we must consider. Now, starting from a perfect lattice, an extended state can be produced simply by inserting a long chain of parallel negative (-J') defects in the lattice. The evidence is that the maximum r value of our extended states increases much more slowly than the size L of the lattice. Let us write its dependence on L as L^q where $q<1$. To obtain similar behaviour with a single chain of defects requires $J'/J \sim L^{q-1}$, which goes to zero as $L \to \infty$. As far as long-range properties are concerned it appears that the extended states have an effect equivalent to breaking up the lattice with chains of weak (approaching zero as $L \to \infty$) links and thus destroying long-range order.

3. Conclusions

The eigenstates discussed here appear to contain the physics of frustration in the ±**J** model. That they express the "wrong bond" free energy and ground state degeneracy is clear. We have argued that they also hold the key to the understanding of the presence or absence of long-range order.

One can apply the formalism to models with a distribution in J values, although degenerate state perturbation theory is not the obvious technique to apply. There is also the possibility of extending the ideas to three dimensions. The Ising model in 3d can be written as an interacting field theory [9], by including a term quartic in the Grassmann variables. At T=0, the quartic term vanishes and we again have a one-to-one correspondence between frustrated plaquettes and a set of basis states. We are currently considering whether the Grassmann variable formalism can offer any new insight into the behaviour of the 3d spin glass in the difficult low T regime.

4. References

1. G. Toulouse : Commun. Phys. 2, 115 (1977)
2. K. Binder and A.P. Young : Rev. Mod. Phys. 58, 801 (1986)
3. T.D. Schultz, D.C. Mattis and E.H. Lieb : Rev. Mod. Phys. 36, 856 (1964)
4. H.S. Green and C.A. Hurst : Order-Disorder Phenomena (Interscience, New York 1964)
5. S. Samuel : J. Math. Phys. 21, 2806 (1980)
6. J.A. Blackman : Phys. Rev. B26, 4987 (1982)
7. J. Poulter and J.A. Blackman : J. Phys. C. 19, 569 (1986)
8. E. Domany : J. Phys. C. 12, L119 (1979)
9. C. Itzykson : Nuclear Physics B210 [FS6], 477 (1982)

Calculation of Static Magnetic Susceptibilities of a Jahn-Teller *E*-Doublet

P. de Vries[1], *A. Lagendijk*[1], *and H. De Raedt*[2]

[1]Natuurkundig Laboratorium, University of Amsterdam, Valckenierstraat 65, NL-1018 XE Amsterdam, The Netherlands
[2]Physics Department, University of Antwerp (U.I.A.), Universiteitsplein 1, B-2610 Wilrijk, Belgium

Here we report on computations of static susceptibilities of a doubly degenerate orbital electronic state having cubic or tetrahedral symmetry. Because of coupling to harmonic modes there is a pronounced dynamic Jahn-Teller effect on the static susceptibilities of the orbital state, e.g. at low temperatures the static magnetic susceptibility is substantially reduced, and approaches asymptotically a Curie-like behavior only at higher temperatures. The calculations are based on a path-integral representation of the partition function of an appropriate model system.

1. INTRODUCTION

The theory of the dynamic Jahn-Teller (JT) effect has been extensively discussed in connection to optical and paramagnetic resonance spectra of paramagnetic ions in solids. The JT-effect is based on a theorem first described by JAHN and TELLER [1], which states that a complex or defect having orbital electronic degeneracy is unstable against asymmetric distortions of nuclei configurations that remove the degeneracy. This can be described by coupling lattice vibrations to a magnetic ion, thus giving the crystal field a dynamic component. The result is a combined (vibronic) system in which the eigenstates have mixed electronic and vibrational components. Matrix elements of electronic operators are modified, because now vibronic eigenstates must be used. The JT-coupling introduces reduction factors in an effective Hamiltonian that describes the electronic state. The reduction factors (also called Ham factors) are a measure of the overlap of the oscillators, which are now displaced from their original positions by the coupling. As a result of this, the JT-coupling has important consequences on the energy splittings of the ion and, consequently, on resonance spectra. For further details in this field we refer to reviews by STURGE [2], HAM [3] and BATES [4].

There is also a considerable dynamic JT-effect on the static susceptibilities of the orbitally degenerate magnetic ion. SASAKI et al [5] were the first to calculate the static magnetic susceptibility of a paramagnetic ion with a triplet ground state interacting with tetragonal vibrational modes. It appears that the temperature dependence deviates greatly from the Curie behavior. Their results are exact since the Hamiltonian they use is already of diagonal form.

We will calculate the static susceptibilities of a system with an orbital doublet electronic state having cubic or tetrahedral symmetry. These quantities will be computed as a function of temperature and ion-lattice coupling strength using path-integral methods. In Sect. 2 we introduce our model Hamiltonian. We define the quantities we wish to compute and formulate the numerical problem. The results are presented in Sect. 3.

2. THEORY

2a. The Hamiltonian and static susceptibilities

A doubly degenerate orbital electronic state in cubic (or tetrahedral) symmetry belongs to the two-dimensional irreducible representation E of the rotation group O of the cube (or the tetrahedral group T_d). Using the states of the doublet, denoted by ψ_θ and ψ_ϵ and transforming respectively as $\frac{1}{2}(3z^2-r^2)$ and $\frac{1}{2}\sqrt{3}(x^2-y^2)$, as a basis we can define four Hermitian electronic operators: the unit operator I, $U_\theta = -\sigma^z$, $U_\epsilon = \sigma^x$ and $A_2 = \sigma^y$, where σ^α are the Pauli matrices. The Hamiltonian we study is given by

$$H = [E_0 + 1/2M(P_\theta^2 + P_\epsilon^2) + \tfrac{1}{2}M\Omega^2(Q_\theta^2 + Q_\epsilon^2)]I + V(Q_\theta U_\theta + Q_\epsilon U_\epsilon) \ , \qquad (2.1)$$

where V is the coupling coefficient for linear JT-coupling. Here we only consider coupling to one pair of vibrational modes (Q_θ, Q_ϵ) transforming by the representation E. Such an approach is obviously more appropriate for a molecule or an isolated complex than for a defect embedded in a crystal, where the electronic doublet interacts with many such modes. In the latter case one mostly uses the "cluster model" [2-5]: the defect and nearest neighbors are considered as an independent cluster. Then M and Ω in Eq. (2.1) are effective quantities describing approximately the actual system. Further, E_0 is the energy of the degenerate electronic state in the symmetrical configuration. The stabilization energy or JT-energy E_{JT} is $V^2/2M\Omega^2$.

The Hamiltonian expressing the presence of external perturbations acting on the electronic doublet (ψ_θ,ψ_ϵ) has the general form [3,4]

$$H_{ext} = g_1 I + g_2 A_2 + g_\theta U_\theta + g_\epsilon U_\epsilon \ , \qquad (2.2)$$

where the g's are functions of the components of the external perturbations. Here we take them to be independent of the oscillator variables. The g's transform respectively as A_1, A_2, E_θ and E_ϵ of the cubic group. Examples of external perturbations are magnetic fields and strains. The static susceptibilities $\chi_{U_\theta U_\theta}$, $\chi_{U_\epsilon U_\epsilon}$ and $\chi_{A_2 A_2}$ determine the susceptibilities measured in an experiment. Because of the JT-interaction these quantities will deviate from the Curie behavior $\chi=\beta=1/T$. We express them in terms of a reduction factor [5], $\gamma(T)$, defined as the ratio of two static susceptibilities with and without JT-coupling:

$$\gamma_{AA}(T) = \chi_{AA}/\chi_0 = \beta^{-1}\int_0^\beta < e^{\lambda H}Ae^{-\lambda H}A > d\lambda \ , \qquad (2.3)$$

where $A=U_\theta$, U_ϵ or A_2 and $\chi_0=\beta$ (independent of A). Because of symmetry the $\theta\theta$- and $\epsilon\epsilon$-reduction factors are equal.

2b. The numerical problem

To derive a path-integral representation for the partition function we use Trotter's formula [6-8] and obtain

$$Z = \mathrm{Tr}(e^{-\beta H}) = \lim_{m\to\infty} \mathrm{Tr}(e^{-\tau H_1}\, e^{-\tau H_2})^m , \tag{2.4}$$

where $\tau=\beta/m$ and where H_1 and H_2 respectively represent the diagonal and the interaction part of (2.1). Inserting complete sets of oscillator states in (2.4) and working out all resulting matrix elements, we get for the m-th approximant Z_m of the partition function

$$Z_m = c\int dQ_\theta \int dQ_\epsilon \exp(-Q_\theta^T K Q_\theta - Q_\epsilon^T K Q_\epsilon)\, \mathrm{tr}(T_1 T_2 \ldots T_m) , \tag{2.5}$$

where tr denotes the trace operation in $(\psi_\theta,\psi_\epsilon)$-space, $Q_{\theta(\epsilon)}$ are vectors with m components and c is an irrelevant constant. The (periodic) matrix K (which can be diagonalized by a Fourier transform) and the matrices T_j are defined as

$$K_{i,j} = M\Omega/2\sinh\tau\Omega\, (2\cosh\tau\Omega\,\delta_{i,j} - \delta_{i,j-1} - \delta_{i,j+1}) , \tag{2.6}$$

$$T_j = \cosh[\tau V(Q_{\theta j}^2+Q_{\epsilon j}^2)^{\frac{1}{2}}]I - \sinh[\tau V(Q_{\theta j}^2+Q_{\epsilon j}^2)^{\frac{1}{2}}](Q_{\theta j}U_\theta+Q_{\epsilon j}U_\epsilon)/(Q_{\theta j}^2+Q_{\epsilon j}^2)^{\frac{1}{2}}. \tag{2.7}$$

Analogously, m-th approximants can be derived for the susceptibilities (2.3). Multi-dimensional integrals can generally be calculated by Monte Carlo techniques, which require a positive probability function. It turns out that, as a function of temperature and coupling constant V, the factor $\mathrm{tr}(T_1..)$ in (2.5) is negative in parts of the phase space of (Q_θ,Q_ϵ). We therefore proceed as follows: we simulate numerically two quantum harmonic modes by generating (Q_θ,Q_ϵ)-configurations, which are distributed according to the exponential factor in (2.5), using a Gaussian probability distribution and an inverse Fourier transform. The resulting trace over the orbital states is then calculated exactly. The advantage of this method over a Monte Carlo approach is that the subsequent configurations are independent of one another. Following this procedure, approximants of expectation values of an operator O are given by

$$\langle O\rangle_m = \langle\langle\, \mathrm{tr}(T_1..T_m O_m(Q_\theta,Q_\epsilon))\,\rangle\rangle / \langle\langle\, \mathrm{tr}(T_1..T_m)\,\rangle\rangle , \tag{2.8}$$

where the double brackets denote averaging with respect to the exponential factor in (2.5).

3. RESULTS

We have set Ω as the energy unit in our calculations. The thermodynamical quantities are then characterized by two parameters: the temperature T/Ω and the normalized JT-energy $D=E_{JT}/\Omega$, which is a measure for the coupling. The values of the 'Trotter'-parameter m are chosen such, that systematic errors due to the approximation of finite m disappear in the statistical noise. The results for the reduction factors $\gamma_{AA}(T)$ ($A=U_\theta$, A_2) are shown in Fig. 1 for several values of JT-energy D as a function of temperature. At intermediate temperatures ($T/\Omega\approx0.2$) typical values of m are 16 or 32 for D=0.125 and 1. For smaller D-values or higher temperatures, when $\tau=\beta\Omega/m$ becomes smaller, the value of m can be decreased. Lowering the temperature (or increasing the "coupling" D) leads to higher values of m. Results are obtained from data of at least two runs of typically 20000 samples each. The relative number of negative

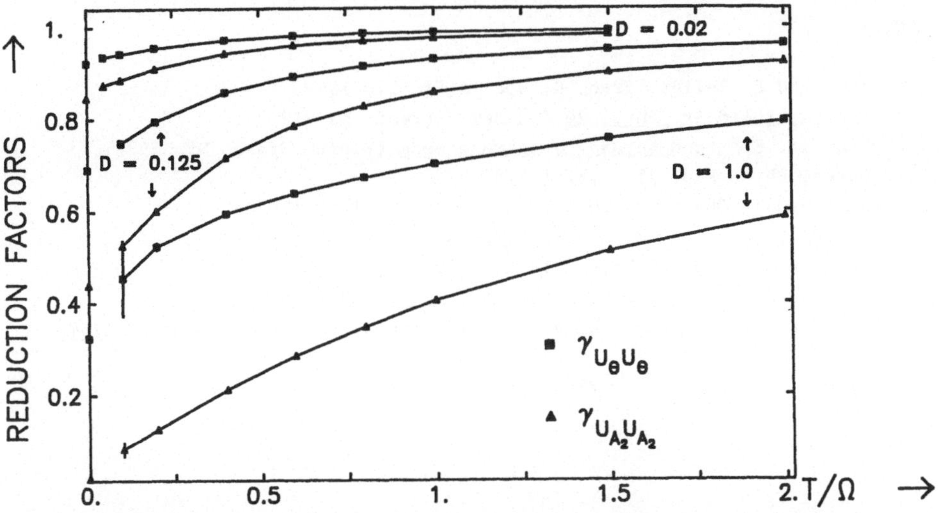

Fig. 1. The reduction factors γ_{AA} ($A=U_\theta$, A_2) as a function of temperature T/Ω for coupling D=0.02, 0.125 and 1.0. Exact results [3,9] for T/Ω=0 are also shown. Error bars not shown have a size smaller than the symbols (average error less than 1%). Drawn lines are guides to the eyes.

contributions to (2.5), or equivalently to the denominator of (2.8), depends on the temperature and the coupling strength; the lower (larger) the temperature (coupling) the larger the relative number of negative contributions. We give some data for D=0.125 and D=1: i) T/Ω=0.2: 0.02% and 9% respectively, ii) T/Ω=0.1: 5% and 32%. At these temperatures we find no negative contributions for D=0.02. The occurrence of negative contributions gives large fluctuations, and in fact frustrates calculations for very low temperatures (or very high coupling values).

We see in Fig. 1 that the susceptibilities are substantially reduced at lower temperatures. At T=0 it can be shown easily that the reduction factors, γ, are equal to the square of the appropriate Ham factors q and p, which are known for T=0 [3,9]. The susceptibilities deviate greatly from the Curie-behavior, the reduction factors show a strong temperature dependence. The m=1 approximants, relevant for the limit of $T/\Omega\to\infty$, can be calculated analytically. We find that the susceptibilities show a Curie-Weiss behavior in the limit $T/\Omega\to\infty$: $\chi_{U_\theta U_\theta}\simeq 1/(T/\Omega+2D/3)$ and $\chi_{A_2A_2}\simeq 1/(T/\Omega+4D/3)$.

Incorporating quadratic and anharmonic coupling terms [2-4] in our numerical procedure is quite straightforward. It is also easy to apply this procedure on a cluster model using two pairs of vibrational modes with frequencies in the optic and acoustic regime [10].

This work is part of the research program of the Dutch "Stichting voor Fundamenteel Onderzoek der Materie (FOM)", which is financially supported by the "Nederlandse Organisatie voor Zuiver Wetenschappelijk Onderzoek (ZWO)". HDR wishes to thank the Belgian National Science Foundation (NFWO) for financial support.

REFERENCES

1. H.A. Jahn and E. Teller, Proc. R. Soc. A161, 220 1937.
2. M.D. Sturge, Solid St. Phys. 20 (Academic Press, NY, 1967).
3. F.S. Ham, in: Electron paramagnetic resonance (Plenum Press, NY, 1972).
4. C.A. Bates, Phys. Rep. 35C, 187 (1978).
5. K. Sasaki and Y. Obata, J. Phys. Soc. Japan 28, 1157 (1970).
6. H.F. Trotter, Proc. Am. Math. Soc. 10, 545 (1959).
7. M. Suzuki, Commun. Math. Phys. 51, 183 (1976).
8. H. De Raedt and A. Lagendijk, Phys. Rep. 127C, 233 (1985).
9. M.S. Child and H.C. Longuet-Higgins, Phil. Trans. Roy. Soc. (London) A254, 259 (1962).
10. B. Halperin and R. Englman, Phys. Rev. B9, 2264 (1974).

Part II

Random Field Problems, Spin Glasses and Chaotic Behaviour

Critical Properties of the Random Field Ising Model

J. Villain[1], *T. Nattermann*[1], *and M. Schwartz*[2]

[1]Institut für Festkörperforschung der Kernforschungsanlage Jülich, Postfach 1913, D-5170 Jülich, Fed. Rep. of Germany
[2]School of Physics and Astronomy, Tel Aviv University, Tel Aviv 69978, Israel

1. Introduction

The purpose of this note is to summarize the ideas of theorists, in June 1987, about what the critical behaviour of the random field Ising model (RFIM) /1,2,3/ might be. A list of possibly exact relations between critical exponents will be given, together with approximate ones which allow a direct comparison with experiment.

"Possibly exact" relations are indeed the only results that theory is able to provide, because they rely on certain assumptions or approximations which will be summarized below. In order to facilitate comparison with experiments in magnetic systems /4-8/, antiferromagnetic interactions will be assumed in contrast with most of the existing theoretical literature. The basic assumptions are listed below.

i) The existence of long-range antiferromagnetic order in dimension $d > 2$, and in particular in $d = 3$, below some transition temperature T_c and if the random field amplitude H is sufficiently smaller than the typical exchange interaction $|J|$. The existence of order in $d = 3$ at $T = 0$ has been derived exactly /9/. At $T \neq 0$ it follows from approximate arguments /10,11/, which also show the absence of order in $d \leq 2$. There are also theories /12-14/, which predict no order for $d = 3$ and $T \neq 0$. In our opinion, they have been refuted /10,11,15,16,17/.

ii) The existence of a continuous phase transition, instead of a first order one as suggested by Monte Carlo simulations /18/, high-temperature expansions /19/, the mean field approximation in certain cases /20/, and some experimental data /5/. Other simulations /21/ are consistent with a continuous transition and experimental data may be blurred by metastability /22/. A crude renormalization group theory /23-25/ described in Section 4 is consistent with a continuous transition too.

iii) Scaling assumptions of the standard sort /23,26,27/ or of a bolder type, yield the equalities listed in the next Section.

2. Critical exponents

The critical behaviour of the random field Ising model is described by the usual exponents α, β, γ, δ, η, ν which have their usual definition. However, it is necessary to introduce an additional exponent $\bar{\eta}$ defined by

$$\overline{\langle m(r')\rangle\langle m(r'+r)\rangle} \sim 1/r^{d-4+\bar{\eta}} \qquad (T = T_c \text{ or } r < \xi), \qquad (1)$$

where brackets denote a thermal average, and the bar an ensemble average. One can also introduce an exponent $\Theta = (d/2)-\sigma$ which describes the renormalisation of the random field. The most physical definition may be that the typical relaxation time τ diverges just above T_c /1/ as $\exp(H\xi^{\Theta} \times \text{Const})$ where ξ is the correlation length, d the space dimensionality, and H the random field amplitude.

The exponents are not independent, but are related by the usual relations

$$\alpha + 2\beta + \gamma = 2, \quad (2) \qquad \gamma = (2 - \eta)\nu. \quad (3)$$

Exponents ρ /23/, Θ /27/, Γ /30/, Y_J, Y_h, Δ /25/ have been introduced by various authors and are related to the present ones by

$$\overline{\eta} = 4-\Gamma = \rho + 2, \qquad Y_J = \Theta = (d/2)-\sigma = (d/2)-\Delta \; , \qquad y_h = d-\beta/\nu \; .$$

Usual "hyperscaling" relations (those which depend on d) do not hold. This is readily seen if one remembers that the upper critical dimension is $d_{cu} = 6$ /28,1,2/. At d = 6 one has the mean field values $\alpha=0$ and $\nu=1/2$ so that $2-\alpha \neq d\nu$ in contrast with the usual hyperscaling relation. Alternative hyperscaling relations will now be proposed.

3. Hyperscaling relations

Using standard scaling assumptions /25,29/, the following relations can be obtained /23,24,27/:

$$\eta = \frac{d}{2} - 2 + \overline{\eta} - \sigma, \quad (4) \qquad\qquad \alpha = 2 - (\frac{d}{2} + \sigma)\nu, \quad (5)$$

$$\delta = (4 + 2\sigma - \overline{\eta})/(d - 4 + \overline{\eta}). \quad (6)$$

In addition to these equations, several authors have suggested the following relations which are equivalent if formulae (2) to (6) hold:

$$\overline{\eta} = 2\eta, \; (7) \qquad 2\sigma = d - 4 + \overline{\eta}, \; (8) \qquad 2 - \alpha = (d - 2 + \eta)\nu. \quad (9)$$

Let the derivation of (7) by Schwartz and Soffer /29/ be summarized. They write

$$\langle m_q \rangle = \chi_q H_q + \psi_q \; , \tag{10}$$

where H_q is the Fourier transform of the random field and

$$\overline{\psi_q H_q} = 0 \; . \tag{11}$$

These relations are just a definition of χ_q which of course depends on H. It can be shown /30/ that χ_q is the differential susceptibility

$$\chi_q = \beta[\; \overline{\langle m_{-q} m_q \rangle} - \overline{\langle m_{-q} \rangle \langle m_q \rangle} \;] \; . \tag{12}$$

Then Schwartz and Soffer assume $\psi_q = \sum_{kk'} A_{kk'} H_k H_{k'} H_{q-k-k'}$ and show that this assumption implies that ψ_q is negligible in (10). Relation (7) follows from (1), (10) and (12).

Thus, Schwartz and Soffer have assumed that the response (11) may be expanded in power of H_q and have shown that their assumption is self-consistent. Self-consistency, however, is no guarantee for correctness. Relations (7) to (9) have also been derived from an equivalent annealing method /31/ and also come out /25/ if one assumes that the random field scales as the staggered magnetisation. Then, the effective random field in a volume L^d should be $H\, L^{-\sigma} \sim \overline{\langle m(0)\; m(L) \rangle}^{1/2}$ and insertion of (1) yields (8).

We close this Section by recalling that certain exact relations /30/ imply, under the only assumption of a continuous transition /23/

$$2\eta > \overline{\eta} \; . \tag{13}$$

4. Renormalisation group theory for ν in isotropic systems

In view of relations (3) to (8) there are 2 independent exponents, e.g. η and ν. ν can be calculated by a renormalisation group method. The principle is to obtain a renormalisation equation for the surface tension g(L) of a domain wall, assumed to depend on the length scale L. This equation will first be given without proof and its consequences will be derived. For small H,g and 1/L one finds

$$\frac{dg}{d(\ln L)} = - H^{\lambda} g^{-\overline{\mu}} L^{-1/\nu}, \qquad (14)$$

where λ, μ, ν are some exponents. We have not shown yet that ν is the usual correlation length exponent but this will follow from the forthcoming proof. Integration of (14) yields $g^{\overline{\mu}+1}(L) - H^{\lambda} L^{-1/\nu} = f(H,T)$. The correlation length ξ (beyond which g does not depend on L below T_c) is given by $\xi^{-1/\nu} = H^{-\lambda} f(H,T)$. The integration constant f(H,T) is a microscopic characteristic and therefore should be analytic in T. Moreover it should vanish at T_c. Thus, $f(H,T) \sim T_c - T$. It follows that ξ diverges at $(T_c - T)^{-\nu}$.

The derivation of (14) is explained in references /23-24/ : $\delta g = dg/d(\ln L)$ can be interpreted as the decrease of g due to bumps of radius about L. The additional energy $\delta W \sim L^{d-1}\,\delta g$ due to creating such a bump consists of two parts: i) an additional surface energy, $W_s = g(L)\delta S$, where $\delta S \approx (h/L)^2 L^{d-1}$ is the change in the surface area of the domain due to creating a bump, and h is the bump height. ii) an energy gain $\delta W_m \approx H(L^{d-1} h^{1-\sigma})^{1/2}$ due to random fields. Here the random field at a distance h from the base of the bump is replaced by an effective field $Hh^{-\sigma}$ in contrast with low-temperature calculations /1,11,12/. Thus

$$\delta W = L^{d-1}\delta g = L^{d-1} dg/d(\ln L) = g(L)\,\frac{h^2}{L^2}\,L^{d-1} - H\sqrt{L^{d-1}\,h^{1-2\sigma}}\,. \qquad (15)$$

The L-dependent surface tension g(L) includes the contribution of degrees of freedom with lengthscale smaller than L. Minimisation of (15) with respect to h yields

$$h \approx (H/g)^{\frac{2}{3+2\sigma}} L^{\frac{5-d}{3+2\sigma}} = h_L \qquad (16)$$

and insertion into (15) yields the free energy per unit area due to fluctuations of length L, namely

$$dg/d(\ln L) \approx -H(H^4/g)^{\frac{1-2\sigma}{3+2\sigma}} L^{\frac{4-2d-4\sigma}{3+2\sigma}}\,. \qquad (17)$$

Identification with (14) yields

$$\frac{1}{\nu} = 2\,\frac{d - 2 + 2\sigma}{3 + 2\sigma}\,. \qquad (18)$$

This result is probably only approximate because the derivation neglects the interaction between different degrees of freedom, free energies are replaced by their root mean square averages, etc. For instance, for d = 5, relation (18) yields $\nu = 1/2$, a relation which should only hold for $d \geq 6$. Well, the difference between 5 and 6 is not that big, and this comparison may even be regarded a favourable test for formula (18).

Bray and Moore /24/ obtain a formula rather different from (18): $1/\nu = d-2$. The contradiction results from the fact that they assume, in contradiction with (15), that domain walls are anisotropic and sensitive to lattice discreteness. Of course, they <u>are</u>. However, it is expected that critical properties are correctly described by a Landau-Ginzburg free energy functional, which is a continuum approximation in which domain walls are isotropic. In fact, even at low temperature, it has been shown by Nattermann /32/ that domain walls can be represented by continuous models in the long wavelength limit even at <u>low</u> temperature for $d \leq 3$.

5. The limit d = 2.

Although there is no order in d = 2, there is in 2 + ε. The limit of σ for $\varepsilon = 0$ appears to be

$$\sigma(d = 2) = 0 . \tag{19}$$

This relation follows, for instance, from inequalities /23/. It implies the absence of domains within domains, and this property implies

$$\bar{\eta} + d - 4 = 0 \qquad (d \to 2) . \tag{20}$$

This follows also from (7), (4) and (19).

According to Bray and Moore (19) and (20) should hold to all orders in ε=d-2, though non-analytic-corrections as exp[-1/(d-2)] are allowed.

Formulae (19) and (18) predict $\nu = \infty$ in 2 dimensions, in agreement with Nattermann /25/ and Bray and Moore /24/.

6. The message to three-dimensional experimentalists

In this last Section we collect the values of the critical exponents which can be expected from the above arguments in d = 3, combined with simulations from Ogielski and Huse /21/.

The exponent σ is equal to 1 for d = 6 according to (8) and to 0 for d = 2 according to (19). From the plausible assumption that σ varies smoothly we expect σ to be rather small in d = 3. This is in agreement with numerical estimates of Ogielski and Huse /21/ who find $\eta = 0.5 \pm 0.1$, because this implies $\sigma < 0.1$ according to (7) and (8). Note that Cheung /34/ finds $\sigma = 0 \pm 0.2$. Since σ reflects the effect of domains within domains we expect $\sigma > 0$ or

$$\sigma = 0.05 \pm 0.05. \tag{21}$$

Then (18) yields $\nu = 1.4 \pm 0.1$.

Other exponents may be derived from relations (2) to (9). $\alpha = -0.2 \pm 0.07$, $\gamma \simeq 2$, $\beta \simeq 0.05$, $\bar{\eta} = 1.1 \pm 0.1$. These values are rather close to those of Cheung /34/,/37/.

Anglés d'Auriac and Rammal /33/, using ground state simulations find the same values of β, γ, but rather different values for η and σ. Their results ($\alpha = \beta = 0$, $\Theta = \nu = \gamma = 2$) are consistent with (2) and (5) but inconsistent with (9) if (8) is accepted. More surprisingly perhaps, their value $\sigma = (d/2) - \Theta = -1/2$ is negative, while domains within domains would be naively expected to renormalize the random field to lower values, thus yielding $\sigma > 0$. Remember that σ=1 for d = 6.

Experimental measurements are very difficult for various reasons: i) the range of validity of the random-field-dominated scaling laws is narrow in weak fields. Above T_c /22/ zero-field scaling laws are valid so long as $\xi < (J/H)^{2/(2-\eta)}$. ii) Random-field effects are blurred with random exchange effects, especially in systems of the Fishman-Aharony type, on which most of systematic studies have been performed. iii) Metastability effects are difficult to overcome. Some critical exponents may be impossible to measure. For instance, η is related, if the Villain-Fisher /22,23,27/ picture is correct, to reversal of large blocks which require a very long time. Since the amplitude of the corresponding fluctuations $\overline{\langle | S_k |^2 \rangle}$ is small compared with that of $\overline{|\langle S_k \rangle|^2}$ (which corresponds to $\bar{\eta}$), η is hardly accessible to experiment. It has also been pointed out by Wong /4/ that indirect measurements of the specific heat (e.g. by birefringence /7/) might be unable to give the equilibrium value. Wong finds $\alpha = -1$, lower than any theoretical expectation.

Since equilibrium is so difficult to reach, the most reasonable idea is to measure dynamical quantities like the dynamical susceptibility /35/. Unfortunately, existing data do not allow to distinguish between the Villain-Fisher picture (with a Vogel-Fulcher-like law for the relaxation time) and the standard picture with a power law.

References

/ 1/ Y. Imry, S. Ma; Phys. Rev. Lett. 35, 1399 (1975)
/ 2/ A.P. Young; J. Phys. C 10: L 257 (1977)
/ 3/ J. Villain in Scaling Phenomena in Disordered Systems, ed. by R. Pynn and A.T. Skjeltorp (Plenum Press, New York, 1986)
/ 4/ Po-zen Wong; Phys. Rev. B 34, 1864 (1986) and references therein .
/ 5/ R.J. Birgeneau, Y. Shapiro, G. Shirane, R.A. Cowley and H. Yoshizawa; Physica 37 B and C, 83 (1986) and references therein.
/ 6/ A.R. King, I.B. Ferreira, V. Jaccarino, D.P. Belanger; Phys. Rev. B (submitted) and references therein.
/ 7/ D.P. Belanger, A.R. King, V. Jaccarino, Phys. Rev. B 31, 4538 (1985)
/ 8/ H. Ikeda, K. Kikuta; J. Phys. C 17, 1221 (1984)
/ 9/ J.Z. Imbrie, Phys. Rev. Lett. 53, 1747 (1984). Commun. Math. Phys. 98, 145 (1985)
/10/ G. Grinstein, S.K. Ma; Phys. Rev. Lett 49, 685 (1982) Phys. Rev. B 28, 2588 (1983)
/11/ J. Villain, J. Physique Lett. 43, 808 (1982); J. Villain, B. Séméria, F. Lançon, L. Billard, J. Phys. C 16, 6153 (1983)
/12/ D. Mukamel, E. Pytte, Phys. Rev. B 24, 6736 (1981)
/13/ H.S. Kogon, D.J. Wallace, J. Phys. A 14, L 527 (1981)
/14/ U. Krey, J. Phys. C 18, 1455 (1985)
/15/ A. Engel, J. Physique Lett. L 409 (1985)
/16/ V. Janiš, Phys. Stat. Sol. (b) 138, 539 (1986)
/17/ D.S. Fisher, Phys. Rev. B 31, 7233 (1985)
/18/ A.P. Young,M. Nauenberg, Phys. Rev. Lett. 54, 2429 (1985)
/19/ A. Houghton, A. Khurana, F.J. Seco, Phys. Rev. Lett. 55, 856 (1985)
/20/ A. Aharony, Phys. Rev. B 18, 3318 (1978)
/21/ A.T. Ogielski, D.A. Huse, Phys. Rev. Lett. 56, 1298 (1986)
/22/ J.Villain, Phys.Rev.Lett. 52, 1543 (1984) and in "Elementary excitations and fluctuations in magnetic systems" ed.by S.W.Lovesey, V.Tognetti and U.Balucani Springer, Heidelberg, 1985) p.142.
/23/ J. Villain, J. Physique 46, 1843 (1985)
/24/ A.J. Bray, M.A. Moore, J. Phys. C, L 927 (1985)
/25/ T. Nattermann, Phys. Stat. Sol. 131, 563 (1985)
/26/ G. Grinstein, Phys. Rev. Lett. 37: 944 (1976)
/27/ D.S. Fisher, Phys. Rev. Lett. 56, 416 (1986)
/28/ A. Aharony, Y. Imry and S.K. Ma, Phys. Rev. Lett. 37: 1364 (1976)
/29/ M. Schwartz, A. Soffer, Phys. Rev. B 33, 2059 (1986)
/30/ M. Schwartz, A. Soffer, Phys. Rev. Lett. 55, 2499 (1985)
/31/ M. Schwartz, J. Phys. C 18, 1455 (1985)
/32/ T. Nattermann, Phys. Stat. Sol. 132, 125 (1985)
/33/ J.C. Anglés d'Auriac, R. Rammal, preprint (1987)
/34/ H.F. Cheung, Phys. Rev. B33 6191 (1986)
/35/ A.R. King, J.A. Mydosh, V.Jaccarino, Phys. Rev. Lett. 56, 2525 (1986)
/36/ In order to derive (19) one needs the inequality $\sigma > 0$, which is questioned by some authors /33/. However, this inequality should hold for $d = 2$. More generally /25,27/, $\sigma > 1-d/2$.
/37/ The value $\overline{\eta} \cong 1$ is consistent with the squared lorentzian lineshape observed experimentally.

Random-Exchange and Random-Field *xy*-Chain for $S=1/2$ *

T. Schneider and A. Politi

IBM Research Division, Zurich Research Laboratory,
CH-8803 Rüschlikon, Switzerland

1. Introduction

Although spin-glass and random-field behavior has been observed in a wide variety of different systems, the identification of universal properties is still disputed [1]. Adopting a quantum statistical point of view, the universal features are hidden in the density of states and the localization properties of the wave functions. In this work, we investigate density of states and localization length of wave functions in an $s=1/2$ xy-chain for random exchange and random field. Using the transformation to spinless fermions [2], the problem is then reduced to a nonlinear map, providing accurate estimates for integrated density of states and exponential localization length of the wave functions. The thermodynamics can be obtained from the density of states since the spinless fermions are free. The results clearly reveal: (i) The appearance of disorder-induced exponential tails in the integrated density of states at the bottom and top of the spectrum. These tails lead to a characteristic field dependence of the zero-temperature magnetization and susceptibility. (ii) Important differences between the random-exchange and random-field models. In the random-exchange case, the state in the middle of the band is found to be extended and the zero-field susceptibility diverges, while in the random-field case, corresponding to the Anderson model [3], all states are exponentially localized and the zero-field susceptibility remains finite. This difference also affects the leading temperature dependence of the zero-field susceptibility and specific heat as $T \to 0$.

The one-dimensional xy-model for $s=1/2$ spins is described by the Hamiltonian

$$\mathscr{H} = \sum_{j=1}^{N-1} \left(2J_j \left(S_j^x S_{j+1}^x + S_j^y S_{j+1}^y\right)\right) - \sum_{j=1}^{N} \left(h_j S_j^z - h S_j^z\right) . \tag{1}$$

We assume J_j and h_j to be independent random variables, while h denotes the external homogeneous field. As in the pure case, Hamiltonian (1) can be transformed to one of noninteracting spinless fermions [2]

$$\mathscr{H} = \frac{1}{2} \sum_{j=1}^{N} h_j + \sum_{m,n} C_m^+ A_{mn} C_n + \frac{1}{2} hN , \quad \text{where} \tag{2}$$

$$A_{mn} = \begin{cases} -h_m - h & \text{for } m = n \\ J_m & \text{for } n = m+1, m = n+1 \\ 0 & \text{otherwise} \end{cases} \tag{3}$$

* Short version of the paper presented at the 31st Annual Conference on Magnetism and Magnetic Materials, Baltimore, Maryland, November 1986 and published in J. Appl. Phys., April 1987.

is a real, random and symmetric matrix. C_m and C_m^+ denote fermion annihilation and creation operators. Thus the problem can be reduced to finding the eigenvalues ω and eigenvectors f of the recursion relation

$$J_{l-1} f_{l-1} - (h_l - \omega) f_l + J_l f_{l+1} = 0 \tag{4}$$

which can be reduced to

$$J_{l-1} \frac{1}{R_l} + J_l R_{l+1} = -\omega + h_l, \quad R_{l+1} = \frac{f_{l+1}}{f_l} . \tag{5}$$

Invoking boundary conditions $f_N = 1$, $f_{N+1} = 0$ and the solution $f_n(\omega)$, the characteristic function of map (5) is

$$\begin{aligned} \Gamma(\Omega) &= -\lim_{N\to\infty} \frac{1}{N} \ln f_{n=0}(\Omega) \\ &= -\lim_{N\to\infty} \left(-\frac{1}{N} \sum_{n=0}^{N-1} \ln J_n + \frac{1}{N} \sum_{n=0}^{N-1} \ln \overline{R}_n \right) , \quad \text{where} \end{aligned} \tag{6}$$

$$\overline{R}_n = J_{n-1} R_n . \tag{7}$$

Thus, it is natural to express map (5) in terms of $\overline{R}_n$'s, yielding

$$J_{n-1}^2 \frac{1}{\overline{R}_n} + \overline{R}_{n+1} = h_n - \omega , \tag{8}$$

to evaluate $\Gamma(\Omega)$. Setting $\Omega = -\omega + i0^+$, from (6) we obtain

$$\begin{aligned} \mathrm{Re}\Gamma(\Omega) = \gamma'(\omega) &= \lim_{N\to\infty} \left(-\frac{1}{N} \sum_{n=0}^{N-1} \ln |J_n| + \frac{1}{N} \sum_{n=0}^{N-1} \ln |\overline{R}_n| \right) \\ \mathrm{Im}\Gamma(\Omega) &= \pi\gamma''(\omega) = \pi N(\omega) . \end{aligned} \tag{9}$$

$\gamma'(\omega)$ is the inverse exponential localization length, and $N(\omega)$ the integrated density of states defined by

$$N(\omega) = \int_{-\infty}^{\omega} \rho(x) dx . \tag{10}$$

$\rho(\omega)$ denotes the density of states and $N(\omega)$ corresponds to the density of negative $\overline{R}_n$'s, i.e., the density of nodes.

From the expression for the free energy,

$$F = \frac{1}{2} h + \lim_{N\to\infty} \frac{1}{2N} \sum_{j=1}^{N} h_j - \frac{1}{\beta} \int_{-\infty}^{+\infty} \rho(\omega) \ln(1 + \exp - \beta(\omega - h)) d\omega , \tag{11}$$

it is seen that the thermodynamic properties are fully determined by the density of states. Moreover, magnetic field h plays the role of the Fermi level. The field dependence of zero-temperature magnetization M and susceptibility χ are then given by

$$M(h, T=0) = -\frac{dF}{dh}\Big|_{T=0} = -\frac{1}{2} + N(h)$$
$$\chi(h, T=0) = \frac{dM(h,T=0)}{dh} = \rho(h) \;, \tag{12}$$

and for the temperature dependence of the zero-field specific heat and susceptibility one finds

$$\left.\begin{aligned} C(h=0,T) &= \frac{\beta^2}{4} \int d\omega\; \omega^2 \rho(\omega) \left(\cosh \frac{\beta\omega}{2}\right)^{-2} \\ \chi(h=0,T) &= \frac{\beta}{4} \int d\omega\; \rho(\omega) \left(\cosh \frac{\beta\omega}{2}\right)^{-2} . \end{aligned}\right\} \tag{13}$$

We are now prepared to discuss random-field and random-exchange models.

2. Random-Field Model

Setting $J_l = J = 1$, from (5) we obtain the random-field model. Here, recursion relation (8) reduces to

$$\frac{1}{R_n} + R_{n+1} = h_n - \omega \;, \tag{14}$$

because $\overline{R}_n = R_n$ for $J_n = J = 1$. It is identical to the nonlinear map resulting from the standard Anderson model for localization [3]. Thus, we have established the connection between the Anderson and the s=1/2 xy-models in a random field. Numerical results for inverse localization length and integrated density of states are shown in Fig. 1, for h_l equally distributed between $\pm\Delta$. These results were obtained from a numerical solution of (14) by invoking (9). The spectrum is bounded by $-2-\Delta \leq \omega \leq 2+\Delta$. Moreover, because matrix A is symmetric, $\gamma'(\omega) = \gamma'(-\omega)$ and $N(\omega) = 1 - N(-\omega)$. From Fig. 1, it is also seen that strong exponential localization sets in by approaching the bottom and top of the spectrum. Here, the integrated density of states exhibits Lifshitz-tail behavior [4,5],

$$N(\omega) \sim \exp\left(-C\,\frac{|\ln(2+\Delta+\omega)|}{(2+\Delta+\omega)^{1/2}}\right) \qquad : \omega \geq -2+\Delta \;, \tag{15}$$

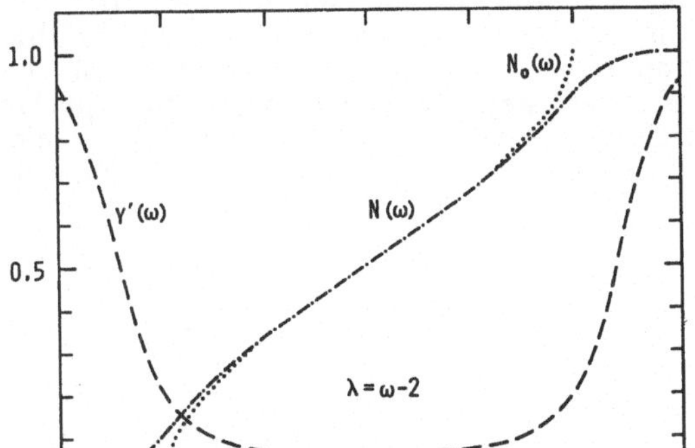

Fig. 1. Numerical results for inverse localization length $\gamma'(\omega)$ and integrated density of states $N(\omega)$ for the random-field model for $\Delta = 1$ and $N = 10^5$. $N_0(\omega)$ is the integrated density of the pure system

and the corresponding expression for $\omega \lesssim 2+\Delta$. As seen in Fig. 1, these tails differ markedly from the power-law behavior in the pure system ($h_l = 0$), where

$$N_0(\omega) = \frac{1}{2} + \frac{1}{\pi} \sin^{-1}(\omega/2) \tag{16}$$

so that

$$N_0(\omega) = \begin{cases} 1/2 + \frac{1}{2\pi}\,\omega & \omega \to 0 \\ 1 - \frac{1}{\pi}\,(2-\omega)^{1/2} & \omega \to 2 \\ \frac{1}{\pi}\,(2+\omega)^{1/2} & \omega \to -2\ . \end{cases} \tag{17}$$

Accordingly, the disorder-induced Lifshitz tails lead to dramatic modifications in the field dependence of magnetization and susceptibility. In the pure system, M and χ are given by [Eqs. (12) and (16)]

$$M(h, T=0) = \begin{cases} -1/2 & :h \geq -2 \\ 1/2 + \frac{1}{\pi} \sin^{-1}(h/2) & :-2 < h < 2 \\ 1/2 & :h > 2 \end{cases} \quad \text{and} \tag{18}$$

$$\chi(h,T=0) = \frac{1}{\pi}\left(1 - \left(\frac{h}{2}\right)^2\right)^{-1/2}, \quad |h| \lesssim 2\ . \tag{19}$$

In the presence of a random field, from (12) and (15) we obtain

$$M(h,T=0) = \ -1/2 + \ldots \exp\left(-C\,\frac{|\ln(2+\Delta+h)|}{(2+\Delta+h)^{1/2}}\right) \quad :h \gtrsim -2-\Delta\ , \tag{20}$$

the corresponding expression for $h \lesssim 2+\Delta$ and

$$\chi(h,T=0) \sim \exp\left(-C\,\frac{|\ln(2+\Delta-|h|)|}{(2+\Delta-|h|)^{1/2}}\right) \tag{21}$$

for $|h| \to 2+\Delta$. Thus, by approaching the top or bottom of the spectrum ($|h| = 2+\Delta$), the power-law behavior of the pure system is replaced by the exponential tail. In particular, the random field removes the divergence of the susceptibility at the top and bottom. At zero field and low temperatures, the specific heat and susceptibility become dominated by $\rho(\omega)$ around $\omega = 0$. Because $\rho(\omega) \to$ const for $\omega \to 0$ (Fig. 1), as in the pure case, the leading asymptotic laws are given by [Eq. (13)]

$$\begin{aligned} C(h=0,T) &\sim T \\ \chi(h=0,T) &\to \text{const}\ . \end{aligned} \tag{22}$$

3. Random-Exchange Model

For $h_j = 0$, Hamiltonian (1) reduces to a random-exchange model, and map (8) reads

$$J_{n-1}^2\,\frac{1}{\overline{R}_n} + \overline{R}_{n+1} = -\omega\ . \tag{23}$$

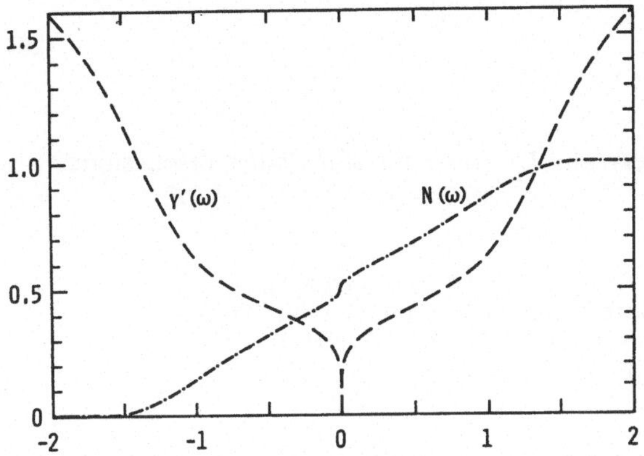

Fig. 2. Numerical results for $\gamma'(\omega)$ and $N(\omega)$ for the random-exchange model for $\Delta = 1$ and $N = 2.10^6$

In contrast to the random-field case [Eq. (14)], randomness enters here as multiplicative noise. Rectangular distributions with $-\Delta \leq J_l \leq \Delta$ are of particular interest in the context of spin glasses. Numerical results for inverse exponential localization length $\gamma'(\omega)$ and integrated density of states $N(\omega)$ as obtained from (9) and (23) are shown in Fig. 2 for $\Delta = 1$. The spectrum is bounded by $-2\Delta \leq \omega \leq 2\Delta$.

Again $\gamma'(\omega) = \gamma'(-\omega)$ and $N(\omega) = 1 - N(-\omega)$, because matrix A is symmetric. Strong localization occurs at the bottom and top of the spectrum, while the zero-energy state is not exponentially localized. In fact, for $\omega = 0$ the leading asymptotic behavior is [6]

$$\gamma'(\omega) = \frac{1}{|\ln\omega|}$$
$$N(\omega) = 1/2 + \frac{1}{|\ln\omega|^2} \, . \tag{24}$$

The appearance of the extended state in the middle of the band differs markedly from the random-field model, where all states are exponentially localized (Fig. 1). Moreover, slow variation of the integrated density of states near $\omega \sim \pm 2\Delta$ also suggests the existence of exponential tails. In fact, Lifshitz behavior [Eq. (15)] also occurs in this model, even though randomness enters as multiplicative noise. As in the random-field model, the disorder-induced Lifshitz tail leads to a characteristic field dependence of magnetization and susceptibility. Here, (20) and (21) also apply. Clearly, $2 + \Delta$ must be replaced by Δ.

Moreover, the appearance of the extended state in the middle of the band will also affect the field dependence of magnetization and susceptibility in terms of the singular behavior of $N(\omega)$ around $\omega = 0$. From (12) and (24), we obtain for $|h| \to 0$,

$$\left.\begin{aligned} M(h,T=0) &= |\ln|h||^2 \\ \chi(h,T=0) &= 2\left(|h|\,|\ln|h||^3\right)^{-1} . \end{aligned}\right\} \tag{25}$$

In the pure system and random-field case, $\chi(h=0, T=0)$ remains finite for $h \to 0$ (Fig. 1). Thus, divergence of the zero-frequency susceptibility appears to be a characteristic feature of the random-exchange model. This singularity also affects the leading temperature dependence of zero-field susceptibility and specific heat. From (13) and (24), we obtain

$$\left.\begin{aligned} C(h=0,T) &\sim |\ln T|^{-3} \\ \chi(h=0,T) &\sim \left(T|\ln T|^3\right)^{-1} . \end{aligned}\right\} \quad (26)$$

It is a pleasure to acknowledge V. Emery and D. Würtz for stimulating discussions.

1. See: Proceedings of the Heidelberg Colloquium on Spin Glasses, ed. by J. L. Van Hemmen and I. Morgenstern (Springer, Berlin 1983)
2. E. Lieb, T. Schultz, D. C. Mattis: Ann. Phys. (NY) **16**, 407 (1961)
3. D. J. Thouless: Phys. Rep. **13**, 93 (1974)
4. B. Simon: J. Stat. Phys. **38**, 65 (1985)
5. A. Politi and T. Schneider: unpublished.
6. T. P. Eggarter, R. Riedinger: Phys. Rev. B **18**, 569 (1978)

Spin-glass-like Behaviour in Spinels

E. Agostinelli and D. Fiorani

I.T.S.E.-C.N.R.-Area della Ricerca di Roma, P.B. 10,
I-00016 Monterotondo Stazione, Italy

Experimental evidence of a spin-glass-like behaviour is reported for two systems with a spinel structure: MeN_2X_4 (Me=Co,Fe; N=Al,Ga,In; X=O,S) and $Zn_xCd_{1-x}Cr_2S_4$.

1 INTRODUCTION

In the field of random magnetism, spin-glasses continue to be the object of considerable interest from both a theoretical and an experimental point of view /1-3/.

However, while the behaviour of dilute metallic alloys is nowadays well understood and their critical properties can be classified within universality classes/4/, some questions still remain open for insulating systems, whose properties are found to vary greatly from one system to another and with the concentration within the same system.

Within this framework, systems with a spinel structure seem to be particularly attractive, as different degrees and types of disorder and frustration may be present. This is due to the fact that in the spinel lattice two crystallographic sublattices, with tetrahedral(A) and octahedral(B) sites, are available for the magnetic ions. As the cations can have different site preferences, by choosing the appropriate atoms it is possible to realize a selective magnetic dilution in one of the two sublattices or a distribution, to a different extent, between the two sublattices. This fact, leading to different degrees of disorder, is responsible for the large variety of magnetic behaviours exhibited by spinels.

For antiferromagnetic interactions only, a qualitative magnetic phase diagram was proposed/5,6/ as a function of the fraction of magnetic ions in the two sublattices. A spin-glass region was predicted over a wide concentration range. This includes a concentration region within the magnetically diluted B-sublattice, which has a topological frustration/7/, and a concentration region with magnetic dilution in both sublattices, where the frustration is due to the concomitant occurrence of A-A, B-B abd A-B interactions. Canted structures are possible in the last concentration range/2,5,8/.

In this paper we shall discuss two additional examples of spin-glass behaviour in spinels. The first one deals with a series of spinels where the magnetic ions are distributed between both sublattices and ferromagnetic(F) and antiferromagnetic(AF) interactions coexist: $MeAl_2O_4$, $MeGa_2O_4$ and $MeIn_2S_4$ (Me=Co,Fe). The second one deals with a very particular case, the non-magnetically diluted spinel $Zn_xCd_{1-x}Cr_2S_4$. In this system the chromium ions occupy only the octahedral sites and the frustration arises from the competition between nn F and nnn AF interactions. The substitutional disorder involves only the non-magnetic ions that occupy the tetrahedral sublattice.

2 EXPERIMENTAL

The samples preparation will be reported elsewhere/9/. X-ray diffraction showed that all the samples are composed of a single phase with spinel structure. The AC susceptibility measurements were performed at ν=198 Hz using a Mutual Inductance Bridge and the DC susceptibility measurements by using a Faraday Balance (H=530 Oe).

3 RESULTS AND DISCUSSION

I - Spinels with magnetic ions distributed on both sublattices.

Percolation curves have been calculated including A-B bonds only/10/ (the A-B interactions are the most important ones) as well as the overall bonds A-A, B-B and A-B/8/. For all our cobalt and iron spinels, $MeAl_2O_4$, $MeGa_2O_4$ and $MeIn_2S_4$(Me=Co,Fe), the relative concentration of magnetic ions between the two sublattices is located above the percolation threshold calculated for A-B bonds.

$CoAl_2O_4/FeAl_2O_4$

In $CoAl_2O_4$ the Co ions occupy almost exclusively the A sites, so predominant AF interactions (J_{AA}) are expected, leading to an AF order as predicted in the magnetic phase diagram/6/. Neutron diffraction experiments/9/ have shown that magnetic reflections appear at T_N=4K. The magnetic structure has been found to be similar to that of Co_3O_4, in agreement with previously published results. However, the magnetic order does not appear completely established and it is accompanied by some short-range order. This fact may be explained as due to the presence of a non detectable degree of inversion in this system.

For $FeAl_2O_4$, where 85% Fe ions are in A sites,a maximum in $\chi_{a.c.}$ was observed at T=13,2±0.5K

$CoGa_2O_4/FeGa_2O_4$

In these spinels 40% Co and 50% Fe ions occupy the A sites, so determining the simultaneous presence of A-B, A-A and B-B interactions. Considering nn interactions only, $J_{AB}<0$, $J_{AA}<0$ while $J_{BB}>0$ (the ferromagnetic B-O-B superexchange contribution prevails over the antiferromagnetic B-B direct exchange). The A-B interactions are expected to be the strongest one.

Neutron diffraction experiments/9/ have shown that no long-range order is stabilized above T=1.6K. For both the samples, a spin-glass-like freezing was observed at T_F=10.0+0.5K and 12.3+0.5K (at ν=198 Hz) for the Co and Fe spinels, respectively. AC susceptibility measurements at different frequencies performed on $CoGa_2O_4$ have shown that there is a very small dependence of T_F on the measuring time ($\Delta T_F/T_F\Delta \log\nu$=0.006), a value comparable to that found in metallic spin-glasses. The static susceptibility measurements for $FeGa_2O_4$ are reported in Fig.1. A spin-glass-like behaviour is observed, as shown by the maximum of the zero-field cooled(ZFC) susceptibility at T_F and the splitting between FC and ZFC susceptibilities below $T \simeq T_F$. The increasing below T_F of the FC susceptibility is similar to the behaviour observed in other insulating spin-glasses with short-range interactions.

$CoIn_2S_4/FeIn_2S_4$

Both these spinels were reported in the literature as being completely inverse, with formula $(In)_A/\{MIn\}_B/S_4$ (M=Co, Fe). Actually, a partial degree of inversion was observed/9/ in both spinels: 26% Co and 5% Fe ions occupy the A sites. In both cases competing ferromagnetic (nn B-B) and antiferromagnetic (nn A-A, A-B) interactions are present. Unlike the former oxispinels, where the electrons are strongly localized and the interactions between nearest-neighbours are dominant, in these two thiospinels the cobalt and the iron ions tend to have σ-bonding 3d electrons delocalized by strong covalent mixing with the anion orbitals. This effect makes

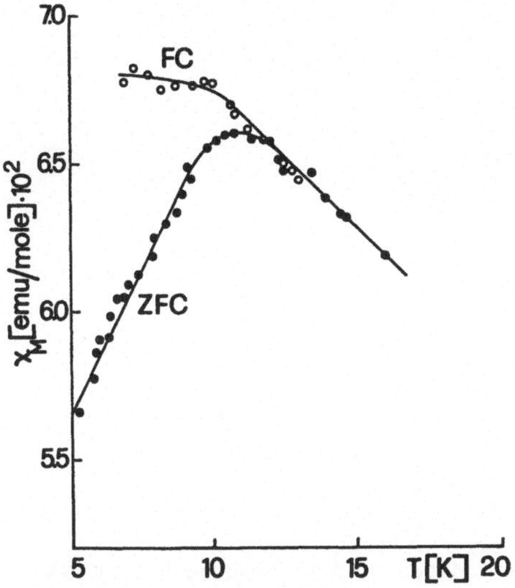

Fig.1 D.C. susceptibility measurements for $FeGa_2O_4$

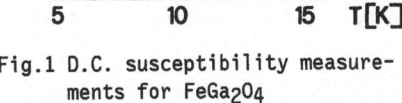

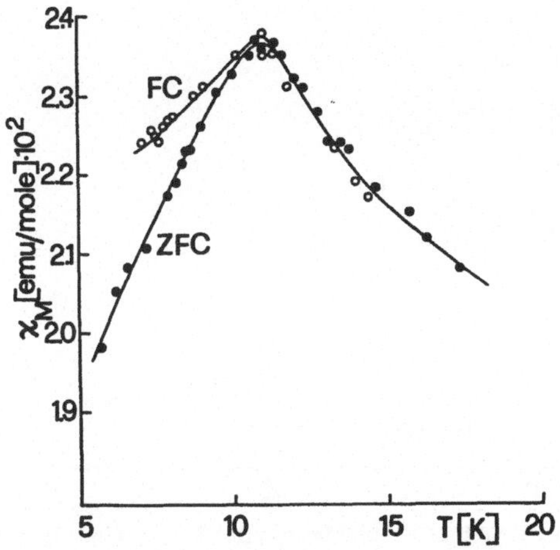

Fig.2 D.C. susceptibility measurements for $FeIn_2S_4$

the superexchange interactions more efficient and consequently the long-range interactions are no longer negligible; therefore the antiferromagnetic contribution coming from nnn A-A and B-B interactions has also to be considered. The delocalization is more important for Co than for Fe ions, as can also be verified observing the very high negative value of θ for $CoIn_2S_4$ (θ=-134K; compare with θ=-76K for $FeIn_2S_4$). Neutron diffraction experiments have shown that no long-range magnetic order is present above 1.6K /9/. AC and DC susceptibility measurements gave evidence for a spin-glass behaviour in both samples, with freezing temperature T_F=12.6+0.2K and T_F=10.8+0.3K for the cobalt and iron spinels, respectively. This is in disagreement with previous susceptibility measurements, which show a paramagnetic behaviour down to 4.2K/11/. We have to recall, however, that in that case the spinels were found to be completely inverse. Moreover, a much higher magnetic field (H = 10kOe) was used. In Fig. 2 the χ_{DC} measurements for $FeIn_2S_4$ are reported. A small maximum was also observed in the FC susceptibility, as shown by other spin-glasses with characteristics of high cooperative freezing.

II - Spinels with magnetically concentrated B-sublattice.

In $Zn_xCd_{1-x}Cr_2S_4$ the magnetic Cr ions exclusively occupy the B-sublattice. Unlike the former class of spinels, no magnetic dilution is present in this case. Competing interactions coexist: nearest-neighbours ferromagnetic (J_1) and higher-order-neigbours antiferromagnetic (J_2) interactions, whose intensity strongly depends on the non-magnetic ions in the A-sublattice (Zn or Cd) which play a role of intermediate in the superexchange path. Actually, the mutual variation of the AF and F interactions (usually indicated by the ratio $R=z_1J_J/z_2J_J$, where z is the neighbours' number) leads to a complex helimagnetic structure in $ZnCr_2S_4$ (R=-0.67) and to a ferromagnetic order in $CdCr_2S_4$ (R=-0.06). In our system the R value is modulated by the substitution between the non-magnetic ions on A sites, leading to a degree of frustration sufficient to stabilize a spin-glass state in the concentration region $0.4 \le x \le 0.7$/12/.

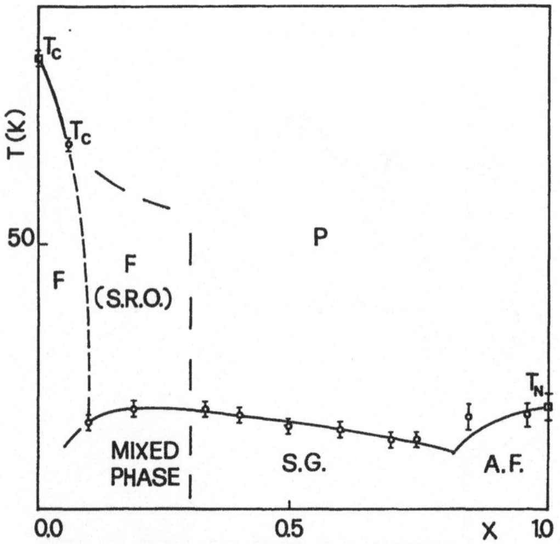

Fig.3 Magnetic phase diagram for $Zn_xCd_{1-x}Cr_2S_4$ deduced by A.C. susceptibility measurements (ν=198Hz)

Thus, a spin-glass state is observed although there is no significant positional disorder on the magnetic sublattice. However.a random distribution of magnetic interaction strengths does exist, because of the disorder on the A sublattice. Only few examples of spin-glass-like properties in concentrated crystalline systems, with competing F and AF interactions, are known/13,14/.

The magnetic phase diagram is reported in Fig.3. The following magnetic regions can be distinguished:

$0.0<x<0.1$: Ferromagnetic region. The susceptibility shows a rapid increase at Tc, below which it decreases very slowly.

$0.1<x<0.3$: Mixed phase. For this composition range, intermediate between the ferromagnetic and the spin-glass region, the A.C. susceptibility curve(χ') shows a peak at low temperature, around 20K, and a shoulder (or a broad maximum, depending on the composition x) at about 60K (Fig. 4). At the same temperatures the out-of-phase component χ'' shows well-defined maxima. The low-temperature peak is frequency dependent (e.g. for x=0.19: T=18.95 at ν = 11Hz and T=20.12K at ν = 11kHz). D.C. measurements show at low temperature a maximum for the F.C. as well as for the Z.F.C. susceptibility. The two curves split in correspondence of the high-temperature shoulder. The results suggest that at the higher temperature some short-range

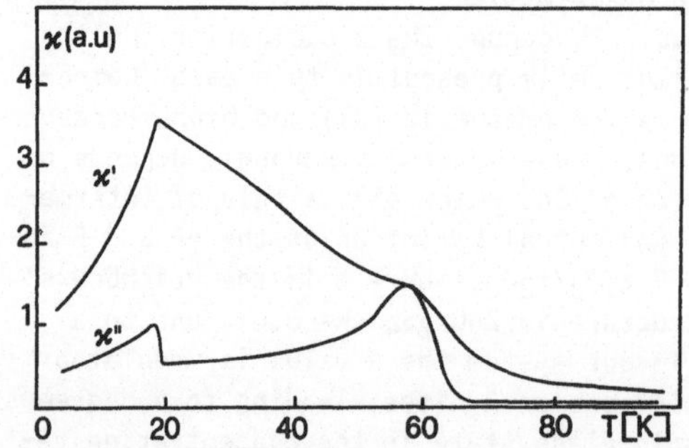

Fig.4 A.C. susceptibility measurements for $Zn_{0.2}Cd_{0.8}Cr_2S_4$ (ν=112Hz)

ferromagnetic order (existence of perturbed ferromagnetic regions) should exist, due to the predominance of ferromagnetic interactions in this concentration range. At the lower temperature the weak long-distance antiferromagnetic interactions become effective stabilizing a non classical spin-glass-like state.

$0.4<x<0.8$: Spin-glass region. With increasing x the strength of the antiferromagnetic interactions increases and the degree of frustration becomes sufficient to stabilize a pure spin-glass state at low temperature. A sharp peak is observed in the A.C. susceptibility. D.C. measurements show for x=0.5 a sharp maximum at T_F in the ZFC susceptibility, while the F.C. susceptibility splits just below T_F, remaining constant at lower temperatures./12/

$0.8<x<1.0$: Antiferromagnetic region. A maximum is observed at low temperature, due to a perturbed antiferromagnetic ordering.

ACKNOWLEDGEMENTS

We thank Drs.J.L.Soubeyroux, J.L.Dormann and M.Nogues for useful discussions. We thank also Mrs.R.Mannocchi for having kindly typed the text.

REFERENCES

1. E.Vincent, J.Hammann and M.Alba, Solid St. Commun. 58,(1986),57
2. R.A.Brand, H.Georges-Bibert, J.Hubsch and J.A.Heller, J.Phys.F:Met.Phys. 15 (1985)1987
3. J.L.Dormann, A.Saifi, V.Cagan and M.Nogues, Phys.Stat.Sol.(b) 131(1985)573
4. N.de Courtenay, H.Bouchiat, H.Hardespint and A.Fert, to be published
5. J.Villain, Z.Physik B 33(1979)31
6. C.P.Poole and H.A.Farach, Z.Phys.B-Condensed Matter, 47(1982)55
7. D.Fiorani, S.Viticoli, J.L.Dormann, J.L.Tholence and A.P.Murani, Phys.Rev.B, 30(1984)2776
8. J.Hubsch and G.Gavoille, J.Magn.Magn.Mat., 36(1983)89
9. J.L.Soubeyroux, E.Agostinelli and D.Fiorani, to be published
10. F.Scholl and K.Binder, Z.Phys.B, 39(1980)239
11. Schlein and A.Wold J.Solid St.Chem. 4(1972)286
12. E.Agostinelli, P.Filaci, D.Fiorani and E.Paparazzo,,Solid St.Commun.56(1985)541
13. K.Westerholt and H.Bach, Phys.Rev.Letters 47(1981)1925
14. J.E.Greedan, M.Sato and Xu Yan, Solid St.Commun. 59(1986)895

Harmonic Magnons in Spin Glasses: $Cd_{0.35}Mn_{0.65}Te$

D.L. Huber[1], *W.Y. Ching*[2], *and T.M. Giebultowicz*[3]

[1]Department of Physics, University of Wisconsin-Madison, Madison, WI 53706, USA

[2]Department of Physics, University of Missouri-Kansas City, Kansas City, MO 64110, USA

[3]Department of Physics, Purdue University, West Lafayette, IN 47907 and National Bureau of Standards, Gaithersburg, MD 20899, USA

The dilute magnetic semiconductor $Cd_{1-x}Mn_xTe$ is a frustrated fcc antiferromagnet which displays spin glass behavior at low temperatures over the range for which samples can be prepared, $x \lesssim 0.7$. Originally, it was believed there was a phase boundary at $x \simeq 0.6$ separating the spin glass phase from an antiferromagnetic phase and a second boundary at $x \simeq 0.2$ separating the spin glass phase from a paramagnetic phase. However, recent neutron scattering measurements show that the spin glass-antiferromagnetic boundary must occur in the physically inacessible region, $x > 0.7$ [1], whereas susceptibility data indicate a paramagnet to spin glass transition down to at least $x = 0.07$ in the magnetically equivalent compound $Zn_{1-x}Mn_xTe$ [2]. Although $Cd_{1-x}Mn_xTe$ exhibits many of the characteristic features of a spin glass [3,4], it is not certain whether it belongs to the same universality class as, for example, <u>Cu</u>Mn.

As with other spin glasses, the characterization of the dynamics of $Cd_{1-x}Mn_xTe$ at low temperatures is a problem of considerable importance. Broadly speaking, one can divide the excitations into two categories. There are relaxational modes involving excitations over barriers between different equilibrium configurations (EC's), and in the case of spin glasses with approximately continuous symmetry (e.g. Heisenberg and XY systems), magnon excitations within an EC. The latter can be subdivided into high-frequency, quasiharmonic excitations and low-frequency, hydrodynamic modes.

In this paper we report on a combined theoretical-experimental investigation of the quasiharmonic magnons in $Cd_{0.35}Mn_{0.65}Te$. The experimental work involved neutron scattering whereas the theoretical studies were based on numerical simulations carried out on finite arrays of spins. Taken together, the two approaches provide the most detailed picture of quasiharmonic magnons in spin glasses available to date. Since the experimental studies have been summarized elsewhere [1,5,6], we will focus attention on the theoretical investigation and its relation to the experimental findings.

The theoretical analysis follows an approach developed by Walker and Walstedt [7] in which EC's are obtained by successively rotating the spins into the directions of their local fields, with the process continued until the energy stabilizes. Having obtained an EC, the equations of motion of the spin operators are then linearized about the equilibrium spin directions. The straightforward but algebraically complex connection between the linearized equations of motion for the spin operators and the green's functions characterizing the scattering is given in Ref. 8.

As noted, $Cd_{1-x}Mn_xTe$ is a frustrated antiferromagnet. Quasielastic scattering studies show that at low temperatures the intensity is peaked around the superlattice points characterizing (short-range) Type III antiferromagnetic order. In a fully occupied fcc magnetic lattice Type III order is present, then the ratio of next-nearest, J_{NNN}, to nearest-neighbor interactions J_{NN}, falls in the range $0<J_{NNN}/J_{NN}<0.5$. In our numerical studies we have taken $J_{NNN}/J_{NN}=0.1$, which is consistent with a variety of experimental findings [2,9], and have neglected more distant interactions which do not play an important role at high Mn concentration.

An interesting result to emerge from the theoretical analysis of the EC's, which is nicely confirmed by experiment, is the asymmetry of the quasielastic scattering about the superlattice points. For example, in the neighborhood of the superlattice point $\vec{Q} = (2\pi/a)(1,1/2,0)$, the intensity is given by [1]

$$I(q) \sim \left[1 + \left((q_x-Q_x)^2 + (q_z-Q_z)^2\right)/\kappa_\perp^2 + (q_y-Q_y)^2/\kappa_{\parallel}^2\right]^{-1}, \qquad (1)$$

with $\kappa_{\parallel}/\kappa_\perp \simeq 1.8$. The value for the ratio of the inverse correlation lengths, $\kappa_{\parallel}/\kappa_\perp$, is indicative of the fact that in perfect Type III order the molecular field arising from interactions between planes of antiferromagnetically coupled spins which are perpendicular to the tetragonal axis of the magnetic unit cell, e.g. (010) planes for $\vec{Q} = (2\pi/a)(1,1/2,0)$, has no contribution from J_{NN}. The spins within these planes are strongly correlated with one another, but there is relatively weak correlation between planes.

The inelastic neutron scattering from $Cd_{0.35}Mn_{0.65}Te$ has been measured at 4.2K and the results compared with the equation-of-motion calculations. In Fig. 1 we show experimental data and theoretical calculations for $\vec{q}$ in the range $(2\pi/a)(2,1/2,0) \rightarrow (2\pi/a)(1,1/2,0) \rightarrow (2\pi/a)(1,1,0)$. The theoretical results were obtained for an array of 6912 sites with a fractional occupancy of 0.65. The open circles and open squares denote the positions of the peaks in the spectrum; the upper and lower bars show the peak width at half-maximum. Since the experimental data are in meV and the theoretical results in units of

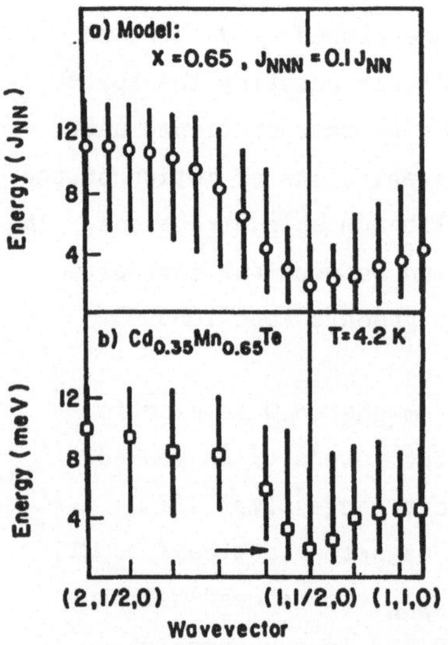

Fig. 1. Results of calculations (upper panel) and measurements (lower panel) of the inelastic magnetic neutron scattering in $Cd_{0.35}Mn_{0.65}Te$ for various momentum transfers. The circles (or squares) indicate the positions of the peaks in the spectra and the bars show the peak width at half maximum [6].

J_{NN} we conclude that the nearest-neighbor interaction in $Cd_{0.35}M_{0.65}Te$ is on the order of 1 meV, a point we shall return to below.

One can understand the relative shifts in peak positions with wave vector from a consideration of an approximate magnon dispersion curve obtained by taking the cubic average of the various terms in the spin wave frequency of a type III antiferromagnet [10]. With nearest and next nearest neighbor interactions we have

$$E(\vec{q}) = 4S[J_{NN}(3A(\vec{q})+1) + J_{NNN}((3/2)B(\vec{q}) - (1/2))]^{1/2}$$

$$\cdot[J_{NN}(1-A(\vec{q})) - 2J_{NNN}(1-B(\vec{q}))]^{1/2}, \tag{2}$$

where S is the spin while

$$A(\vec{q}) = (1/3)[\cos(q_xa/2)\cos(q_ya/2) + \cos(q_xa/2)\cos(q_za/2)$$

$$+ \cos(q_ya/2)\cos(q_za/2)], \text{ and} \tag{3}$$

$$B(\vec{q}) = (1/3)[\cos(q_xa) + \cos(q_ya) + \cos(q_za)]. \tag{4}$$

For $J_{NNN} = 0.1J_{NN}$, $E(\vec{q})$ has the values $10.3J_{NN}$, 0, and $3.7J_{NN}$ for $\vec{q} = (2\pi/a)(2,1/2,0)$, $(2\pi/a)(1,1/2,0)$, and $(2\pi/a)(1,1,0)$, respectively.

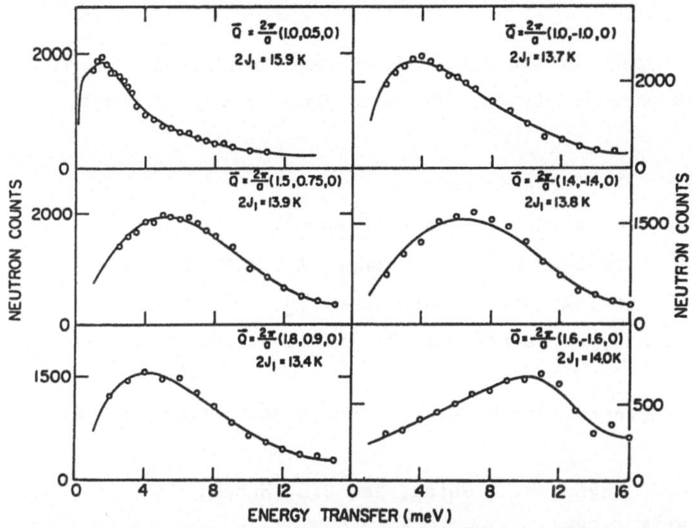

Fig. 2. Comparison of the experimental and theoretical inelastic magnetic neutron scattering for $Cd_{0.35}Mn_{0.65}Te$. The circles show the measured intensities at various q values. The lines show fits of the theoretical lineshapes (convoluted with the instrumental resolution) to these data. The values of the nearest-neighbor exchange constant $J_{NN} = 2J_1$ (the only adjustable parameter) obtained from each fit are displayed on the plots [12].

Not only is there overall agreement between experiment and theory with respect to peak positions and half-widths, the numerical simulations also reproduce the shape of the inelastic peaks. This is shown in Fig. 2 for a variety of points in reciprocal space. In fitting the data the nearest-neighbor exchange interaction $J_{NN} = 2J_1$ was used as an adjustable parameter. With the exception of the superlattice point, $\vec{q} = (2\pi/a)(1,1/2,0)$, a good fit was obtained with $J_{NN} = (13.8\pm0.2)$K in excellent agreement with the value $J_{NN} = (13.8\pm0.3)$K obtained from high-temperature susceptibility data [11].

It is not surprising that the agreement between experiment and theory is not as good at $\vec{q} = (2\pi/a)(1,1/2,0)$. Since the magnon frequencies in a Heisenberg antiferromagnet with long-range Type III order vanish at the superlattice points [cf. Eq. (2)], the spectrum in the immediate neighborhood $(2\pi/a)(1,1/2,0)$ particularly sensitive to the presence of anisotropic, e.g. dipolar and Dzyaloshinskii-Moriya, interactions, which have not been included in the theoretical calculations.

In summary, very good agreement between inelastic neutron spectra measured at 4.2K and theoretical calculations in the harmonic magnon approximation have been obtained for $Cd_{0.35}Mn_{0.65}Te$. Additional data at higher temperatures and lower Mn concentrations will be presented elsewhere.

Acknowledgments

Research supported in part by the National Science Foundation. Computer time on the Cray X-MP was provided by the Office of Basic Energy Science, Department of Energy.

References

1. T.M. Giebultowicz and T.M. Holden (to be published).
2. A. Twardowski, C.j.M. Denissen, W.J.M. de Jonge, A.T.A.M. de Waele, M. Demianiuk and R. Triboulet, Solid State Comm. 59, 199 (1986).
3. P. Nordblad, P. Svedlindh, J. Ferre', and M. Ayadi, J. Mag. Mag. Met. 59, 250 (1986).
4. M. Saint-Paul, J.L. Tholence, and W. Giriat, Solid State Comm. 60, 621 (1986).
5. T.M. Giebultowicz, J.J. Rhyne, W.Y. Ching, and D.L. Huber, J. Appl. Phys. 57, 3415 (1985).
6. T.M. Giebultowicz, J.J. Rhyne, W.Y. Ching, D.L. Huber, and R.R. Galazka, J. Mag. Mag. Met. 54-57, 1149 (1986).
7. L.R. Walker and R.E. Walstedt, Phys. Rev. B22, 3816 (1980).
8. W.Y. Ching, D.L. Huber, and K.M. Leung, Phys. Rev. B21, 3708 (1980).
9. M. Escorne and A. Mauger, Phys. Rev. B25, 4674 (1982).
10. D. ter Haar and M.E. Lines, Philos. Trans. R. Soc. London, Ser. A 254, 521 (1962); 255, 1 (1962).
11. A. Lewicki, J. Spalek, J.K. Furdyna, and R.R. Galazka, J. Mag. Mag. Met. 54-57, 1221 (1986).
12. T.M. Giebultowicz, J.J. Rhyne, W.Y. Ching, B. Lebech and R.R. Galązka (to be published).

On a Symmetric $(2S+1)$-State System Coupled to Its Environment: Tunneling Splitting and Quantum Coherence

T. Tsuzuki

Department of Physics, Tohoku University, Sendai 980, Japan

1. Introduction

We study quantum coherence [1,2,3] in a spin system [4,3] coupled to its environment. The Hamiltonian of the system is assumed to be $H = H_S + H_B + H_{S-B}$ and

$$H_S = -DS_z^2 - AS_x , \tag{1a}$$

$$H_B = \Sigma_j \hbar\omega_j (b_j^+ b_j + \frac{1}{2}) , \tag{1b}$$

$$H_{S-B} = S_z u , \quad u = \Sigma_j \lambda_j (b_j^+ + b_j) , \tag{1c}$$

where S_α is the α-component of the spin operator with magnitude S, and D and A are assumed to be positive. The environment H_B is assumed to be described by a heat bath composed of an ensemble of bosons with characteristic energies $\{\hbar\omega_j\}$. The coupling between the systems is denoted by H_{S-B}. Degeneracies of every doubly degenerate state of $-DS_z^2$ are lifted by the transfer coupling $-AS_x$, if one considers an isolated system described by H_S. So the probability $P_0(t;M \to -M)$ of finding a state $|S_z=-M\rangle$ at time t oscillates with a characteristic frequency $2\Delta_M/\hbar$ essentially, where $2\Delta_M$ is the level splitting between the two states that come from the bare states $|\pm M\rangle$, provided the spin is set in a state $|M\rangle$ initially and if the transfer coupling is weak, $AS \ll DS^2$. When $S = 1/2$, it is given exactly by

$$P_0(t;1/2 \to -1/2) = \{1 - \cos(2\Delta t/\hbar)\}/2 , \tag{2}$$

where $\Delta = AS$. This oscillation is one example of quantum coherence caused by quantum tunneling.

In the present report we first study tunneling splitting of an isolated system H_S, and discuss the effect of H_{S-B} on energy spectra of the whole system H. Then we examine the effect of dissipation caused by H_{S-B} on quantum coherence at finite temperatures.

2. Tunneling Splitting of an Isolated System H_S

Let us first study quantum mechanics of an isolated system described by H_S, by proposing a method of decomposing the contribution from the transfer coupling to the eigenvalue into the anharmonic part and the tunneling splitting. This is a generalization of the familiar method of computing the tunneling splitting of the energy eigenvalues of a particle moving in one dimension in a potential with a couple of symmetrically degenerate minima, say in a double-well potential. The anharmonic contribution is found by solving an eigenvalue problem $H_S\Psi^{(\pm)} = \varepsilon\Psi^{(\pm)}$ in the right ($M>0$) and left ($M<0$) half-spaces by turning off the coupling between both half-spaces, that is, by setting $\langle S-[S]-1|S_x|S-[S]\rangle = \langle S-[S]|S_x|S-[S]-1\rangle = 0$, where $S_z|M\rangle = M|M\rangle$ and [S] is the maximum integer not larger than S. We obtain in total

$([S]+1)$ eigenvalues $\{\varepsilon_\ell; \ell = 0,1,\cdots,[S]\}$ by

$$D_{S-[S]-1}(\varepsilon) = 0 \ , \tag{3}$$

where, writing $E_0(M) = -DM^2$, $D_S(\varepsilon) = 1$, $D_{S-1}(\varepsilon) = E_0(S)-\varepsilon$ and

$$D_{S-M}(\varepsilon) = [\prod_{\ell=0}^{M-1} \{E_0(S-\ell)-\varepsilon\}]\cdot[1+\sum_{p=1}^{[M/2]} (-)^p \sum_{\ell_1=0}^{M-2p} \sum_{\ell_2=\ell_1+2}^{M-2(p+1)} \cdots \sum_{\ell_p=\ell_{p-1}+2}^{M-2} \{\prod_{j=1}^{p} U_{\ell_j}\}] \ , \tag{4}$$

$$U_\ell(\varepsilon) = Q_{S-\ell}^2[\{E_0(S-\ell)-\varepsilon\}\{E_0(S-\ell-1)-\varepsilon\}]^{-1} \ ,$$

$$Q_{S-\ell} = \langle S-\ell|H_S|S-\ell-1\rangle = \langle S-\ell-1|H_S|S-\ell\rangle = -AS[(1-\frac{\ell}{2S})\frac{\ell+1}{2S}]^{1/2} \ .$$

The eigenfunctions $\Psi_\ell^{(\pm)}$ are not written down to save space, but it is easy to find them.

Next let us find the eigenvalues of the full system by recovering the correct value of $Q_{S-[S]}$ in order to fulfill the continuity condition between $\Psi^{(+)}$ and $\Psi^{(-)}$. In the case of even 2S we find (S+1) eigenvalues $\{E_\ell^s; \ell = 0,1,\cdots,S\}$ with symmetric eigenfunctions and S eigenvalues $\{E_\ell^a; \ell = 1,2,\cdots,S\}$ with antisymmetric eigenfunctions by

$$D_{-1}(E^s) = Q_0Q_1D_1(E^s) \quad \text{and} \quad D_0(E^a) = 0 \ , \tag{5}$$

$$\Psi_\ell^{(s)} = C_{\ell 0}^{(s)}|0\rangle + \frac{1}{\sqrt{2}} \sum_{M=1}^{S} C_{\ell M}^{(s)} \{|M\rangle + |-M\rangle\} \ , \tag{6s}$$

$$\Psi_\ell^{(a)} = \frac{1}{\sqrt{2}} \sum_{M=1}^{S} C_{\ell M}^{(a)} \{|M\rangle - |-M\rangle\} \ . \tag{6a}$$

In the case of odd 2S we get ([S]+1) symmetric (+) and antisymmetric (-) eigenstates by

$$D_{-1/2}(E) \pm Q_{1/2}D_{1/2}(E) = 0 \ , \tag{7}$$

$$\Psi_\ell^{(s,a)} = \frac{1}{\sqrt{2}} \sum_{M=S-[S]}^{S} C_{\ell M}^{(s,a)} \{|M\rangle \pm |-M\rangle\} \ . \tag{8s,a}$$

We do not write coefficients in (6s,a) and (8s,a) to save space again. In both cases E_ℓ^s and E_ℓ^a tend to $E_0([\ell+1/2])$ in zero transfer coupling.

Thus we find the shifts of energies $\Delta_\ell^{s,a}$ and the tunneling splitting Δ_ℓ due to the quantum tunneling by $\Delta_\ell^s = \varepsilon_\ell - E_\ell^s$, $\Delta_\ell^a = E_\ell^a - \varepsilon_\ell$, and $\Delta_\ell = \Delta_\ell^s + \Delta_\ell^a$. A few examples are as follows. We denote $d = DS^2$ and $\Delta = AS$. For $S = 1/2$, $\varepsilon_1 = -d$, $\Delta_1^s = \Delta_1^a = \Delta$ which implies a symmetric splitting, and $\Delta_1 = 2\Delta$. For S=1, $\varepsilon_0 = \{\sqrt{d^2+2\Delta^2} - d\}/2$, $\varepsilon_1 = -\{\sqrt{d^2+2\Delta^2} + d\}/2$, $-\Delta_0^s = \Delta_1^s = \{\sqrt{d^2+4\Delta^2} - \sqrt{d^2+2\Delta^2}\}/2$, and $\Delta_1^a = \{\sqrt{d^2+2\Delta^2} - d\}/2$. The splitting is not symmetric ($\Delta_1^s \neq \Delta_1^a$) due to the existence of the third level ε_0. Our formulas are convenient for computing the effects of the transfer coupling in powers of Δ^2 perturbationally for arbitrary magnitude of S.

3. H_{S-B} and Tunneling Splitting

Let us study the effect of coupling H_{S-B} on tunneling splitting and low-lying excitations of the whole system by expanding the wave function Ψ in terms of eigenfunctions of H_S and H_B as

$$\Psi = \sum_{\ell} \sum_{\vec{n}} \{X_\ell(\vec{n})\psi_\ell^{(s)} + Y_\ell(\vec{n})\psi_\ell^{(a)}\}\cdot\phi_{\vec{n}} \ , \tag{9}$$

where $\psi_\ell^{(s,a)}$ have been given by (6a,b) or (8a,b). and $H_B\phi_{\vec{n}} = \omega(\vec{n})\phi_{\vec{n}}$ with notations of $\omega(\vec{n}) = \Sigma_j \hbar\omega_j(n_j^\ell+1/2)$ and $\vec{n} = (n_1,n_2,\cdots)$. From Schrödinger equation $H\Psi = \Omega\Psi$ we get the following set of equations for $X_\ell(\vec{n})$ and $Y_\ell(\vec{n})$:

$$\{E_\ell^s + \omega(\vec{n}) - \Omega\}X_\ell(\vec{n}) + \sum_{\ell'}\sum_{\vec{n}'} K_{\ell\ell'}u(\vec{n},\vec{n}')Y_{\ell'}(\vec{n}') = 0 \ , \tag{10a}$$

$$\{E_\ell^a + \omega(\vec{n}) - \Omega\}Y_\ell(\vec{n}) + \sum_{\ell'}\sum_{\vec{n}'} K_{\ell'\ell}u(\vec{n},\vec{n}')X_{\ell'}(\vec{n}) = 0 \ , \tag{10b}$$

where $K_{\ell\ell'} \equiv \Sigma_{M>0}MC_{\ell M}^{(s)}C_{\ell' M}^{(a)}$ and $u(\vec{n},\vec{n}') \equiv \langle\phi_{\vec{n}}|u|\phi_{\vec{n}'}\rangle$.

Let us consider the case of $S = 1/2$, where X_ℓ and Y_ℓ have only a single component respectively, $K_{11} = 1/2$, and $-E_1^s = E_1^a = \Delta$. Then we find

$$\{-\Delta + \omega(\vec{n}) - \Omega - Z_{\vec{n}}(\Omega;\Delta)\}X(\vec{n}) = 0 \ , \tag{11a}$$

$$\{\Delta + \omega(\vec{n}) - \Omega - Z_{\vec{n}}(\Omega;-\Delta)\}Y(\vec{n}) = 0 \ , \tag{11b}$$

$$Z_{\vec{n}}(\Omega;\Delta) = \lim_{p\to\infty} Z_{\vec{n}}^{(p)}(\Omega;\Delta) \ , \tag{12a}$$

$$Z_{\vec{n}}^{(p)}(\Omega;\Delta) = Z_{\vec{n}}^{(p-1)}(\Omega;\Delta) + F_{\vec{n},\vec{n}}^{(p)}(\Omega;\Delta) \ , \tag{12b}$$

$$Z_{\vec{n}}^{(1)}(\Omega;\Delta) = F_{\vec{n},\vec{n}}^{(1)}(\Omega;\Delta) \ , \tag{12c}$$

$$F_{n,n''}^{(p+1)}(\Omega;\Delta) = \sum_{\vec{n}''\neq\vec{n},\vec{n}'} \frac{F_{\vec{n},\vec{n}''}^{(p)}(\Omega;\Delta)F_{\vec{n}'',\vec{n}'}^{(p)}(\Omega;\Delta)}{-\Delta + \omega(\vec{n}'') - \Omega - Z_{\vec{n}}^{(p)}(\Omega;\Delta)} \ , \tag{13a}$$

$$F_{\vec{n},\vec{n}'}^{(1)}(\Omega;\Delta) = \frac{1}{4}\sum_{n''} \frac{u(\vec{n},\vec{n}'')u(\vec{n}'',\vec{n}')}{\Delta + \omega(\vec{n}'') - \Omega} \ . \tag{13b}$$

We set the origin of energy at the zero-point energy of bosons $\Sigma_j\hbar\omega_j/2$. If $u=0$, two independent sets of eigenvalues are obtained : $\Omega_X(\vec{n}) = -\Delta + \omega(\vec{n})$ for $\psi^{(s)}$ and $\Omega_Y(\vec{n}) = \Delta + \omega(\vec{n})$ for $\psi^{(a)}$. Let us consider the weak coupling limit where $Z_{\vec{n}}$ is replaced by $Z_{\vec{n}}^{(1)}$. By introducing the spectral distribution function of bosons $J(\omega)$ by $\Sigma_j\lambda_j^2\delta(\omega_j-\omega) = 2\hbar^2\alpha J(\omega)$ and restricting ourselves to the ohmic case where $J(\omega) = \omega$ for $\omega_c \geq \omega \geq 0$ and zero otherwise [1,2], we obtain the following low-lying excitations composed of the spin and a single boson with quantum number j :

$$\Omega_X(j) = -\Delta^{(1)} + \hbar\omega_j \quad \text{and} \quad \Omega_Y(j) = \Delta^{(1)} + \hbar\omega_j \ , \tag{14}$$

where $\Delta^{(1)} = \Delta\{1 - \alpha\cdot\ln(\hbar\omega_c/2\Delta) + O(\alpha^2)\}$. These are the bottoms of continua of excitations with the definite symmetries (denoted by X and Y) of the whole system. The spin dressed by bosons forms a pair of tunneling eigenstates split symmetrically with reduced tunneling splitting $2\Delta^{(1)}$. Note that the energy balance between the spin and the boson system is even. These are the essential points of the effect of H_{S-B}. Due to the symmetric dependence of eigenvalue equations (11a and b) our observations are valid for arbitrary strength of u unless the coupling destroys the tunneling splitting [2]. A pair of degenerate spin states $|1/2\rangle$ and $|-1/2\rangle$ are available as a set of eigenstates above the critical strength of coupling. Consideration similar to the above is applicable to the case of arbitrary S, resulting in the qualitatively same description of the lowest part of excitations of H.

4. Quantum Coherence

We consider the following probability function by restricting ourselves to S = 1/2 :

$$P(t,\beta;1/2 \to b) \equiv Z_B^{-1} \sum_{\vec{n}} \exp[-\beta \sum_j \hbar\omega_j n_j]\cdot|\langle b|U(t)|1/2,\vec{n}\rangle|^2 , \tag{15}$$

where $U(t) = \exp[-iHt/\hbar]$ and $Z_B = \mathrm{Tr}\exp[-\beta H_B]$. This describes the probability of finding the spin in a state $|b\rangle$ at time t when H_{S-B} is turned on at t=0, provided the spin is set in an initial state $|1/2\rangle$ and the boson system is in equilibrium. This function is reduced to (2) for $|b\rangle = |-1/2\rangle$ when $H_{S-B} = 0$. At zero temperature this function still exhibits coherent oscillation with frequency defined by reduced tunneling splitting for $\alpha < \alpha_c$, where α_c is a critical value [1,2]. We want to study the effect of thermal bosons on quantum coherence in the weak coupling limit.

In doing so we notice that the leading effect of the coupling is to renormalize tunneling splitting, leaving the boson excitation unchanged. Then we can use the following approximation to U(t) :

$$U(t) = \exp[-iH_B t/\hbar]\cdot\exp[-it\{H_S + S_z G(t)\}/\hbar] , \tag{16}$$

$$G(t) = \sum_j \lambda_j\{g(\omega_j t)b_j^+ + g^*(\omega_j t)b_j\} , \tag{17}$$

where $g(z) = \{\exp[iz] - 1\}/iz$ is a characteristic function of propagation of free bosons. By the use of (16) we neglect $[u(t), G(t)]$ which is proportional to α and give up the renormalization of tunneling splitting, but we can study the intrinsic effect of thermal distribution of bosons. We can get

$$P(t,\beta;1/2 \to -1/2) = \frac{1}{\sqrt{2\pi}}\int_{-\infty}^{\infty} dx \, \frac{\Delta^2 e^{-x^2/2}}{2(\Delta^2 + \kappa x^2)} \{1 - \cos\frac{2t}{\hbar}[\Delta^2 + \kappa x^2]^{1/2}\} , \tag{18}$$

$$\kappa = \frac{1}{2}\hbar^2\alpha\int_0^{\infty} d\omega\, J(\omega)\cdot\coth(\beta\hbar\omega/2)\cdot\{\frac{\sin(\omega t/2)}{(\omega t/2)}\}^2 , \tag{19}$$

without introducing further approximation. This may be interpreted as follows. (1) The coupling to the environment produces the effective magnetic field by which the effective tunneling splitting of spin is enlarged. (2) The quantum coherence in the case of a particular value of the effective field is described by the elementary process of quantum tunneling. (3) The probability function is an average taken over a Gaussian distribution of the effective field with a width $\sqrt{\kappa}$.

Let us examine the long-time behaviour of P_2 by assuming the ohmic case of $J(\omega)$. For $\pi t/\beta\hbar \gg 1$, $\kappa \to \pi\hbar\alpha/\beta t$, so that, if $\kappa \ll \Delta^2$, that is, if $\tau \equiv (2\Delta\cdot t/\hbar)/2\pi \gg \tau_c \equiv 2\alpha(K_B T/2\Delta)$,

$$P(t,\beta;1/2 \to -1/2) = \frac{1}{2}\,[(1-A_1) + A_1\{1 - \cos(\frac{2\Delta}{\hbar}\,t + \theta_1)\}]\ , \tag{20a}$$

$$P(t,\beta;1/2 \to \ 1/2) = \frac{1}{2}\,[(1-A_1) + A_1\{1 + \cos(\frac{2\Delta}{\hbar}\,t + \theta_1)\}]\ , \tag{20b}$$

where $A_1 = [1 + \delta^2]^{-1/4}$, $\theta_1 = 2^{-1}\tan^{-1}\delta$ and $\delta = 2\pi\alpha/\beta\Delta$. These reach the same oscillatory state with frequency 2Δ, which should be replaced by $2\Delta^{(1)}$ as seen in the preceding section, and get a decreased amplitude and a phase shift which are determined by a single parameter δ. Note that this final state is reached in the extremely long time limit because κ tends to zero very slowly like $1/t$ as $t \to \infty$. P recovers P_0 as $T \to 0$ and tends to 1/2 at the high temperature limit. At intermediate temperatures the coupling makes the quantum tunneling hard, resulting in a reduction of the amplitude of oscillation from one to A_1. This oscillatory part remembers the initial quantum coherence still. The factor $(1-A_1)/2$, which is zero initially, is considered as the incoherent part of probability. We are now extending the present study to the case of finite coupling.

In conclusion, the interplay of the coupling and the thermal distribution of bosons weakens the quantum coherence in the amplitude of oscillation in the long-time limit.

Acknowledgements – The author thanks Drs. H. De Raedt, H. Thomas, P. Talkner and A.J. Leggett for their critical comments and discussions. This work was supported in part by the Yoshida Foundation for Science and Technology, and in part by the Grant-in-Aid for Fundamental Scientific Research by Ministry of Education, Science and Culture of Japan.

References

1. A.O. Caldeira and A.J. Leggett : Ann. Phys. 149, 374 (1983), and references quoted therein.
   A.J. Leggett : Prog. Theor. Phys. Suppl. 69, 80 (1980).
2. A.J. Leggett, S. Chakravarty, A.T. Dorsey, Mathew P.A. Fisher, Anupan Garg and W. Zwerger : Rev. Mod. Phys. 59, 1 (1987), and references quoted therein.
3. S. Chakravarty : Physica 126B, 385 (1984).
4. S. Chakravarty and A.J. Leggett : Phys. Rev. Lett. 52, 5 (1984).
   A.J. Bray and M.A. Moore : Phys. Rev. Lett. 49, 1545 (1982).

Regular and Chaotic Dynamics of Classical Spin Systems

N. Srivastava[1], C. Kaufman[1], G. Müller[1], R. Weber[2], and H. Thomas[2]

[1]Department of Physics, University of Rhode Island, Kingston, RI02881, USA

[2]Institut für Physik, Universität Basel, Klingelbergstraße 82, CH-4056 Basel, Switzerland

1. Introduction

In the context of the present discussion of the nature of quantum chaos /1/, the dynamics of spin clusters has been the subject of several investigations /2-6/ recently. The interest is focussed on the characteristic properties of quantum spin systems whose classical counterparts are nonintegrable. As a basis for the investigation of regular and chaotic spin motion, we have examined the problem of integrability of classical spin clusters /7,8/.

We consider a system of N identical classical three-component spins $\vec{S}_\ell$, $\ell=1,\dots,N$, of constant length $|\vec{S}_\ell| = S$ specified by a spin Hamiltonian $H(\vec{S}_1,\dots\vec{S}_N)$. The time evolution of this system is governed by the equations of motion

$$d\vec{S}_\ell/dt = -\vec{S}_\ell \times (\partial H/\partial \vec{S}_\ell) \ . \quad (1)$$

They coincide with the classical limit ($\hbar\to 0$, $|\vec{S}| \to \infty$, $\hbar|\vec{S}|$ finite) of the Heisenberg equations of motion of the spin operators S_ℓ of the corresponding quantum system. Each spin $\vec{S}_\ell$ is confined to a sphere $|\vec{S}_\ell| = S$, and may be expressed in spherical coordinates (θ_ℓ,ϕ_ℓ),

$$\vec{S}_\ell = S(\sin\theta_\ell\cos\phi_\ell,\ \sin\theta_\ell\sin\phi_\ell,\ \cos\theta_\ell) \ . \quad (2)$$

In terms of the variables $q_\ell = \phi_\ell$, $p_\ell = S\cos\theta_\ell$, the equations of motion (1) assume Hamiltonian form

$$\dot{q}_\ell = \partial H(\{q_\ell,\ p_\ell\})/\partial p_\ell,\ \dot{p}_\ell = -\partial H(\{q_\ell,\ p_\ell\})/\partial q_\ell \ . \quad (3)$$

Therefore, an N-spin system represents a Hamiltonian system with N degrees of freedom, with a compact 2N-dimensional phase space consisting of the product of N spheres $|\vec{S}_\ell| = S$. According to Liouville's theorem, such a system is completely integrable if there exist N independent constants of motion $I_k(\vec{S}_1,\dots,\vec{S}_N)$, $k=1,\dots,N$, which are mutually in involution. A completely integrable system is characterized by the property that phase space is foliated into invariant N-tori which are obtained as the intersections of the N hypersurfaces $I_k(\vec{S}_1,\dots,\vec{S}_N)$=const. Each trajectory is confined to one of these N-tori (regular motion). If fewer than N independent integrals of motion exist, the foliation into invariant tori is incomplete, leaving room for the occurrence of chaotic trajectories whose course through phase space is erratic and extremely sensitive to slight changes in initial conditions.

2. Integrability of Two-Spin Clusters

We have studied the integrability of spin motion in two-spin clusters specified by a Hamiltonian

$$H = -\sum_\alpha J_\alpha S_1^\alpha S_2^\alpha + \frac{1}{2}\sum_\alpha A_\alpha [(S_1^\alpha)^2 + (S_2^\alpha)^2], \quad \alpha = x,y,z . \tag{4}$$

Since H is not explicitly time-dependent, it is itself conserved, and complete integrability requires the existence of one additional constant of motion $I(\vec{S}_1,\vec{S}_2)$ We have searched for integrals of the form

$$I = \sum_\alpha \{ -g_\alpha S_1^\alpha S_2^\alpha + \frac{1}{2} K_\alpha [(S_1^\alpha)^2 + (S_2^\alpha)^2]\} \tag{5}$$

and have found the following results:
i) For pure exchange anisotropy, i.e. $A_x=A_y=A_z=0$, the two-spin cluster is always integrable, and the second constant of motion is given by

$$I = -\sum_{cycl} J_\alpha J_\beta S_1^\gamma S_2^\gamma + \frac{1}{2}\sum_\alpha J_\alpha^2 [(S_1^\alpha)^2 + (S_2^\alpha)^2]. \tag{6}$$

ii) For nonzero single-site anisotropy (not all A_α equal), a second integral of the form (5) exists only if the parameters of the Hamiltonian (4) satisfy the condition

$$(A_x-A_y)(A_y-A_z)(A_z-A_x) + \sum_{cycl} J_\alpha^2(A_\beta-A_\gamma) = 0 . \tag{7}$$

It is then given by

$$I = -\sum_\alpha g_\alpha S_1^\alpha S_2^\alpha , \qquad \text{where} \tag{8}$$

$$g_\alpha = (J_x+J_y+J_z)J_\alpha + (A_\alpha-A_\beta)J_\gamma + (A_\alpha-A_\gamma)J_\beta - (A_\alpha-A_\beta)(A_\alpha-A_\gamma) . \tag{9}$$

As an example, we consider an XY-type model with

$$J_x = J(1+\gamma),\ J_y = J(1-\gamma),\ J_z = 0,\ A_x = -A_y = J\alpha,\ A_z = 0 \tag{10}$$

for which the condition (7) takes the form

$$\alpha^2 - \gamma^2 = 1 . \tag{11}$$

We focus here on the general case of biaxial symmetry, where the existence of a second constant $I(\vec{S}_1,\vec{S}_2)$ is nontrivial. In the two limits $\alpha=0$, $\gamma=0$ (isotropic XY limit) and $\alpha=0$, $\gamma=\pm1$ (Ising limit), the system has rotational symmetry, and Noether's theorem guarantees the conservation of the total spin component M along the rotation axis. In fact, the combination $I-4\gamma H$ of the two invariants H and I reduces to a quadratic function of M in these limits.

Our numerical calculations provide strong indication that a violation of the condition (7) or (11) renders the system nonintegrable. The behaviour of trajectories may be visualized by means of Poincaré surfaces of section. On such a surface, invariant tori are represented as closed curves, whereas chaotic trajectories fill two-dimensional areas. As an example, Fig. 1 shows the projection of the Poincaré surface of section defined by $\theta_2=\pi/2$ onto the (θ_1,ϕ_1)-plane of the system (10) with $\gamma=0$, $\alpha=-1/2$, and fixed energy for various initial conditions. One notices the coexistence of regular and chaotic regions. The chaotic trajectory acts as separatrix between two types of regular motion: precession of spin $\vec{S}_1$ about the z-axis (top and bottom), both accompanied by a considerable amount of nutation, and quasiperiodic oscillations of various complexity without precession about the z-axis (center).
The occurrence of chaotic trajectories for this model implies the nonexistence of a second analytic invariant. However, the abundance of invariant tori observed in the phase flow suggests that fragments of the second invariant I survive in some form.

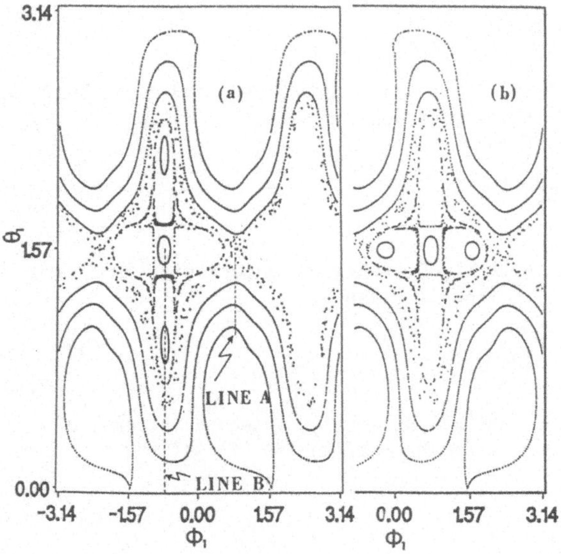

Fig. 1

Projection of the Poincaré surfaces of section defined by (a) $\theta_2 = \pi/2$, $\dot{\theta}_2 < 0$ and (b) $\theta_2 = \pi/2$, $\dot{\theta}_2 > 0$ onto the (θ_1, ϕ_1)-plane for the non-integrable two-spin model (10) with $\gamma=0$, $\alpha=-0.5$ and $H=-0.09957501$

Let us first consider a completely integrable case. If the second integral I were not known explicitly, it could be reconstructed numerically as follows: Pick any dynamical variable X which is independent of H. The time average of X over any trajectory is by construction time-independent, i.e. a constant of motion. It is in fact an analytic function of the initial conditions $(\vec{S}_1, \vec{S}_2)$, and may therefore be identified as the second integral $\langle X \rangle = I(\vec{S}_1, \vec{S}_2)$. In nonintegrable cases, according to Birkhoff's theorem, the time average still exists for almost all trajectories, even chaotic ones. However, it is no longer an analytic function of the initial conditions: Its values for regular trajectories will represent the remains of the analytic invariant, whereas in regions filled densely by a chaotic trajectory, one may expect ergodic behaviour leading to a constant value of the average. Thus, as a function of the initial coordinates on a given energy shell, the time average would display a step-like behaviour, consisting of horizontal pieces in chaotic regions and isolated points on regular trajectories. In those parts where the chaotic regions are very thin, the graph of $\langle X \rangle$ will look like a smooth curve.

The calculation of time averages for a chaotic trajectory represents a highly nontrivial problem. Because of the positive Lyapunov exponent, numerical errors grow exponentially, i.e. two different integration procedures starting from the same initial condition will produce numerical orbits ("itineraries") which separate exponentially. Empirically, we find, however, that time averages calculated for different itineraries coincide within certain error bars. To obtain a measure of the convergence, we have calculated time averages over each of ten successive long time periods ($T = 10^4 J^{-1}$) and determined the standard deviation.

In Fig. 2a, the time average $\langle (S_1^y)^2 \rangle$ is shown for a system with $\gamma=0$, $\alpha=-0.5$ as a function of the initial value θ_1 along the line B marked in Fig. 1. In regions where invariant tori dominate, $\langle (S_1^y)^2 \rangle$ is found to converge rapidly, and looks like a well-behaved function of θ_1. In chaotic regions, on the other hand, the time average converges much more slowly, as indicated by the large error bars. This slow convergence has been shown to be caused by the existence of fragments of tori ("cantori") which act as barriers in phase space. In Fig. 2b, calculated for a system with $\gamma=0$, $\alpha=-0.7$ along a corresponding line, such fragments are much rarer, and the convergence is better even in the chaotic region. Within numerical accuracy, the system appears to be ergodic in the chaotic region.

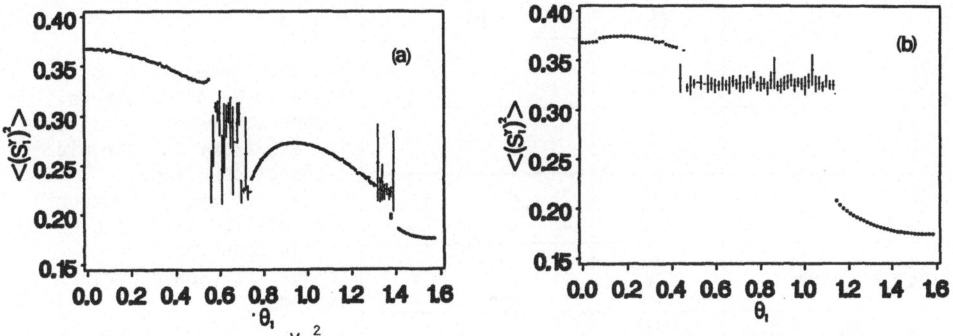

Fig. 2: Time average $<(S_1^y)^2>$ over single trajectories as a function of initial value θ_1 for $\phi_1(t=0)=-\pi/4$ (the line B of Fig. 1), for the model (10) with $\gamma=0$, $\alpha=-0.5$ in (a) and $\gamma=0$, $\alpha=-0.7$ in (b)

3. Integrable N-Spin-Clusters

As a contribution to the question of integrability of spin systems with $N>2$, we list here a number of results (see Fig. 3). We have established complete integrability for a special class of N-spin systems which may be described as follows: The system consists of two arrays A and B of N_A and N_B spins, respectively, such that every spin of array A is coupled to every spin of array B by a constant anisotropic exchange interaction J , but spins belonging to the same array do not interact directly (first column of Fig. 3). This class includes a two-sublattice model of an antiferromagnet with constant inter-sublattice and zero intra-sublattice coupling.

Such a system is described by a Hamiltonian

$$H = -\sum_\alpha J_\alpha T_A^\alpha T_B^\alpha \tag{12}$$

where $\vec{T}_{A,B}$ are the total spins of the arrays A,B

$$\vec{T}_A = \sum_{\ell\varepsilon A} \vec{S}_\ell \ , \quad \vec{T}_B = \sum_{\ell\varepsilon B} \vec{S}_\ell \ . \tag{13}$$

The motion of the effective two-spin system $(\vec{T}_A, \vec{T}_B)$ is governed by the equations

$$d\vec{T}_A/dt = -\vec{T}_A \times (\partial H/\partial \vec{T}_A) \ , \quad d\vec{T}_B/dt = -\vec{T}_B \times (\partial H/\partial \vec{T}_B) \ . \tag{14}$$

It is completely integrable for arbitrary anisotropic J_α , since it is equivalent to the system described by Eq. (4) with $A_x=A_y=A_z=0$, except for the fact that in general $|\vec{T}_A| \neq |\vec{T}_B|$. The motion of the individual spins in the two arrays A, B follows the motion of $\vec{T}_{A,B}$ in a rigid manner, such that all scalar products $\vec{S}_\ell \cdot \vec{S}_{\ell'}$, with ℓ and ℓ' belonging to the same array are constants of motion.

Another class of completely integrable N-spin systems are clusters in which every spin is coupled to every other spin by a constant isotropic exchange interaction (second column of Fig. 3). The Hamiltonian has the form

$$H = -\frac{1}{2} J \sum_{\ell \neq \ell'} \vec{S}_\ell \cdot \vec{S}_{\ell'} \tag{15}$$

(Kittel-Shore model). In this case, the total spin

$$\vec{T} = \sum_\ell \vec{S}_\ell \tag{16}$$

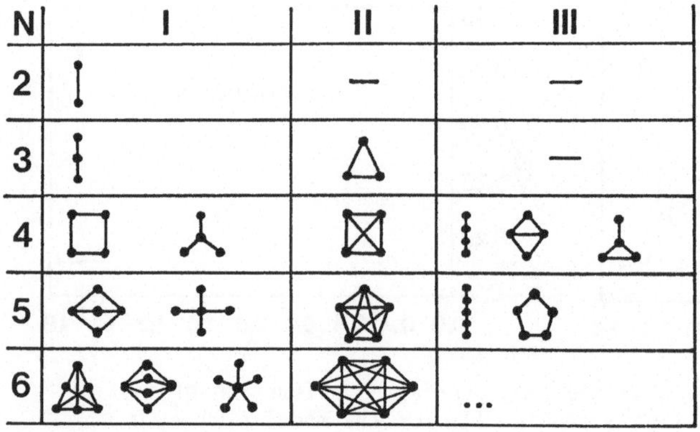

Fig. 3

Table of integrable and non-integrable N-spin clusters. Column I: Systems integrable for arbitrary anisotropic exchange coupling. Column II: Systems integrable only for isotropic exchange coupling. Column III: Nonintegrable systems

is a constant of motion. The individual spins precess about $\vec{T}$ with constant frequency, such that all components parallel to $\vec{T}$ and all relative azimuthal angles are conserved.

In the last column of Fig. 3, we have listed some spin clusters for which the numerical evidence indicates non-integrability.

The work at University of Basel was supported by Schweizerischer Nationalfonds, and the work at URI by a grant from Research Corporation, and by the National Science Foundation, Grant Number DMR-86-03036.

References

1. Müller, G.: Phys.Rev. A34, 3345 (1986)
2. Feingold, M., Peres, A.: Physica D9, 433 (1983)
3. Feingold, M., Moiseyev, N., Peres, A.: Phys.Rev. A30, 509 (1984)
4. Nakamura, K., Nakahara, Y., Bishop, A.R.: Phys.Rev.Lett. 54, 861 (1985)
5. Nakamura, K., Bishop, A.R.: Phys.Rev. B33, 1963 (1986)
6. Frahm, H., Mikeska, H.J.: Z.Phys. B60, 117 (1985)
7. Magyari, E., Thomas, H., Weber, R., Kaufman, C., Müller, G.: Z.Phys. B65, 363 (1987)
8. Srivastava, N., Kaufman, C., Müller, G., Magyari, E., Weber, R., Thomas, H.: J.Appl.Phys. 61, 4438 (1987)

The Kicked Quantum Spin: A Model System for Quantum Chaos

H.J. Mikeska and H. Frahm

Institut für Theoretische Physik, Universität Hannover, D-3000 Hannover 1, Fed. Rep. of Germany

1. Introduction

We have studied a simple quantum spin model, which shows chaotic behaviour in the classical limit. The essential feature of this model is the nonpredictability already on the classical level (i.e. the strong dependence on initial conditions). Our aim was to contribute to the understanding of how the corresponding quantum system approaches the classical limit and which properties of the quantum system reflect the stochastic character of its classical counterpart. A particularly interesting aspect in this context is the relation between the two sources of unpredictability, classical stochasticity and quantum uncertainty - leading to the question whether a unified description of these phenomena exists.

For an investigation of these questions a quantum spin system is particularly appropriate owing to its finite Hilbert space with dimension 2S + 1, where S is the spin length. Our system is defined by the Hamiltonian

$$H = A(S^z)^2 - \mu B f(t) S^x, \tag{1}$$

where f is a sequence of δ-functions with period τ, and we will use the length of spin S to measure the strength of quantum effects. Invoking Floquet's theorem, it is immediately clear that the dynamics of this system is governed by a finite number of frequencies, implying a quasiperiodic time dependence. The main interest, however, is in the approach of the classical limit and in the discussion of remnants of the classically chaotic behaviour in the quantum system. In this article we will restrict to discussing the main results of our work and we refer the reader to our original publications /1-3/ for details, as well as to the contributions of other groups who have studied the same Hamiltonian /4,5/.

Previous investigations of similar questions have been done using different models, in particular the model of the kicked rotator /6-9/, the quantum cat map /10/ and the highly excited hydrogen atom in an external electric field /11,12/.

2. Phase Space Approach

For a discussion of the approach to the classical limit in the spirit of the correspondence principle a description of our system in terms of wave packets in a semiclassical phase space appears to be the most direct approach. Owing to the finite size of cells in phase space for a quantum system, $\Delta p \Delta q \sim \hbar$, the uncertainty from stochastic behaviour in a classical system as measured by the Liapunov exponent λ (inverse rate of exponential growth of small deviations), develops a conflict with the quantum uncertainty relation for $t > t_0 \cong (\ln \hbar)/2\lambda$. Thus our problem should be characterised by a crossover time t_0, beyond which classical concepts are no more useful.

For a quantitative investigation of this approach we have followed the time development of a wave packet of the spin system, starting essentially from a spin coherent state /13/, and allowing the complex width of this state to vary with time. Considering the mean square fluctuations of the canonical variables to $O(1/S)$ we find the following unified picture: The classical equations for stochastic deviations are recovered in the limit $S \to \infty$ (implying that we find the correct value for the classical Liapunov exponent from this approach) whereas for finite values of $1/S$ a quantum Liapunov exponent can be extracted considering the trajectories up to the crossover time t_0 /1/. The fact that this quantum Liapunov exponent decreases with S illustrates the statement that quantum mechanics tends to stabilise the stochastic effects in classical systems.

3. Levelstatistics Approach

An independent approach to the stochastic properties of classically chaotic quantum systems is possible following the conjecture that a classical chaotic system is characterised by a well-defined energy level statistics when considered as a quantum system in its classical limit. The investigations usually concentrate on two quantities, the distribution of spacings between neighbouring energy levels $P(x)$, and the rigidity $\Delta_3(L)$, which describes the deviations from a smooth level sequence in some energy interval L (in units of the average level spacing). These quantities are conjectured to behave significantly different in integrable and nonintegrable systems:

Integrable systems (Poisson distribution):

$$P(x) = e^{-x} \qquad \Delta_3(L) = L/15\,. \tag{2}$$

Nonintegrable systems (Wigner distribution):

$$P(x) = \frac{\pi}{2}\, x\, e^{-\frac{\pi}{4}x^2} \qquad \Delta_3(L) \cong \ln L - 0.007\,. \tag{3}$$

This conjecture has been verified numerically for various systems /14/, and has also been supported by analytical arguments /15/. In particular an interpolation formula has been developed /16/, which interpolates between the limiting cases (2) and (3) and allows to extract a quantity ρ, $0 \le \rho \le 1$, which measures the degree of nonintegrability from the numerically found level statistics.

For the system (1) these concepts can be applied after replacing energy levels by Floquet frequencies or quasi-energy levels owing to the periodic time dependence of the external field. An analysis of the quasi-energy levelstatistics for different values of the spin length S shows /2/ that the interpolation parameter ρ decreases towards 0, its value for a regular system, with increasing strength of quantum effects. In close analogy to the results for the Liapunov exponent as described in the previous paragraph, this is a manifestation of how quantum effects suppress classical stochasticity.

4. Wave Functions and Quantum Predictability

The methods described in the previous two paragraphs provide a discussion of the quantum system, which largely relies on classical concepts and which is useful for $t < t_0$, resp. for large values of the spin length S, i.e. in the semiclassical region. It is an open question to what degree remnants of classical stochasticity can be pinned down in the genuine quantum region. As a contribution to this problem we discuss in the following the consequences of different initial conditions for the time development of a typical expectation value in our model (1).

In Fig. 1 we show the time dependence (n is the number of kicks of the external field in (1)) of the quantity $\langle S^z\rangle_1 - \langle S^z\rangle_2$, where the subscript characterises slightly different initial conditions, realised by coherent spin states. The example

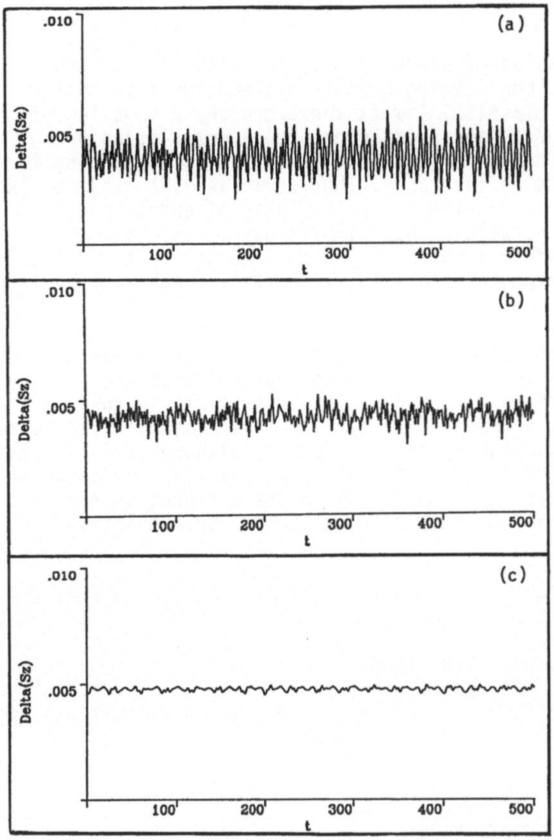

Fig. 1 Time development of an initial uncertainty in S^z for S=10 and different values of the stochasticity parameter s: (a) s=2, (b) s=1, (c) s=0.1

shown is for S = 10 and results are given for different values of the "stochasticity parameter" $s = 2(AS\mu B)^{1/2}$. s = 2 corresponds to complete irregularity in the classical system, s = 0.1 to a regular classical system and s = 1 is an intermediate value. There is no indication of a dramatic increase in uncertainty, such as the exponential rise for the (semi)classical system at s = 2; however, it is clear that the stochastic character of the classical s = 2 system also adds to the non-predictability of the genuine quantum system. This added nonpredictability appears as bounded in time and is thus in qualitative contrast to the exponential increase of the classical case.

Thus it appears that in order to locate stochasticity effects in genuine quantum systems, wave functions have to be discussed in detail. In fact the results given in Fig. 1 can be related to the observation that a typical eigenfunction of the unitary time evolution operator for one period of our kicked system, when discussed in terms of its Wigner distribution, covers an increasing portion of phase space with increasing degree of stochasticity of the corresponding classical system /5/. Thus with increasing stochasticity parameter an increasing number of Fourier components will be present and the maximum amplitude of $\Delta\langle S^z\rangle$ will be increased. We expect that on the basis of these observations it will be possible to develop also a more quantitative theory of stochasticity effects in truly quantum mechanical systems.

5. Conclusions

The mathematical rigour of the discussion of stochasticity in classical systems is based on considering the limit $t \to \infty$. Considering quantum systems, we note that this limit does not commute with the classical limit; therefore any discussion of stochasticity in quantum systems is of an approximate nature. A description using classical concepts is only possible for rather small values of $\hbar$, corresponding to rather large values of the spin length in our model. Thus semiclassical trajectories can only be followed for finite time and a statistical analysis of energy levels can only be done with finite precision. Within these limits our investigations clearly show that quantum effects tend to suppress classical stochasticity, as indicated by a decrease of the Liapunov exponent and a shift in statistical distributions from Wigner to Poisson.

On the other hand there are remnants of classical stochastic behaviour in genuine quantum systems, i.e. beyond the crossover time t_0, which characterises the onset of typically quantum mechanical interference effects: The uncertainty in quantum mechanical expectation values, i.e. the result of a typical measurement, is increased if the classical counterpart of the quantum system is chaotic, although this increase is rather mild and by no means exponential as in the classical system. The crossover time t_0 thus can also be interpreted as the time for which the dominant source of unpredictability changes from classical stochasticity to quantum uncertainty.

References

1. H. Frahm, H.J. Mikeska: Z. Phys. B60, 117 (1985)
2. H. Frahm, H.J. Mikeska: Z. Phys. B65, 249 (1986)
3. H.J. Mikeska, H. Frahm: In Fractals and Chaos, ed. by R. Pike (Adam Hilger, 1987)
4. F. Haake, M. Kus, R. Scharf: Z. Phys. B65, 381 (1986)
5. K. Nakamura, Y. Okazaki, A.R. Bishop: Phys. Rev. Lett. 57, 5 (1986)
6. G. Casati, B.V. Chirikov, F.M. Izrailev, J. Ford: In Stochastic Behaviour in Classical and Quantum Hamiltonian Systems, ed. by G. Casati, J. Ford, Lecture Notes in Physics 93 (Springer, Berlin, 1979)
7. S. Fishman, D.R. Grempel, R.E. Prange: Phys. Rev. Lett. 49, 509 (1982)
8. F.M. Izrailev: Phys. Rev. Lett. 56, 541 (1986)
9. T. Geisel, G. Radons, J. Rubner: Phys. Rev. Lett. 57, 2883 (1986)
10. M. Toda, K. Ikeda: Preprint, Kyoto University (1986)
11. J.N. Bardsley, B. Sundaram, L.A. Pinnaduwage, J.E. Bayfield: Phys. Rev. Lett. 56, 1007 (1986)
12. G. Casati, B.V. Chirikov, I. Guarneri, D.L. Shepelyanski: Phys. Rev. Lett. 56, 2437 (1986)
13. J.M. Radcliffe: J. Phys. A4, 313 (1971)
14. O. Bohigas, M.J. Giannoni, C. Schmit: Phys. Rev. Lett. 52, 1 (1984)
    T.H. Seligman, J.J.M. Verbaarschot, M.R. Zirnbauer: Phys. Rev. Lett. 53, 215 (1984)
15. P. Pechukas: Phys. Rev. Lett. 51, 943 (1983)
    T. Yukawa: Phys. Rev. Lett. 54, 1883 (1985)
16. M.V. Berry, M. Robnik: J. Phys. A17, 2413 (1984)

Chaotic Dynamics of Spin Wave Instabilities

S.M. Rezende, F.M. de Aguiar, and A. Azevedo

Departamento de Fisica, Universidade Federal de Pernambuco,
50739 Recife, Brazil

1. Introduction

Recently the investigation of spin wave phenomena has gained renewed interest due to observations of chaotic dynamics similar to those in other nonlinear physical systems such as turbulent fluids, lasers, plasmas and several solid state systems. The existence of low-frequency oscillations, self-pulsation and turbulence in microwave-pumped magnetic samples has been known since the early days of the spin wave parametric pumping experiments [1]. For years the origin of these self-oscillations was unclear, until several Soviet authors [2] showed that they resulted from the nonlinear dynamics of spin waves when pumped above the threshold for parametric excitation. However their extensive work fell short of connecting the spin wave turbulence phenomena with the modern ideas of chaotic dynamics of nonlinear dissipative systems.

NAKAMURA et al [3] first predicted that two nonlinearly interacting magnon modes driven by a microwave field parallel to the static field should display chaotic behavior when pumped above the threshold intensity. Furthermore they showed that the route to chaos was a universal cascade of period doublings in the low-frequency self-oscillations. This stimulated the experimental search for chaotic dynamics in spin wave instabilities. WALDNER et al [4] observed a route to chaos by irregular periods in $(NH_3)_2(CH_2)_2CuCl_4$ under parallel pumping at low temperatures. YAMAZAKI [5] did low-temperature work in $CuCl_2 2H_2O$ and found one period doubling but no chaotic behavior. A complete period-doubling cascade route to chaos was first observed by GIBSON and JEFFRIES [6] in the prototype "ferromagnet" $Y_3Fe_5O_{12}$ - yttrium iron garnet (YIG) in the second-order Suhl perpendicular pumping, not in the parallel pumping process. The first observation of subharmonic routes to chaos in parallel pumping was made by the present authors [7,8] in YIG at room temperature, but the predicted universal cascade was not always observed. Recently YAMAZAKI and coworkers [9,10] made a series of measurements to characterize the fractal dimension of the strange attractor in several systems under parallel pumping. In this paper we review briefly our earlier work on the parallel pumping and present new results on chaotic dynamics in the first-order Suhl perpendicular pumping in YIG.

2. Experiments

The experiments were performed with spherical and disk-shaped samples of pure YIG at room temperature. The samples were held at the center of a TE_{102} microwave cavity (Q=3 000, $f_p=\omega_p/2\pi=9.4$ GHz) with the magnetic rf field $\vec{h}$ either parallel or perpendicular to the applied dc field $\vec{H}_o$. The cavity was critically coupled and the power was provided by a 3 Watt TWT amplifier fed by a klystron tube, frequency stabilized with an external crystal oscillator. Cavity detunning due to spin wave excitation was manually corrected by adjusting the frequency to the center of the cavity resonance. When the driving field reaches a critical amplitude h_c, one spin wave mode is driven unstable, resulting in an abrupt increase in the reflected microwave power. h_c is a function of H_0 and its value depends on the wavector $\vec{k}$ and the relaxation rate of the unstable mode. In a narrow range of intensity h above h_c the reflected power shows no amplitude modulation. As h increases in the post-threshold region a low-frequency (50-300kHz) modulation develops, with an amplitude and spectrum which depend on the sample shape and orientation and the values of H_0 and h.

We describe initially the results obtained in the parallel pumping configuration. In this process pairs of spin waves with opposite wavevectors $\vec{k}$ and $-\vec{k}$ are driven directly by the microwave photons. The critical field h_c above which one spin wave pair with frequency $\omega_k=\omega_p/2$ grows unstably is $h_c=\omega_k\gamma_k/\pi\gamma M\sin^2\Theta_k$, where γ_k is the relaxation rate of the unstable mode, Θ_k is the angle between the wavevector $\vec{k}$ and the field $\vec{H}_0$, M is the saturation magnetization and γ is gyromagnetic ratio (=2.8 GHz/kOe). Since $\omega_k=\omega_p/2$ is fixed, the threshold h_c at each value of H_0 corresponds to a different $\vec{k}$. For a spherical sample oriented with the [111] axis along $\vec{H}_0$ the minimum threshold occurs at H_0=1540 Oe corresponding to the $\Theta_k=\pi/2$, $k\approx0$ mode which is the one with lowest relaxation rate. This is the situation at which we observe chaotic behavior with a unique route to chaos. Figure 1 shows some spectra of the modulation of the reflected microwave signal for a 1mm diameter sphere of pure YIG with $\vec{H}_0$//[111], H_0=1540 Oe. The pumping field is characterized by a control parameter defined by $R\equiv h/h_c$. A nearly sinusoidal modulation with frequency $f\simeq100$ kHz was first observed at $R\simeq1.50$. As R increases the fundamental frequency increases linearly with R and various cascades of bifurcations are observed

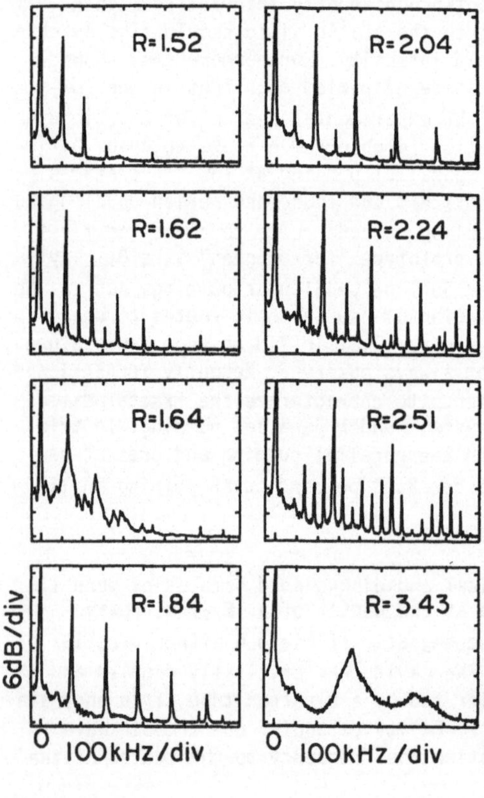

FIG.1 Power spectra of the experimental scenario in the parallel pumping experiment described in the text: oscillation at $f_1\simeq$ 100kHz and harmonics $2f_1$ and $3f_1$ at R=1.52; bifurcation to $f_1/2$ at R=1.62; spin wave turbulence at R=1.64; oscillation at $f_2\simeq$ 160kHz and harmonics $2f_2$, $3f_2$, $4f_2$ and $5f_2$, at R=1.84; bifurcation to $f_2/2$ at R=2.04; bifurcation to $f_2/4$ at R=2.24; power spectrum for a period-6 oscillation at R=2.51; chaos at R=3.43

within the range of available power. In the first cascade a period doubling bifurcation occurs at R=1.62 but there is no evidence of an infinite sequence between this value and R=1.64 where chaotic behavior develops. A coherent modulation reappears at R=1.84 and another sequence of bifurcations develops distinctly from a window of the Feigenbaum scenario. Periods 2T and 4T are observed for $R\doteq 2.04$ and 2.24, but this cascade of period doubling bifurcations is interrupted and a series of periods 3T, 6T, etc, sets in leading to chaos at R=3.43. A similar route to chaos in which a period-doubling cascade crosses over to an odd-period bifurcation series has been observed recently in a Rayleigh-Bénard experiment [11]. We also observed chaotic behavior with a typical Feigenbaum scenario for other values of H_0 and crystal orientation.

If the microwave field is applied perpendicular to $\vec{H}_0$, in the usual ferromagnetic resonance configuration, the radiation couples only to the k=0 uniform mode. This in turn can excite nonlinearly other $k\neq 0$ pair modes through the magnon-magnon interaction. In the second-order Suhl process two k=0 modes drive a degenerate $\vec{k}$, $-\vec{k}$ pair through the 4-magnon interaction,resulting in the premature saturation of the main resonance at a frequency $\omega_0=\gamma H_0$. Chaotic dynamics associated with this process has been previously investigated [6,12,13]. The first-order Suhl process is one by which a k=0 mode drives a $\vec{k}$, $-\vec{k}$ pair via the 3-magnon interaction. Since the magnons have frequency $\omega_k \simeq \omega_p/2$ this process gives rise to a broad peak in the absorbed power near half the field for resonance. This is called the subsidiary resonance, observed only above a threshold field h_c. Figure 2 shows oscilloscope traces and spectra of auto-oscillation in the amplitude of the reflected microwave signal in the subsidiary resonance region. This result was obtained with a 1mm YIG sphere with $\vec{H}_0$//[100], H_0=1950 Oe. The shape and frequency of the modulation have a strong dependence on sample orientation and pump power. The self-oscillation in Figure 2a has frequency 450 kHz for a pumping field $R=h/h_c=2.12$. The other photographs show periods 2T, 4T and chaos for R=2.19, 2.29 and 2.54 respectively. Period 8T was also observed,indicating a typical Feigenbaum route to chaos.

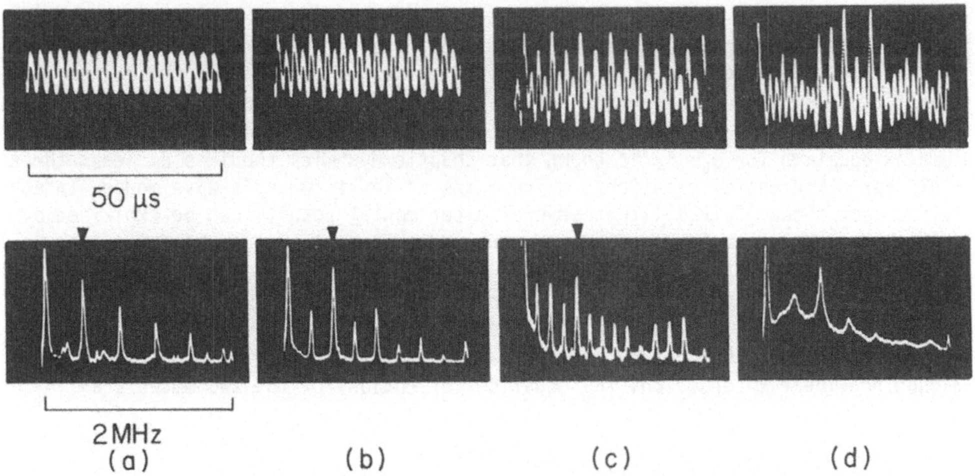

FIG.2 Oscilloscope traces showing (a) period-T, (b) period-2T, (c) period-4T,and (d) chaotic oscillations (top) and respective spectra (bottom) observed in the subsidiary resonance experiments described in the text.

3. Theoretical Interpretation

The low-frequency modulation in the microwave signal arises when more than one spin wave pair mode is driven parametrically. Due to the nonlinear magnon interaction, a dynamic competition between the modes can produce the growth of one mode at the expense of the others in an alternating manner, resulting in auto-oscillations. The equations that describe this process can be obtained from the following Hamiltonian:

$$H = \sum_k \hbar\omega_k c_k^\dagger c_k + H^{(3)} + H^{(4)} + H'(t), \tag{1}$$

where $c_k^\dagger$ and c_k are the magnon creation and destruction operators and

$$H^{(3)} = \sum_{k_1k_2k_3} V_{123}\; c_1^\dagger c_2 c_3 \Delta(\vec{k}_1 - \vec{k}_2 - \vec{k}_3) + \text{term in } c_1c_2c_3 + \text{h.c.}, \tag{2}$$

$$H^{(4)} = \sum_{k_1k_2k_3k_4} V_{1234}\; c_1^\dagger c_2 c_3 c_4 \Delta(\vec{k}_1 + \vec{k}_2 - \vec{k}_3 - \vec{k}_4) + \text{other terms} \tag{3}$$

represent the magnon-magnon interactions and $H'(t)$ represents the interaction with the external driving fields. In the parallel pumping case

$$H'(t) = \frac{1}{2}\hbar \sum_k h\rho_k\, e^{-i\omega_p t}\; c_k^\dagger c_{-k}^\dagger + \text{h.c.}, \tag{4}$$

where h is the microwave field and $\rho_k = \gamma\pi M\sin^2\Theta_k/\omega_k$ represents the coupling of the field with the $\vec{k}, -\vec{k}$ magnon pair. If the microwave field $H_1 \cos \omega_p t$ is applied perpendicular to H_0, in the usual ferromagnetic resonance configuration

$$H'(t) = \hbar\gamma(SN/2)^{1/2}\; H_1(c_0 e^{-i\omega_p t} + \text{h.c.}). \tag{5}$$

In this case the radiation field couples only to the k=0 uniform mode.

Since the parametric spin waves are driven in pairs the modes are best described by the ensemble averages $n_k(t) = \langle c_k c_k^\dagger \rangle$ and $\sigma_k = \langle c_k c \rangle_k \exp(i\omega_p t)$, the magnon occupation number and a Cooper-pair correlation respectively. In the stationary state, after a time $t \sim \gamma_k^{-1}$, $n_k \approx |\sigma_k|$, so the evolution of each mode is governed by one complex equation for σ_k. It is known that chaotic dynamics requires at least three nonlinear differential equations, which means at least two spin wave modes. Indeed it has been shown [8,12,13] that several experimental results can be explained by a two-mode model, but the presence of more modes leads to richer behavior.

In the parallel pumping process we can consider a set of two parametric pair modes with wavevectors $\vec{k}_1$ and $\vec{k}_2$. They could be for instance the mode with lowest threshold and any other nearly degenerate mode coupled to the first by the magnon-magnon nonlinear interaction. The equations of motion for the two modes are [14]:

$$\frac{1}{2}\frac{d\sigma_1}{dt} = -\,[\gamma_1 + i\Delta\omega_1 + i2(S_1+2T_1)n_1 + i4T_{12}n_2]\sigma_1 - i2S_{12}\sigma_2 n_1 - i\rho_1 h n_1,$$
$$\frac{1}{2}\frac{d\sigma_2}{dt} = -\,[\gamma_2 + i\Delta\omega_2 + i2(S_2+2T_2)n_2 + i4T_{12}n_1]\sigma_2 - i2S_{12}\sigma_1 n_2 - i\rho_2 h n_2, \tag{6}$$

where $\Delta\omega_k = \omega_k - \omega_p/2$, the S's and T's are nonlinear parameters related to the microscopic interaction constants [14] and depend strongly on the wavevectors of the two modes and the orientation of the sample with respect to $\vec{H}_0$. Since in YIG the nonlinear parameters S and T are of the order of 10^{-12} sec^{-1} and $\gamma_k \sim 10^5 - 10^6$ sec^{-1}, we divide Eq.(6) by γ_1 and work with normalized variables and parameters $t' = \gamma_1 t$, $n' = Fn$, $\sigma' = F\sigma$, $T' = F^{-1}T$ and $S' = F^{-1}S$, where $F \sim S/\gamma_1 \sim 10^{-17} - 10^{-18}$ sec^{-1}. Thus n', T'/γ_1 and S'/γ_1 are of the order of unity. The set of equations (6) describing the nonlinear dynamics of a four-dimensional phase space can be solved in a computer to give the time evolution of n_1 and n_2. The solutions display a rich variety of periodic and chaotic trajectories depending on the values of the parameters.

In the first-order perpendicular pumping process the simplest situation is that where only the off-resonance k=0 mode and a $\vec{k}, -\vec{k}$ pair are present. In this case the equations of motion are

$$\frac{dc_0}{dt} = (-i\Delta\omega_0 - \gamma_0 - i2T_0 n_0)c_0 - iS_0\sigma_0 c_0^* - i\frac{F_k}{2}\sigma_k - i\gamma(SN/2)^{1/2} H_1 , \tag{7}$$

$$\frac{d\sigma_k}{dt} = [-i2\Delta\omega_k - 2\gamma_k - i4(S_k + 2T_k)n_k]\sigma_k - iF_k c_0 n_k , \tag{8}$$

where $c_0 = \langle c_0 \rangle \exp(i\omega_p t)$ is a slowly varying variable.

Numerical solution of Eqs.(7) - (8) yields a wide variety of behavior depending on the values of the nonlinear parameters and dissipation rates. Here we consider only the parallel pumping equation (6). For some sets of parameters and $h > h_c$, the trajectories in phase space $n_1 \times n_2$ can be attracted into one of the nontrivial asymetric fixed points $n_1=0$, $n_2\neq0$ or $n_1\neq0$, $n_2=0$. In this case only one pair mode is driven and there is no self-oscillation. For other sets the trajectories may be attracted to limit cycles with a frequency that depends on the parameter values and pumping intensity. In this case a low-frequency modulation arises in the power absorbed by the sample due to the dynamic competition of the two parametric spin wave modes. As the pumping increases the trajectories may bifurcate and eventually lead to a strange attractor corresponding to chaos. NAKAMURA et al [3] considered two symmetric modes with the fictitious parameters $\Delta\omega_1 = \Delta\omega = -1.0\gamma_1$, $\gamma_1 = \gamma_2$, $\alpha = \rho_2/\rho_1 = 1.0$, $S_1 + 2T_1 = S_2 + 2T_2 = 1.0\gamma_1$, $S_{12} = -0.4\gamma_1$ and $T_{12} = 0.75\gamma_1$ and obtained a Feigenbaum route to chaos. However, other unique routes indicative of the higher dimensional nature of the system were found with other sets of parameter values. A particularly interesting route that is qualitative similar to the experimental results was obtained with the values: $\Delta\omega_1 = \Delta\omega_2 = 0$, $\gamma_2/\gamma_1 = 5.0$, $\alpha = 0.7$, $S_1 + 2T_1 = -1.0\gamma_1$, $S_2 + 2T_2 = 0.5\gamma_1$, $S_{12} = 2.5\gamma_1$ and $T_{12} = 1.125\gamma_1$. Figure 3a shows the phase-space $n_1 \times n_2$ trajectory and the the Fourier spectrum of n_1 for a control parameter $\Gamma = h\rho_1/\gamma_1 = 6.0$. In this situation we have self-oscillation with frequency $f_0 \simeq \gamma$. Period doubling is observed in Fig. 3b where the driving field has been increased to $\Gamma = 7.7587$. A noisy period-2 attractor is shown in Fig. 3c for $\Gamma = 7.759$ and fully developed chaos appears in Fig. 3d for $\Gamma = 7.83$. We varied Γ up to the fifth decimal place in the range 7.50-7.76 and saw no trace of periods 4 or 8, just like the experimental results. By increasing Γ further in the chaotic region we observe a period-4 window about the value $\Gamma = 7.925$. Another self-oscillation with period-1 develops at $\Gamma = 9.5$ and period-2 occurs at 10.95, but contrary to the experimental results this sequence does not lead to another chaotic state. A return map constructed at $\Gamma = 8.20$ [14] bears resemblance to the one observed by HUNT and ROLLINS for a nonlinear electronic circuit [15].

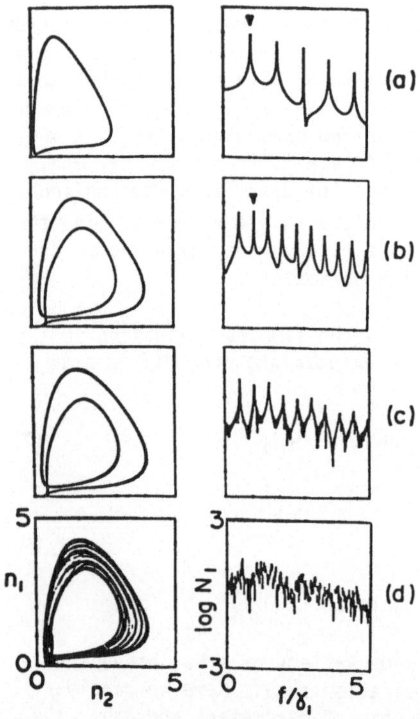

FIG.3 Phase space $n_1 x n_2$ trajectories and power spectra $N_1(f)$ obtained with the numerical solution of the asymmetric two-mode model for parameters given in the text. The arrow in (a) and (b) indicates the fundamental frequency.

According to our numerical results the self-oscillation frequency is $f_0 \simeq \gamma_1$. Since the dissipation rate measured by the value of the threshold field (1.4×10^6 sec^{-1}) is an order of magnitude larger than the observed f_o(~100kHz) we suggest that γ_k at large magnon numbers is smaller than it is near or below threshold. This is also consistent with the predictions for the driving intensities. We can see this by writing $\Gamma = \gamma_{th} h/h_c \gamma_1$ where γ_{th} is the relaxation rate near or below threshold, and noting that theory predicts $\Gamma \simeq 7.75$ for period doubling, whereas the observed value for h/h_c is 1.62. A smaller dissipation rate at high power levels has also been suggested [12] for the perpendicular pumping configuration. A possible source for this is the heating of selected modes in the magnon bath due to a bottleneck effect similar to that observed in nonequilibrium phonon systems.

This work has been supported by FINEP, CNPq and CAPES.

References

1. See for example T.S. Hartwick, E.R. Peressini, and M.T. Weiss, J. Appl. Phys. 32, 223S (1961)
2. See for example V.E. Zakharov, V.S. L'vov and S.S. Starobinets, Usp. Fiz. Nauk 114, 609 (1974) [Sov. Phys. Usp. 17, 896 (1975)]
3. K. Nakamura, S. Ohta and K. Kawasaki, J. Phys. C15, L143 (1982)
4. F. Waldner, D.R. Barberis and H. Yamazaki, Phys. Rev. A31, 420 (1985)
5. H. Yamazaki, J. Phys. Soc. Japan 53, 1155 (1984)
6. G. Gibson and C. Jeffries, Phys. Rev. A29, 811 (1984)
7. S.M. Rezende, F.M. de Aguiar and O.F. de Alcântara Bonfim, J. Mag. Mag. Materials 54-57, 1127 (1986)
8. F.M. de Aguiar and S.M. Rezende, Phys. Rev. Letters 56, 1070 (1986)

9. M. Mino and H. Yamazaki, J. Phys. Soc. Japan 55, 4168 (1986)
10. H. Yamazaki and M. Warden, J. Phys. Soc. Japan 55, 4477 (1986)
11. A. Arneodo, P. Coullet, C. Tresser, A. Libchaber, J. Maurer and D. d'Humières, Physica 6D, 385 (1983)
12. X.Y. Zang and H. Suhl, Phys. Rev. A32, 2530 (1985)
13. S.M. Rezende, O.F. de Alcântara Bonfim, and F.M. de Aguiar, Phys. Rev. B33, 5153 (1986)
14. S.M. Rezende and F.M. de Aguiar, Rev. Bras. Física 16, 324 (1986)
15. E.R. Hunt and R.W. Rollins, Phys. Rev. A29, 1000 (1984)

Part III

Theory of Spin Excitations

Spin 1 vs Spin 1/2 Antiferromagnetic Chains: The Staggered Magnetization

D.C. Mattis[1] *and C.Y. Pan*[2]

[1]Physics Department, University of Utah, Salt Lake City, UT 84112, USA
[2]Physics Department, Utah State University, Logan, UT 84322, USA

1. Introduction

F.D.M.Haldane [1] has proposed that the properties of Heisenberg antiferromagnetic linear chains depend in a major way on whether the spins have integer (s=1,2,...) or half-integer (s=1/2,3/2,...) magnitudes. One important consequence concerns the existence of an energy gap at the isotropy point ($J_x=J_y=J_z$), with all that this entails for the correlation functions and the low-temperature properties.

Haldane's conjecture was motivated by analogy with the nonlinear O(3) sigma model in one dimension. But other exactly soluble one-dimensional models of interacting particles, also known to have some bearing on the Heisenberg antiferromagnet in one dimension for s≥1, are entirely devoid of energy gaps [2]. Indeed, in view of the striking similarity between exact solutions of isotropic s=1/2 and classical (s→∞) linear-chain antiferromagnets, prior to Haldane's remarks [1] there was general concensus that none of the interesting properties of the Heisenberg antiferromagnet depended on s in any crucial manner, no matter what the number of dimensions.

Thus, when confirmation of Haldane's gap was first announced by Botet and Jullien [3] on the basis of a numerical study of rather short chains (which was accompanied by a rather elaborate and novel phase diagram in parameter space), a vigorous controversy ensued. A number of articles argued one side of the issue [4] or the other [5]. This is one of those unfortunate instances when even numerical analyses can be deemed inconclusive, due to convergence difficulties and possible misinterpretation of numerical trends.

In the lecture titled "Energy gap in isotropic antiferromagnets: is Haldane's conjecture correct?" which one of us (D.C.M) presented at this conference, an attempt was made to resolve the controversy analytically by means of standard spin-wave analysis. It was noted that in the limit $s>>1$ the "transverse" operators $S_i^{\pm}$ are insensitive to whether the individual spin is integer or half-integer, whereas the "longitudinal" operators S_i^Z differ in an essential way. In particular, the integer spins can have $S_i^Z=0$ while the half-integer spins always have nonvanishing longitudinal interactions. According to the by-now standard arguments of Lieb, Schultz and Mattis [6], all antiferromagnets of half-integer spins are devoid of an energy gap [7] whereas, aside from Haldane's conjecture and the disputed numerical work, little is known about the integer spins. Thus, as it is possible to show that the integer spin antiferromagnet is similar to the half-integer antiferromagnet in an external field, one has a convincing argument that the isotropy point has shifted; for if the energy gap for half-integer spins vanishes at the isotropy point, then it can not do so for the integer spins.

Because our claims suffered from a certain qualitative flavor (aside from the no-gap theorem [6], [7], all the properties of half-integer spin $s\geq3/2$ antiferromagnets have to be obtained approximately, e.g. using spin-wave theory) we undertook to supplement them with some numerical calculations. With the procedures outlined below, it has not been necessary to proceed to such long chains as had heretofore been thought necessary, obviating the need for Monte Carlo.

2. The Ground States

The first question concerns the ground state energy per site, e_0, for the isotropic antiferromagnet. With exchange parameter $J=1$, classically $e_0=-s^2$ whereas quantum mechanically the energy is lower; conventionally, one writes [8] $e_0 = -s^2(1+ \gamma_S/s)$ with $\gamma_S \rightarrow \gamma$ (constant) for large spins $s>>1$. Because the interesting cases $s=1/2$ or 1 are not large, individual calculations must be undertaken. In Figs.1 and 2 we illustrate upper/lower bounds for $s=1/2$ and $s=1$, obtained as follows.

Upper bound for $s=1/2$: obtained variationally using real-space renormalization. We solve for the doublet ground eigenstates of

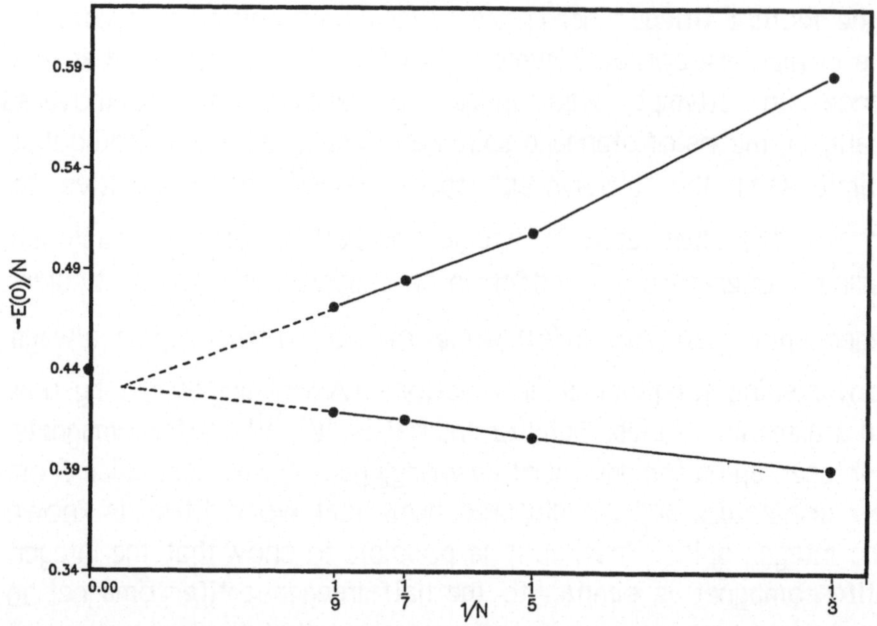

Fig.1. Upper and lower bounds to the ground state energy e_0 calculated for various segments of length N and extrapolated to $N \to \infty$ (i.e. $1/N = 0$). We plot $-e_0(N)$ vs. $1/N$ for a linear chain antiferromagnet, $J=1$, $s=1/2$.

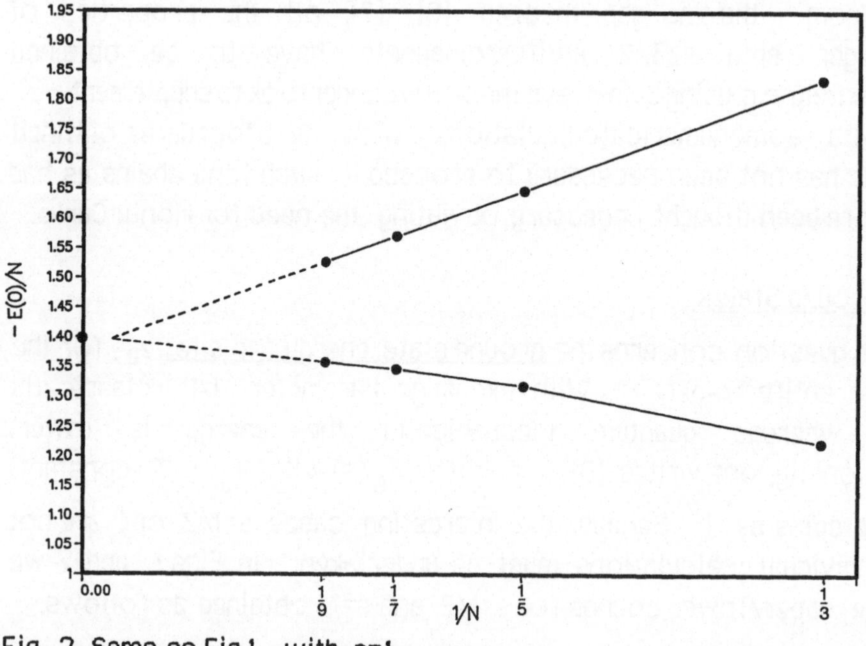

Fig. 2. Same as Fig.1, with $s=1$.

chainlets (S_1,S_2,S_3), (S_4,S_5,S_6), (S_7,S_8,S_9), ... Within the ground state subspace, we calculate the new bonds J' connecting the chainlets (S_3 to S_4, S_6 to S_7, etc.,) then construct new chainlets out of 3 primitive chainlets using the new coupling constant J'. Infinite iteration of this procedure yields a geometric series which is easily summed to yield $e_0(3)=-0.391...$, which is already within 10% of the exact answer. We now repeat with chainlets of N=5, obtaining $e_0(5)=-0.407...$then with N=7, and 9. Of course, here we know the exact limit thanks to the well known [8] Bethe-Hulthen result, $e_0(\infty)= -0.443...$. Although convergence to better than 1% is not numerically achievable at any reasonably small value of the odd integer N, the plot of $e_0(N)$ vs. 1/N does extrapolate accurately to the exact value as seen in Fig.1. What is more, extrapolation of a lower bound serves to pinpoint the correct result.

Lower bound for s=1/2: obtained by taking the energy of overlapping chainlets (S_1,S_2,S_3), (S_3,S_4,S_5), (S_5,S_6,S_7), ... and estimating $e_0(3)$ therefrom, then constructing chainlets such as $(S_1,...,S_5)$, $(S_5,...,S_9)$, $(S_9,...)$ ... to estimate $e_0(5)$, and proceeding similarly to N=7,9. (These numbers are of course already known from the previous calculation.) Bonds are correctly counted at each value of N although end spins (e.g. S_3,S_5,S_7 for the chainlets of N=3, $S_5,S_9,...$ for chainlets of N=5, etc.) are double-counted. This calculation thus results in a lower bound to the exact ground state energy. But as the error is localized at the ends, the error per spin diminishes roughly as 1/N. We can therefore expect the linear extrapolation of the calculated $e_0(N)$ to $1/N \to 0$ to yield an accurate $e_0(\infty)$. And indeed, it does.

Upper/lower bounds for s=1 : We proceed as for s=1/2. As long as N is chosen to be an odd integer, the ground states are always triplets (as we have explicitly verified) and the renormalization procedure works as simply as in the preceding case, with the extrapolation to $1/N \to 0$ falling within the same bounds of accuracy.

3. Ground State Antiferromagnetic Correlations

We now turn to the vexing questions of the existence/ nonexistence of an energy gap, and of long range order. The ground state of an ordinary antiferromagnet should be highly responsive to an external staggered field just as the ground state of a ferromagnet is highly responsive to an homogeneous field. While the staggered susceptibility χ indeed diverges for a three-dimensional antiferromagnet at all $T<T_N$, it is not expected to do so in one dimension because long-wavelength fluctuations prohibit the institution of long-range order. Nevertheless, vestigial (short-range ordered) Néel correlations, if present, will reveal themselves in an external staggered field, by exhibiting a large value of χ. For chains of s=1/2 the lack of an energy gap, the finite density of excited states near zero energy, and the relatively slow decay of correlation functions do conspire to yield a substantial χ. Conversely, the Haldane ground state is supposed to be separated from the spectrum of elementary excitations by an energy gap, and should consequently exhibit little, if any, of the Néel type correlations (even of the short-range variety.) This is precisely what we now find.

To test these ideas, we have added to the Heisenberg antiferromagnetic Hamiltonian the perturbing Hamiltonian,

$$H' = h\sum(-1)^n s_n^z \ . \tag{1}$$

Assuming the ground state energy to be lowered by an amount $-(N/2)\chi h^2$ for small h, with $\chi \propto s(s+1)$, we have defined $X_s=\chi/s(s+1)$ as the scaled response parameter <u>per spin</u>, and have plotted $X_{1/2}$ as well as X_1 for chains of lengths N=2,4,6,8. Odd length chains are inconvenient here, because the low field response is dominated by the odd spin. Nor is the renormalization procedure used, for in a field J' becomes anisotropic $(J_z' \neq J_x'=J_y')$, which complicates the iteration procedure. The results we have found seem to justify the relatively crude approximations made to obtain them. Fig.3 shows the response functions $X_{1/2}$, X_1, calculated at various N=2,4,6,8. When extrapolated to $1/N \to 0$ they clearly show the response of spins 1 to the staggered field to be some 20 times smaller than that of spins 1/2, as expected if antiferromagnetic correlations are indeed absent from the s=1 ground state.

4. Conclusions

While the present results are not an "acid test" of Haldane's conjecture, which a direct calculation of the eigenvalue spectrum would have been, neither do they suffer from the convergence difficulties which have

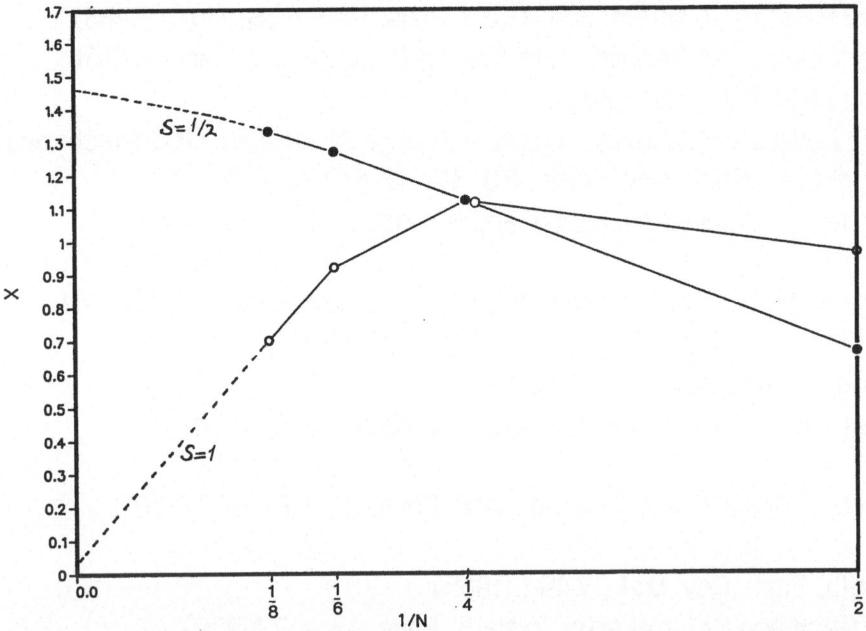

Fig.3. Ground state susceptibility **X**, per site, in units of Js(s+1), measuring the response to an external staggered field as calculated for various length segments N. Upper curve, s=1/2. Lower curve, s=1 . Note the significant difference in the magnitude of the response in the extrapolated limit of an infinite chain (i.e. 1/N=0).

plagued the more ambitious enterprises [4], [5]. An improved calculation of X using the paraphernalia of renormalization group calculations is evidently quite within reach. It should be interesting to see whether it will support the preliminary conclusions presented here.

References

1. F.D.M.Haldane, Phys. Lett. **93A**, 464 (1983), Phys. Rev. Lett. **50**, 1153 (1983)

2. F.D.M.Haldane, J.Phys. **C15**, L1309 (1982)
L.A.Takhtajan, Phys. Lett. **87A**, 479 (1982)
H.M.Babujian, Phys. Lett. **90A**, 479 (1982)
N.Andrei and H.Johannesson, Phys. Lett. **104A**, 370 (1984)

3. R.Botet and R.Jullien, Phys. Rev. **B27**, 613 (1983)

4. *For*: R.Botet, R.Jullien and M.Kolb Phys. Rev. **B28**, 3914 (1983)
J.Parkinson, J.C.Bonner, G.Müller, M.P.Nightingale and H.Blöte, J. Appl. Phys. **57**, 3319 (1985)
W.J.L.Buyers, R.M.Morra, R.L.Armstrong, M.J.Hogan, P.Gerlach and H.Hirakawa, Phys. Rev. Letts. **56**, 371 (1986)
I.Affleck, T.Kennedy, E.H.Lieb and H.Tasaki, to be published

5. *Against*: J.C. Bonner and H.Müller, Phys. Rev. **B29**, 5216 (1984)
S.T.Chui and K.B.Ma, Phys. Rev. **B29**, 1287 (1984)
K.Sogo, Phys. Lett. **104A**, 51 (1984)
J.Solyom and T.A.L.Ziman, Phys. Rev. **B30**, 3980 (1984)

6. E.Lieb, T.Schultz and D.Mattis, Ann. Phys.(N.Y.) **16** ,407 (1961)

7. M.Kolb, Phys. Rev. **D31** , 7494 (1985)
I.Affleck and E.Lieb, Letts. in Math. Phys. **12** , 57 (1986)
M.Kohmoto and D.Mattis, unpublished (also dealing with the extension to d>1 in the manner of ref.6)

8. D.Mattis "The Theory of Magnetism I : Statics and Dynamics", Springer, Berlin, New York, 1981; see chapter 5

Excitations and Critical Behavior in Generalized Heisenberg Spin Chains

J.C. Bonner[1], G. Müller[1], and J.B. Parkinson[2]

[1]Physics Department, University of Rhode Island, Kingston, RI 02881, USA

[2]Department of Applied Mathematics, UMIST, Manchester M60 1QD, United Kingdom

1. Introduction

Heisenberg spin chains which represent the simplest realistic models for magnetic insulators were thought to be well understood and generically similar for any spin-value s. This is expressed in the spin-wave approach to Heisenberg spin chains [1]. Consequently surprise and some degree of controversy resulted from recent work of Haldane [2,3], who proposed a dramatically different picture. Consider the spin-s XXZ Hamiltonian with anisotropy parameter Δ:

$$H = J \sum_{\ell=1}^{N} (S_\ell^x S_{\ell+1}^x + S_\ell^y S_{\ell+1}^y + \Delta S_\ell^z S_{\ell+1}^z). \quad (1)$$

For half-integer s, the region $0 \leq \Delta < 1$ is a gapless phase with power-law decay of the two-spin correlation functions, terminating in an essential singularity at $\Delta = 1$. For $\Delta > 1$, the ground state consists of two degenerate singlet states associated with long-range order and a gap to an excitation continuum. For integer s, on the other hand, the gapless phase associated with planar anisotropy extends only over a range $0 \leq \Delta \leq \Delta_1$ ($\Delta_1 < 1$) and the phase with gap and ordered ground state extends over $\Delta \geq \Delta_2$ ($\Delta_2 > 1$). A new phase (called hereafter the Haldane phase) appears in the region $\Delta_1 < \Delta < \Delta_2$ encompassing the Heisenberg point at $\Delta = 1$. The Haldane phase ground state is a non-ordered singlet with exponentially decaying spin correlation functions, and there is a gap to an excitation continuum, which has its maximum value at $\Delta = 1$. The spin-dependent gap at $\Delta = 1$ is given by $\Delta E/J \sim s^2 \exp(-\pi s)$.

Generalized Heisenberg spin chains are not exactly solvable (Bethe Ansatz integrable) for $s > 1/2$, except in very special situations. Hence Haldane's conjecture has, perforce, been investigated with a variety of numerical techniques, including scaled-gap and finite-size scaling calculations, finite-chain extrapolations, variational approaches and various correlation function calculations [4]. Obtaining reliable numerical results turns out to be quite difficult and great care must be taken. For example, while the first numerical, finite-size scaling calculation on spin-1 XXZ chains [5] revealed the predicted Haldane phenomena, it was subsequently demonstrated that a "pseudo-Haldane picture" is obtained also for the spin-1/2 XXZ model [6] in contradiction to exact analytic results. However, it is now the consensus of a large body of numerical work that the Haldane picture is, nevertheless, correct, and experimental support for this conclusion is starting to appear [7].

On the basis of a comprehensive numerical study and survey we have concluded that the various classes of excitations predicted by Haldane for the spin-1 XXZ model are all present and behave as conjectured [8]. However, we observe additional interesting features, in particular, classes of excitations which have the potential for modifying somewhat the basic Haldane picture, as we will discuss.

The addition to the basic Heisenberg Hamiltonian of exchange anisotropy, single-ion anisotropy, biquadratic exchange, or a magnetic field, generates a rich and complicated phase diagram for chains with $s > 1/2$ [4]. Unusual spectral features in generalized spin-1 chains will be discussed in the context of nonintegrability effects and quantum chaos [9,10].

2. Spin-1 Heisenberg Antiferromagnetic Chain

Since the predicted singlet-triplet Haldane gap is a maximum at the isotropic Heisenberg point, $\Delta = 1.0$, numerical attention has focussed on this limit. While several numerical techniques have been devised to study this problem [8], the most direct approach is to examine the behavior of the Haldane gap for a sequence of finite systems of increasing size and examine the trend as $N \to \infty$. Fig. 1 shows finite-N gaps for up to $N = 20$ spins with $s = 1/2$ as a function of $1/N$. The gaps extrapolate convincingly to a value very close to zero, in agreement with the exact result that this system should be gapless in the limit $N \to \infty$. Exact results up to $N = 14$ for the $s = 1$ case, on the other hand, show concave upwards curvature, consistent with a nonzero gap in the limit $N \to \infty$. However, since earlier studies [6] have pointed to the importance of obtaining data for very long chains to be sure of observing a reliable large N trend, a quantum Monte Carlo approach was developed to obtain data out to $N = 32$ spins [11]. The Monte Carlo data continue the concave upward trend of the exact finite-N data and predict a limiting singlet-triplet gap of magnitude

$$\Delta E/J \sim 0.41. \tag{2}$$

This gap occurs at the Brillouin zone boundary, since the excited triplet is at $k = \pi$ and the ground state is always a $k = 0$ state. The Haldane conjecture implies that a gap of equal magnitude should occur at the zone center. In Fig. 1, the finite-N triplet gaps for the $k = 2\pi/N$ mode are shown, for $N \leq 14$. The convergence is quite regular, and the extrapolated limit is in reasonable agreement with Eq. 2.

There does, however, appear to be one remarkable feature associated with the spectral excitations in the Heisenberg limit and vicinity. Numerical studies have revealed a crossover in the character of the spectral excitations as a function of field [12]. At high fields, the low-lying dispersion spectra are qualitatively similar to those for $s = 1/2$, i.e. have quantum character. At low fields, on the other hand, the dispersion spectra display notable classical character, with one complication. An additional set of modes occurs which appears unrelated to the classical spectra. When extrapolated as a function of $1/N$, these anomalous modes project <u>below</u> all other excitations with the same value of S_T^z. For this reason these states have been termed "supersoft" modes [12]. In particular the $S_T^z = 1$ mode lies at $k = \pi/N$, and its excitation energy is included in Fig. 1. Extrapolating below all other excited states it appears to extrapolate <u>below</u> the triplet at $k = \pi$ which has been used to determine the Haldane gap! The curvature is consistently concave downwards and increases with increasing N, making it unlikely that the curve could develop an inflection point for larger N and tend to the value (2).

Detailed information on the $T = 0$ phase behavior of quantum spin chains may be inferred from the integrated intensity

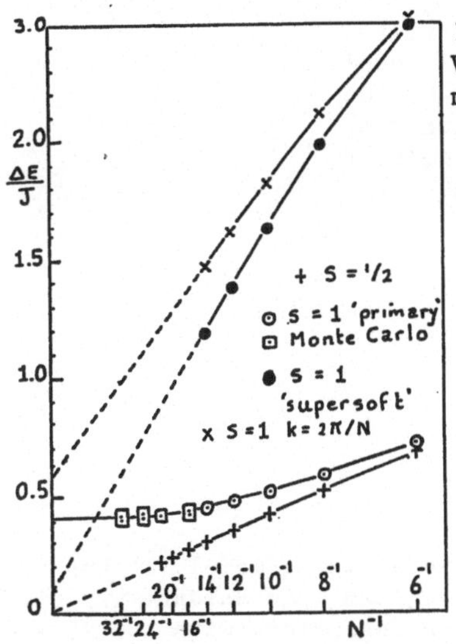

Fig. 1.
Various excited modes for the s = 1 antiferromagnetic Heisenberg chain as discussed in the text.

$$I(k) = \sum_{R=-\infty}^{\infty} e^{ikR} \langle \vec{S}_\ell \cdot \vec{S}_{\ell+R} \rangle, \tag{3}$$

At zero temperature, I(k) is a property of the ground state solely, and yet also contains information on the excited states, at wave number k. It can be shown [8] that in the presence of a gap ΔE between the ground state and the lower edge of the excitation continuum, the integrated intensity must satisfy the following inequality:

$$I(k) \leq 2|E_G| (1 - \cos k)/3\Delta E, \tag{4}$$

where E_G is the ground-state energy per spin. For the Heisenberg spin-1 chain calculations of Blöte and Nightingale [11] give $E_G/J \sim -1.4015$. Together with (3) this yields the following lower bound for the inverse integrated intensity: $I^{-1}(\pi) \geq 0.219$. A plot of $1/I(\pi)$ versus $1/N$ is consistent with this bound in the limit $N \to \infty$, but a plot of the inverse intensity $1/I(\pi-2\pi/N)$, which is dominated by the $S_T^z = 1$ supersoft mode, fails appreciably to satisfy this criterion and is not consistent with an energy gap (2).

Hence present numerical evidence out to N = 16 spins is consistent with the presence of a special class of excitations not predicted by Haldane. These isolated excitations probably have insufficient thermodynamic weight to affect the thermal properties, but since there exist N/2 such modes (corresponding to $S_T^z = 1,2,\ldots,N/2$), they should be experimentally observable in the T = 0 magnetization isotherm at low magnetic fields. In particular, the T = 0 magnetization isotherm should become zero at a much lower field than that corresponding to the Haldane gap (2).

3. Spin-1 XXZ Model

Of all the excitation phenomena predicted by Haldane to occur for the s = 1 XXZ model, the one which is most difficult to confirm numerically has been the behavior in the vicinity of the critical point Δ_2 [13]. In fact, doubt has even been expressed concerning the validity of this particular aspect of the conjecture. Haldane predicts that the transition at $\Delta = \Delta_2$ should be in

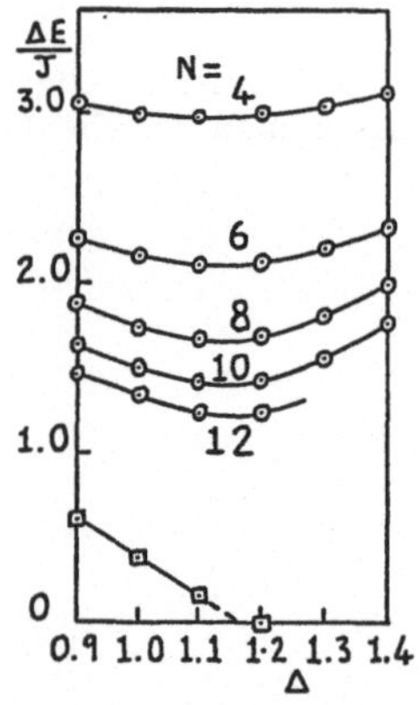

Fig. 2. $S_T^z = 0$ "scaling" states for the s = 1 XXZ model for N = 4,6,8,10 and 12 (denoted by ⊙). The excitation denoted by ⊡ is the extrapolated limit of the upper component of the $\Delta \geq \Delta_2$ ground state.

the universality class of the spin-1/2 transverse Ising model. It has already been established [4,8] that at $\Delta_2 \sim 1.18 - 1.20$, the Haldane gap disappears and an excited $S_T^z = 0$ state becomes degenerate with the $S_T^z = 0$ ground state for $\Delta > \Delta_2$. The mapping to the transverse Ising model implies the existence in the limit $N \to \infty$ of an infinite continuum of scaling states quasi-degenerate with the ground state (states) at. and only at, $\Delta = \Delta_2$. The development of this scaling continuum is not very apparent in the spectra for small finite systems. A search for the scaling states implied by the Haldane conjecture revealed a class of high-lying $S_T^z = 0$ excitations at k=0. Plotted as a function of Δ, these excitations are shown in Fig. 2. A minimum develops in the vicinity of $\Delta \sim 1.1 - 1.2$ which intensifies with increasing N. Fitting a polynomial through data in the vicinity of the minimum and plotting the excitation gap at the polynomial minimum versus 1/N yields an extrapolated value well below the lower edge of the triplet ($S_T^z = 1$) continuum, and even consistent with a value zero. We conclude these are the Haldane scaling states at Δ_2. This conclusion is reinforced by a detailed study of corresponding excitations for the spin-1/2 transverse Ising model near the critical field. The transverse Ising model picture shows a striking resemblance to Fig. 2.

A surprising feature of the Haldane prediction [3] is that the zz-correlation function decays as a power law, $|\langle S_\ell^z S_{\ell+1}^z \rangle| \sim R^{-1/4}$, whereas the correlation function $\langle S_\ell^x S_{\ell+1}^x \rangle$ decays exponentially for $R \to \infty$. This unusual prediction implies that the fluctuations are critical in the longitudinal (z) direction but not in the transverse (x) direction. Such a situation has not been observed previously. Hence we examined also the $S_T^z = 1$ excitations at k = 0 as a function of Δ, and again observed the development of minima with increasing N in the vicinity of Δ_2. These minima also display a potential for extrapolating to zero, analogous to the $S_T^z = 0$ minima, and in contrast to k = 0 excitations in the same class corresponding to $S_T^z = 2$, etc. If gapless $S_T^z = 1$ excitations occur at $\Delta = \Delta_2$, the transverse xx-correlations will also display power-law decay, in disagreement with Haldane's specific predictions for the behavior of the two-spin correlation functions [2].

4. Nonintegrability Aspects of the Spectra

The above numerical studies of the spin-1 XXZ model, including the Heisenberg point, show that the Haldane prediction represents a remarkably successful mapping. All classes of states predicted by Haldane appear to be present and behave generally as predicted. It appears, however, that we have discovered additional classes of states whose presence have the potential of playing an important role in the extended T = 0 phase diagram. The $|S_T^z| = 1$ "scaling" states at $\Delta \sim \Delta_2$ are perhaps less surprising since analogous states occur for the transverse Ising model near the critical field. The "supersoft" modes at $\Delta = 1$, on the other hand, are a remarkable feature not encountered previously in any integrable model. Here we discuss in more detail the possible occurrence in generalized $s \geq 1$ Heisenberg spin chains of spectral features which are characteristic of nonintegrable models.

Consider the spin-1 Heisenberg antiferromagnetic chain generalized by addition of biquadratic exchange:

$$H/J = \sum_{\ell=1}^{N} \vec{S}_\ell \cdot \vec{S}_{\ell+1} - \beta \sum_{\ell=1}^{N} (\vec{S}_\ell \cdot \vec{S}_{\ell+1})^2. \tag{5}$$

Special limits of this bilinear-biquadratic exchange Hamiltonian are:

$\beta = 0$ - Heisenberg model (nonintegrable)
$\beta = 1$ - "Russian" model [14] (integrable)
$\beta = \infty$ - pure biquadratic (nonintegrable)

Hamiltonian (8) at the special point $\beta = 1$ is an example of an $s > 1/2$ model which is Bethe Ansatz integrable and gapless [14,15]. Hence it is of interest to investigate the parameter range $0 \leq \beta \leq \infty$ to determine the extent of the gapless region. A prediction of Affleck [16] is that only the point $\beta = 1$ is gapless and that a gap opens on either side of the Russian point as $\Delta E \sim |1 - \beta|$. Numerical studies including scaled-gap [17] and other finite-size scaling calculations [18,19] are consistent with Affleck in predicting the opening of a singlet-triplet gap at $\beta = 1$, but differ in predicting that the gap opens up more slowly than linearly. Furthermore, numerical calculations revealed a very curious phenomenon in the Russian-biquadratic regime ($0 \leq 1/\beta \leq 1$) [18]. This phenomenon is illustrated in Fig. 3. Close to the biquadratic limit, the first excited state is no longer the triplet at $k = \pi$ for sufficiently large N, but instead becomes a singlet at π. A crossover effect occurs in this regime, illustrated in Fig. 3, which implies that as $N \to \infty$, the lowest excited state is a singlet over the whole regime. Traditional finite-size scaling approaches which assume, as an act of faith, that the dominant excited states for small finite N remain the dominant states as

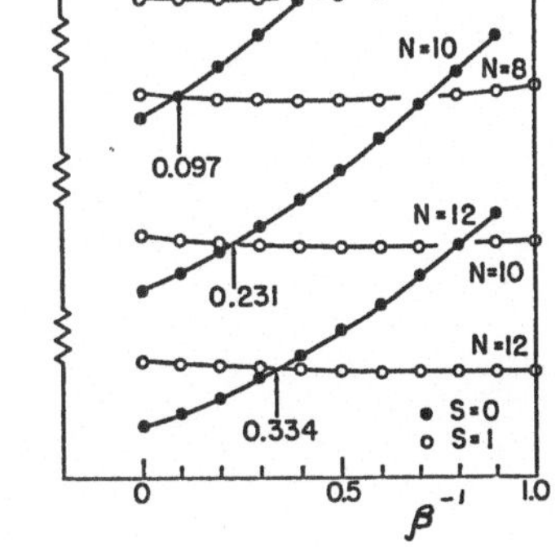

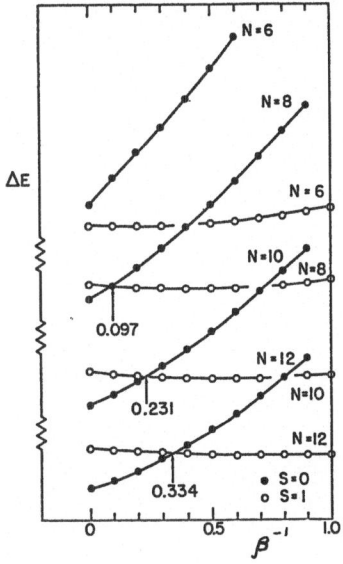

Fig. 3. A plot of the values of β^{-1} at which the lowest excited state in the range $0 \leqslant \beta^{-1} \lesssim 1$ changes from a singlet to triplet versus N^{-1}. The rough extrapolated limit of the crossing points is consistent with the value unity, implying that the singlet is ultimately the dominant excited state in this range.

as $N \to \infty$, can be seriously misled by a situation such as the above. Just as in the case of the supersoft modes, a new class of important modes appears at sufficiently large N. This triplet-to-singlet crossover phenomenon does not occur for integrable systems (e.g. the spin-1/2 XXZ model), and is presumably a spectral nonintegrability effect.

We acknowledge contributions by J. Oitmaa. Support has been provided by the US NSF grant DMR 86-03036 and the Research Corporation.

References

1. See, e.g., R. Kubo, Phys. Rev. 87, 586 (1952).
2. F. D. M. Haldane, Phys. Lett. 93A, 464 (1983); F. D. M. Haldane, Phys. Rev. Lett. 50, 1153 (1983).
3. F. D. M. Haldane, ILL. Report No. SP81/95 (unpublished).
4. J. C. Bonner, J. Appl. Phys. 61, 3941 (1987) (invited review).
5. R. Botet and R. Jullien, Phys. Rev. B 27, 613 (1983).
6. J. C. Bonner and G. Müller, Phys. Rev. B 29, 5216 (1984).
7. J. P. Renard et al., Europhys. Lett. In press. See also W. J. L. Buyers et al., Phys. Rev. Lett. 56, 371 (1986).
8. J. C. Bonner, J. B. Parkinson and G. Müller, unpublished work.
9. G. Müller, J. Bonner and J. Parkinson, J. Appl. Phys. 61, 3950 (1987).
10. G. Müller, Phys. Rev. A 34, 3345 (1986).
11. M. P. Nightingale and H. W. J. Blöte, Phys. Rev. B 33, 659 (1986).
12. J. B. Parkinson and J. C. Bonner, Phys. Rev. B 32, 4703 (1985) and J. Phys. C (in press).
13. See, e.g., K. Kubo and S. Takada, J. Phys. Soc. Jpn. 55, 438 (1986).
14. L. A. Takhtajan, Phys. Lett. 87A, 479 (1982); H. M. Babujian, Phys. Lett. 90A, 479 (1982).
15. J. B. Parkinson and J. C. Bonner, J. Phys. C 19, 6063 (1986).
16. I. Affleck, Nucl. Phys. B265, 409 (1986).
17. J. Oitmaa, J. B. Parkinson and J. C. Bonner, J. Phys. C 19, L595 (1986).
18. J. C. Bonner, J. B. Parkinson, J. Oitmaa and H. W. J. Blöte, J. Appl. Phys. 61, 4432 (1987).
19. H. W. J. Blöte and H. W. Capel, Physica 139A, 387 (1986).

Exact Results for the Out of Plane Dynamical Correlation Function in a Classical Easy Plane Ferromagnet at Low Temperatures

G. Reiter

University of Houston, Houston, TX77004, USA

Projection operator methods[1] reduce the problem of calculating the spin density dynamical correlation function of spin systems to that of calculating the spin current relation function $\gamma_q^\alpha(\omega)$, by means of the exact result

$$R_q^\alpha(z) = \int_0^\infty d^{izt} \langle S_q^\alpha(t)\, S_{-q}^\alpha \rangle dt = \frac{i\langle S_q^\alpha\, S_{-q}^\alpha\rangle}{z - \omega_q^{\alpha,2}/(z + \gamma_q^z(z))} , \tag{1}$$

where $\omega_q^{\alpha,2}$ is the exact second moment $\langle \dot{S}_q^\alpha\, \dot{S}_{-q}^\alpha\rangle / \langle S_q^\alpha S_{-q}^\alpha\rangle$, and $\langle\ \rangle$ denotes an equilibrium average. For classical systems below their lower critical dimension, the coherence length diverges as $T \to 0$. One expects to observe well defined spin waves at all wave vectors at $T = 0$, which implies that $\gamma_q^\alpha(\omega) \to 0$. One can expand $\gamma_q^\alpha(\omega)$ as a power series in the temperature,

$$\gamma_q^\alpha(\omega) = KT\gamma_{1,q}^\alpha(\omega) + (KT)^2\gamma_{2,q}^\alpha(\omega) + \ldots \tag{2}$$

and, barring singularities in the higher order terms, the dynamics will be determined by $\gamma_{1,q}^\alpha(\omega)$ or perhaps if this vanishes, $\gamma_{2,q}^\alpha(\omega)$. These, however, can be calculated at $T \doteq 0$, where the spin wave theory gives an exact picture of the small amplitude dynamics. The first calculation of this kind made use of an equation of motion approach for the dynamics of the operators appearing in the definition of $\gamma_q^\alpha(\omega)$., i.e.,

$$\gamma_q^\alpha(\omega) = \langle \ddot{S}_q^\alpha\, Q(z - QLQ)^{-1} Q\ddot{S}_{-q}^\alpha\rangle / \langle \dot{S}_q \dot{S}_{-q}\rangle , \tag{3}$$

where Q is a projection operator that projects out the part of an operator that contains S_q^α or $\dot{S}_q^\alpha$, and L is the Liouville operator for the system. It was subsequently realized that a straightforward spin wave calculation led to the same results for the Heisenberg model.[2] The spin wave theory is applicable, I believe, to any system below its lower critical dimension. A treatment has been given for the two dimensional Heisenberg[2] model and for the in plane motion of the easy plane ferromagnet,[3]

$$H = -J \sum_i (\vec{S}_i \cdot \vec{S}_{i+1} + D(S_i^z)^2) . \tag{4}$$

The results of this theory agree with that of the isotropic model as $D \to 0$, and the theories of Villain,[4] and Nelson and Fisher[5] at long wavelengths. They disagree with calculations by Cieplak and Sjolander[6] based upon the equation of motion method, although the two results should agree in principle. I attribute this to numerical errors in the final evaluation of their results.

The results also disagree with numerical simulations of Loveluck et al.[7], again presumably due to numerical error in their codes.

We present here results for the out of plane correlation function for temperature $\ll \sqrt{8DJ}$, the crossover temperature. The out of plane linewidth, which one might naively expect to be $\gamma_q^z(\omega_q)$, where ω_q is the non-interacting spin wave frequency

$$\omega_q = S((J_0 - J_q)(J_0 - J_q + 2D))^{1/2} , \tag{5}$$

would actually be infinite if that expression were used, as $\gamma_q^z(\omega)$ will be shown to be logarithmically divergent at $\omega = \omega_q$. The results of Cieplak and Sjolander for these functions at ω_q, which are finite, are therefore incorrect.

A remarkable feature of $\gamma_q^z(\omega)$ is the appearance of a discontinuity as a function of ω. This arises as a result of a catastrophe, in the mathematical sense, that occurs when a region of phase space that contributes to the damping ceases to be accessible. The discontinuity should be observable in the out of plane linewidth as a function of temperature or wave vector.

The second moment can be calculated using the identity $\langle \dot{S}_q^z \dot{S}_{-q}^z \rangle = KT\langle [\dot{S}_q^z, S_{-q}^z] \rangle$ and the spin wave theory to evaluate the nearest neighbor pair correlation function and is

$$\omega_q^{2,z} = \omega_q^2(1 - K_\perp a/r)(1 - K_\perp a) , \tag{6}$$

where $r = (\bar{D}(\bar{D} + 2))^{1/2}$, and $\bar{D} = D/J$. $K_\perp a = 1/2KT/JS^2$ is in the plane inverse coherence length. a is the lattice parameter.

The quantity $\ddot{S}_q$ can be calculated straightforwardly from the equation of motion, and the usual spin wave approximation, followed by a Bogolyubov transformation. The effect of Q is to subtract the pairings of operators on the same side of the bracket in (3). The Q operator appearing in conjunction with the Liouville operator in (3) can be ignored to lowest order in T.

This leads to an expression for $\gamma_q^z(\omega + i\varepsilon)$, the imaginary part of $\gamma_q^z(z)$, given in (7)

$$\begin{aligned}\gamma_q^{z''}(\omega) = {} & \frac{\pi(KT)^2}{8S^2} \sum \frac{(\Gamma'^5_{123})^2 [\delta(\omega-\omega_1+\omega_2+\omega_3) + \delta(\omega+\omega_1-\omega_2-\omega_3)]}{(1-\cos q_1+\bar{D})(1-\cos q_2+\bar{D})(1-\cos q_3+\bar{D})} \delta(q-q_1-q_2-q_3) \\ & + \frac{\pi(KT)^2}{24S^2} \sum \frac{(\Gamma'^4_{123})^2 [\delta(\omega-\omega_1-\omega_2-\omega_3) + \delta(\omega+\omega_1+\omega_2+\omega_3)]}{(1-\cos q_1+\bar{D})(1-\cos q_2+\bar{D})(1-\cos q_3+\bar{D})} \delta(q-q_1-q_2-q_3) ,\end{aligned} \tag{7}$$

where the vertices $\Gamma^{4,5}$ are obtained from the equations of motion by the methods discussed in ref. (2).

The second δ function in the first term in (7) is responsible for the singularity in γ^z. The singularity occurs in the region $q_1 \simeq -q$, $q_2 \simeq q$, $q_3 \simeq q$, and is due to a divergence in the three spin wave density of states. To see this, let $q_1 = -q - \rho - \Delta$, $q_2 = q + \rho$, $q_3 = q + \delta$. Then for small ρ, Δ, the integral is proportional to

$$\int \delta(\delta\omega + a\rho\delta - b(\rho^2\delta + \delta^2\rho)) , \tag{8}$$

where $a = \partial^2\omega/\partial q^2$, $b = (1/2)\, \partial^3\omega/\partial q^3$, and $\delta\omega = \omega - \omega_q$.

It is the vanishing of the linear term in (8) that produces the singularity. With a straightforward change of variables, and doing one integration, the integral becomes

$$\sum_i \int_{x_a^i}^{x_b^i} \frac{dx}{((\delta\omega + ax)^2 - 4b^2x^3)^{1/2}} , \tag{9}$$

where the limits of integration are either the roots of the cubic polynomial in (9) or correspond to the boundary of the domain of integration in (7). The origin of the logarithmic singularity is now clear. As long as $a \neq 0$, the behavior near $\delta\omega = 0$ arises from small values of x and the cubic term can be neglected (if $a = 0$, which occurs for small $\bar{D}$ near $qa = .5\pi$, the divergence is stronger), leading to a logarithmic singularity. One branch that contributes to the total corresponds to the case that the denominator has three real roots, and the upper and lower arguments of the integral in (9) correspond to the roots x_2 and x_3 that come together and become imaginary when $\delta\omega$ increases to $-\delta\omega = (1/27)\, a^3/b^2$. The contribution from this branch is finite at the frequency at which it disappears, since

$$\lim_{x_2 \to x_3} \int_{x_2}^{x_3} \frac{dx}{((x-x_1)(x-x_2)(X_3-x))^{1/2}} = \frac{1}{(x_2-x_1)^{1/2}} \int_{x_2}^{x_3} \frac{dx}{((x-x_2)(x_2-x_3))^{1/2}} = \frac{\pi}{(x_2-x_1)^{1/2}} \,. \quad (10)$$

There is, then, a discontinuity in $\gamma_q^z(\omega)$ when this branch disappears. This is shown in Fig. 1, where we have plotted the singular part of $\gamma_q^z(\omega)$ for several values of q. The position of the discontinuity is determined by the sign of a, and corresponds to $\delta\omega < 0$ when $a > 0$ and vice versa. Since a is not quite zero at $q = 0.5$, there is still a discontinuity there, and in fact, the discontinuity is proportional to $1/a$, but it is too close to $\delta\omega = 0$ to appear in the graph. The singularity when $a = 0$ is proportional to $(\delta\omega)^{-1/3}$.

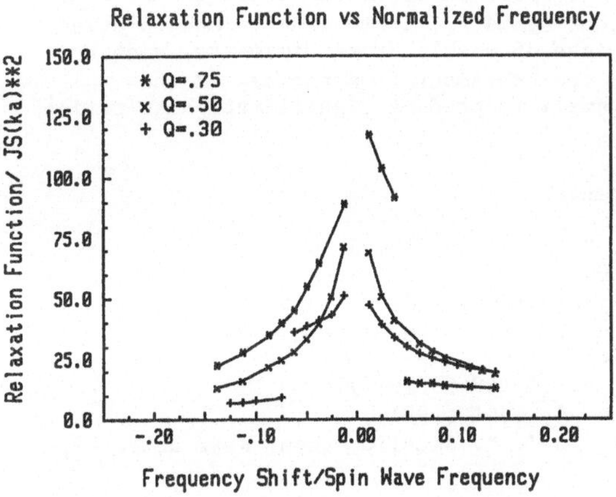

Fig. 1 The singular part of $\gamma_q^z(\omega)$ for several values of q and ω near the resonance frequency, vs $\delta\omega/\omega_q$. $D = 0.21$, the value appropriate for C_2NiF_3.

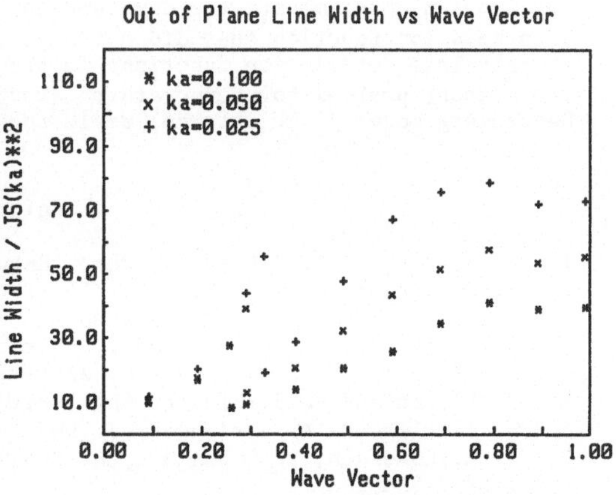

Fig. 2. The linewidth $\gamma_q^z((\omega_q^{z,2})^{1/2})$ as a function of q for several values of temperature. The overall factor of $(K_\perp a)^2 JS$ has been normalized away. The actual linewidth decreases as $K_\perp a$ decreases. $D = 0.21$.

In order to compute the linewidth at low temperatures, we must account for the fact that the resonant frequency is shifted to lower frequency at any finite temperature. There is a static shift, due to the change in the second moment, which is of order $K_\perp a$ and a dynamic shift, coming from the real part of γ^z that is of order $(K_\perp a)^2$ and can be neglected. Putting in the shift, we find that the linewidth is proportional to $(K_\perp a)^2 |\ln K_\perp a|$. We show in Fig. 2, the linewidth for several values of the temperature. The linewidth at $K_\perp a = 0.1$ and $q = 0.3$ is about 0.6 MeV for C_3NiF_3, and should be observable. There will, of course, be some rounding due to the fact that the delta function in (7) will have a width at finite temperatures. We estimate this to be of order $K_\perp$.

At long wavelengths, the behavior is diffusive, with the diffusion coefficient being given by $c^2/\gamma_0^z(0)$, where $\omega = cq$ for long wavelengths and $c^2 = 2DJS^2$.
As a function of $\overline{D}$, $\gamma_0(0)$ can be shown to be logarithmically divergent,

$$\gamma_0^z(0) = (K_\perp a)^2 JS \, \frac{4}{\pi} \ln(\overline{D}/\overline{D}_0) \,, \quad (11)$$

where $\overline{D}_0$ is evaluated numerically to be 1.75.

If we note that the coherence length along the z axis is $K_\parallel = (2\overline{D})^{1/2}$, we see that the diffusion coefficient is of the form

$$\left(\frac{K_\parallel}{K_\perp}\right)^2 |\ln K_\parallel a|^{-1} \tag{12}$$

This should be compared with the diffusion coefficient in the isotropic case, which is proportional to $|\ln Ka|^{-1}$.[1] In the latter case, the results have been corroborated numerically[1], and violate the dynamical scaling hypothesis. It appears that a functional form such as $(K_\parallel^2 + K_\perp^2)/K_\perp^2 \ln \sqrt{K_\parallel^2 + K_\perp^2}a$ would provide an interpolation between the two limits, although at the moment, a theory does not exist for temperatures at or above the easy plane crossover temperature.

There is a value of D ($\approx$ 5.5) above which the leading term in the temperature expansion of $\gamma_0^2(0)$ vanishes, and it is presumably then proportional to $(K_\perp a)^3$.

The methods used here can be trivially extended to the case that the anisotropy is of the exchange form, and hence can treat the x-y model. Indeed, as we have already mentioned, they should be useful for the dynamics of any system at or below its lower critical dimension. Since they rely heavily on the use of the symmetry of the problem to eliminate the divergences that would occur if one tried to calculate a correlation function that did not have that symmetry, it remains an open problem to calculate correlation functions in the case of weakly broken symmetry, such as the present problem for temperatures above the crossover temperature, or the Heisenberg model in an external field too small to produce significant polarization.

References

1. G. Reiter and A. Sjolander, Phys. Rev. B. 21, 5356 (1980).
2. G. Reiter, Phys. Rev. B. 21, 5356 (1980).
3. G. Reiter, "Magnetic Excitations and Fluctuations", S. W. Lovesey, U. Balucani, F. Borsa and V. Tognetti, eds., (Springer, Berlin, Heidelberg 1984).
4. J. Villain, J. de Physique 35, 27 (1974).
5. D. Nelson and D. S. Fisher, Phys. Rev. B. 16, 4945 (1977).
6. M. Cieplak and A. Sjolander, J. Phys. C 14, 4861 (1981).
7. J. M. Loveluck, T. Schneider, E. Stoll and J. R. Jauslin, Phys. Rev. Lett. 45, 1505 (1980).

A Comparison of the Bethe Ansatz and Green Function Approaches to Two-Magnon States of the Ferromagnetic Chain

P.D. Loly and W.K. Scott

Department of Physics, University of Manitoba, Winnipeg, Manitoba, Canada, R3T 2N2

1. Introduction

Until recently investigations of magnetic excitations in the generalized Heisenberg chain by the Bethe Ansatz method [1] were largely limited to S=1/2. The exception being for special models with particular coefficients, e.g. the completely integrable "Russian" models [2], which have recently been studied by TAKHTAJAN [3] and BABUJIAN [4] for general S, and PARKINSON and BONNER for S=1 [5]. The lack of complete integrability in the general case means that the Bethe Ansatz fails in general for n-magnon excitations. However MATTIS [6] has shown that the Bethe Ansatz could be used for two magnon excitations (n=2) from the ferromagnetic ground state for S>1/2. Two groups have recently given a fuller analysis of two-magnon states. The possibility of higher spin opens the way for inclusion of uniaxial anisotropy and higher order exchange.

First, HODGSON and PARKINSON [7] derived the basic equations, now three for S>1/2 rather than the two for S=1/2, since two excitations on a single site are now allowed. Their formulation included Ising anisotropy and they studied both the bound and continuum states, tabulating the phase shift information for all states of an N=12 chain for S=1 with no anisotropy. In addition they calculated the probabilities for pairs of spins at various separations in the periodically closed chain.

Second, PAPANICOLAOU and PSALTAKIS [8] derived the equivalent equations for the Heisenberg model supplemented by uniaxial anisotropy. They also considered an easy-plane model, and the case of biquadratic exchange. However their applications focussed only on the bound states via the large N limit of the infinite chain. They mainly recovered results already known from application of earlier bound state investigations, particularly those obtained from the Green function approach introduced by WORTIS [9], and subsequently used by many others. However, almost all applications have been restricted to nearest neighbour (NN) interactions. Of those a number discuss the 1D chain: TONEGAWA [10] studied the effect of Ising and uniaxial anisotropies on two-magnon bound states; PINK et al [11] gave extensive discussions of bound states including biquadratic exchange; CHIU-TSAO et al [12] studied the S=1 biquadratic exchange problem finding an SU(3) symmetric Goldstone mode exhausting the continuum for the Schrodinger case; and SCHNEIDER et al [13] studied dynamic form factors for the Heisenberg-Ising chain to study solitons and magnon bound states. BAHURMUZ and LOLY [14] studied the next nearest neighbour (NNN) chain, with analytic lattice Green functions [15], finding a very rich structure in the continuum spectra.

Our interest in the Bethe Ansatz approach arose from curiosity about its relationship to the Green function method which LOLY et al [16] have used extensively in 1-, 2- and 3-D. While the Green function method is only exact for n=2 excitations from ferromagnetic ground states (and of course at absolute zero), <u>it can be applied in any dimension and for any range of interactions</u> (<u>i.e.</u> <u>beyond nearest neighbours</u>). All this can be carried out for a very general Hamiltonian with various kinds of anisotropies and higher order exchange. Thus the Green function method has a greater breadth of application than one can ever hope to attain with the Bethe Ansatz. However, the advantage of the Bethe Ansatz is that it gives the wavefunctions for both the bound and continuum states [7]. Such information cannot be obtained from the Green function approach which only yields spectral information. The present contribution extends the discussion of HODGSON and PARKINSON [7] to include uniaxial anisotropy and includes a similar numerical analysis for N=13 chains. As a result we can abbreviate some of the discussion that may be found in their paper. The results are, however, presented in a somewhat different and more conventional way.

2. Bethe Ansatz for general S with Ising and Uniaxial Anisotropies

We consider the following Hamiltonian with $0 \leq \rho \leq 1$ and $D \geq 0$:

$$H = -J \sum_{i=1}^{N} [S_i^z S_{i+1}^z + \rho(S_{i+1}^x S_i^x + S_i^y S_{i+1}^y)] - D(S_i^z)^2 \tag{1}$$

and look for a wavefunction of the form:

$$\sum_{i=0}^{N-2} \sum_{j=i+1}^{N-1} f(i,j)[i,j\rangle + \sum_{i=0}^{N-1} g(i,i)[i,i\rangle \tag{2}$$

where $[i,j\rangle$ are unnormalised basis states for deviations on sites i and j. Substituting this into the Schrödinger equation we obtain three equations:

$$Ef(i,j)=J((4S+\mu)f(i,j)-\rho S[f(i-1,j)+f(i+1,j)+f(i,j-1)+f(i,j+1)]); \quad i+2\leq j\leq N-1 \tag{3a}$$

$$Ef(i,i+1)=J((4S-1+\mu)f(i,i+1)-\rho S[f(i-1,i+1)+f(i,i+2)]-\alpha[g(i,i)+g(i+1,i+1)]) \tag{3b}$$

$$Eg(i,i)=J((4S+\mu-2d)g(i,i)-\alpha[f(i-1,i)+f(i,i+1)]) \tag{3c}$$

with $\mu=2d(2S-1)$; $\alpha^2=\rho^2 S(2S-1)$; $d=D/J$

where E is the energy relative to the ground state of aligned spins.

Taking the usual Bethe Ansatz for (3a) of the form

$$f(i,j)=\exp(ik_1R_i+ik_2R_j+\phi/2) + \exp(ik_1R_j+ik_2R_i-\phi/2) \tag{4}$$

and applying the cyclic boundary conditions for a chain of N atoms, i.e. $f(i,j)=f(i+N,j)$, with the symmetry $f(i,j)=f(j,i)$, we find the phase-shifted wavevectors

$$Nk_1=2\pi\lambda_1+\phi \text{ and } Nk_2=2\pi\lambda_2-\phi \tag{5}$$

where λ_1, λ_2 are integers and $\mathrm{mod}(\phi)\leq\pi$.

We can then satisfy (3a,b,c) for the phase angle ϕ for a given λ_1, λ_2 provided that

$$\cot(\phi/2)=\tan(q/2)(2\rho S\cos(k/2)\cos(q/2)[\cos(q/2)+\rho(2S-1)\cos(k/2)]-d\cos(q/2)) \\ /(2\rho S\cos(k/2)\cos(q/2)[\rho\cos(k/2)-\cos(q/2)]+d[\cos(q/2)-2\rho S\cos(k/2)]) \tag{6}$$

and

$$E=J(4S+\mu)-2\rho JS[\cos(k_1)+\cos(k_2)] \tag{7}$$

where K is the total wavevector (k_1+k_2) and q is the difference wavevector (k_1-k_2). We may also obtain an expression for g(i,i) by the use of (3c,4):

$$g(i,i)=\exp(ikR_i)[(2S-1)/S]^{1/2}\, 2\rho S\cos(q/2)\cos(k/2)\,[\cos(\phi/2)+\tan(q/2)\sin(\phi/2)] \\ /(2\rho S\cos(q/2)\cos(k/2)-d) \tag{8}$$

where the phase angle ϕ has now been defined by (6).

The probabilities, P_n, of having two spin deviations separated by n sites can be evaluated to give

$$P_0=2G/(2G+F), \text{ and } P_n=2\,[\mathrm{mod}(f_n)]^2/(2G+F)\ (n\neq 0) \tag{9}$$

where

$$F = \sum_{p=1}^{N-1} [\mathrm{mod}(f_n)]^2 = 2N-4\cos(\phi/2)[\cos(\phi/2)-\cot(q/2)\sin(\phi/2)] \tag{10}$$

with $\mathrm{mod}(f_p)=\mathrm{mod}[f(i,i+p)]$ for any i, and

$$G=(\mathrm{mod}[g(i,i)])^2=(\mathrm{mod}[g(0,0)])^2$$

$$=[2\rho\cos(q/2)\cos(K/2)]^2[\cos(\phi/2)+\tan(q/2)\sin(\phi/2)]^2$$

$$/[2\rho\cos(q/2)\cos(K/2)-d]^2, \text{ for } S=1. \tag{11}$$

There are two sum rules of interest, first, for each allowed value of the total wavevector K:

$$\sum_{E_i} P_n(E_i) = 1 \text{ for any } n, \tag{12}$$

and second, for each K and E_i combination:

$$P_0 + 1/2\sum_{n=1}^{N-1} P_n(E_i) = 1 \text{ for any solution } E_i. \tag{13}$$

3. Results for N=13, S=1 and D>0 (ρ=1)

The numerical solution of (6) takes a pair of λ_1, λ_2, giving $k_1(\phi)$, $k_2(\phi)$ so that (6) yields a set of solutions ϕ_ℓ, of which those with real wavevectors are scattering states in the continuum, and those with complex pairs of wavevectors are bound states.

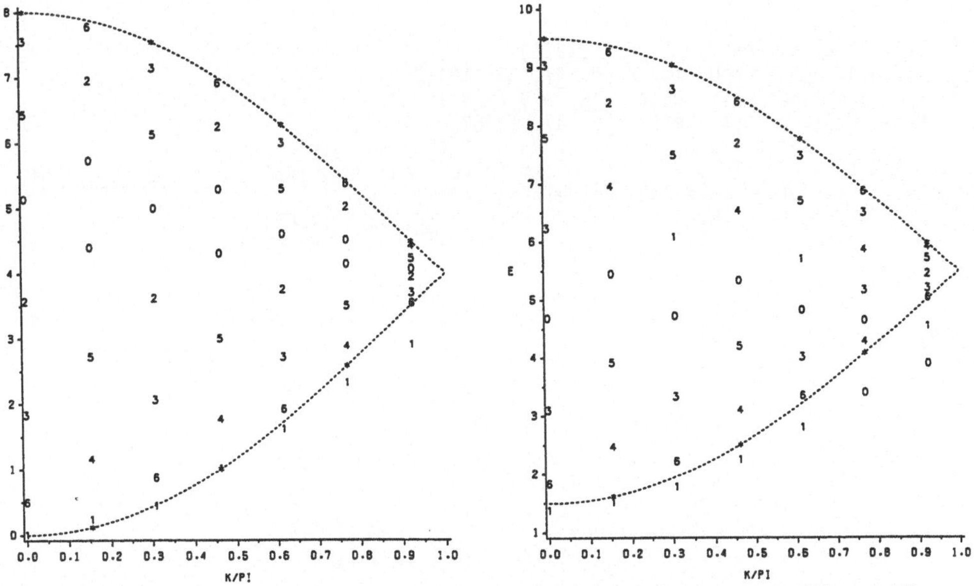

Fig. 1. Two-magnon states as a function of K for N=13, (a) d=0 and (b) d=0.75. The dashed lines give the boundaries of the two-magnon continuum and the indices within the axes indicate which of the P_n are largest for a given (E,K) location. The bound states appear below the continuum in these cases. Asterisks denote states of zero total probability. (The energy scale uses J=1.)

Figure 1 shows results for bound and continuum states for the isotropic case (d=0) and also for d=0.75 when the single-ion and exchange bound states are not degenerate at the zone boundary. The binding energy of exchange and single-ion bound states at the zone boundary are JS and 2D respectively (below the middle of the continuum), giving degeneracy when d=0.5 for S=1. We refer the reader to diagrams given by SILBERGLITT and TORRANCE [17] and PAPANICOLAOU and PSALTAKIS [8] for various values of d which are greater or smaller than that case. In those works the infinite chain limit allows the bound states to be drawn as continuous curves rather than the few discrete points available here for N=13. In Fig. 1 we have shown which P_n is greatest for a given E,K by the appropriate index (some others may of course be close in value and require tabulations to be given in a fuller paper). Comparison of the data in Fig. 1 for d=0 and d=0.75 clearly shows a shift of a single-ion continuum resonance in the presence of $D \geq 0$ to lower energies, as well as hybridizing interactions between the single-ion bound state, which pops out of the continuum for D>0, and the original bound state of the D=0 case. Figure 1(b) shows clearly the character change of the lower bound state from exchange type at low K to single-ion type at high K. These bound state results are in accord with those obtainable from the Green function method [17,18]. For the Green function method in the past, the spectral intensities were usually only evaluated the single-ion (n=0), NN exchange (n=1) and NNN exchange (n=2) processes [15,18,19]. These correspond to the excitation processes likely in most materials and there was little interest in larger values of n in a chain of infinitely many sites. By contrast, in finite chain calculations it is natural to look at the entire structure through results for all n.

Acknowledgements

This work has been supported by the Natural Sciences and Engineering Research Council of Canada. Discussions with B.W.Southern and S.C.Bell have been helpful.

References

1. H.A.Bethe: Z. Physik 71, 205 (1931)
2. E.Schrodinger: Proc. Ir. Acad. 47, 39 (1941)
3. L.A.Takhtajan: Phys. Lett. 87A, 479 (1982)
4. H.M.Babujian: Phys. Lett. 90A, 479 (1982)
5. J.B.Parkinson and J.C.Bonner: J. Phys. C 19, 6063 (1986)
6. D.C.Mattis: The Theory of Magnetism (Harper and Row, 1965), and The Theory of Magnetism I (Springer-Verlag, Heidelberg, 1981)
7. R.P.Hodgson and J.B.Parkinson: J. Phys. C 18, 6385 (1985)
8. N.Papanicolaou and G.C.Psaltakis: Phys. Rev. B 35, 342 (1987)
9. M.Wortis: Phys. Rev. 132, 85 (1963)
10. T.Tonegawa: Prog. Th. Phys. Supp. 46, 61 (1970)
11. D.A.Pink and R.Tremblay: Can. J. Phys. 50, 1728 (1972), D.A.Pink and R.Ballard: Can. J. Phys. 52, 34 (1973)
12. S.T.Chiu-Tsao, P.M.Levy and C.Paulson: Phys. Rev. B12, 1819 (1975)
13. T.Schneider: Phys. Rev. B24, 5327 (1981), T.Schneider and E.Stoll: Phys. Rev. Lett. 47, 377 (1981)
14. A.A.Bahurmuz and P.D.Loly: J. Phys. C 19, 2241 (1986)
15. A.A.Bahurmuz and P.D.Loly: J. Math. Phys. 22 ,564 (1981)
16. P.D.Loly: (a review) - Can. J. Phys. - to be published
17. R.Silberglitt and J.B.Torrance, Jr.: Phys. Rev. B2, 772 (1970)
18. A.A.Bahurmuz and P.D.Loly: Phys. Rev. B22, 1294 (1981)
19. A.A.Bahurmuz and P.D.Loly: Phys. Rev. B21, 1924 (1980)

Nonlinear Effects in Low-Dimensional Magnetism: Solitons and Vortices

A.R. Bishop[1,*], *C. Kawabata*[2], *F.G. Mertens*[3], *and G.M. Wysin*[4]

[1]ICTP, P.O. Box 586, I-34100 Trieste, Italy
[2]Okayama University Computing Center, Okayama 700, Japan
[3]Physics Institute, University of Bayreuth, D-8580 Bayreuth, Fed. Rep. of Germany
[4]Los Alamos National Laboratory, Los Alamos, NM 87545, USA

The primary purpose of this report is to outline our recent results [1,2] on the dynamics of easy-plane classical ferromagnetic spins in two spatial dimensions - particularly signatures of unbound vortices above a Kosterlitz-Thouless topological phase transition [3]. However, for completeness we briefly place this in the context of our wider studies of low-dimensional magnetism since the last meeting in this series. These may be summarized as:

(i) 0-dimensional systems. Here we include single spins [4] and small spin clusters [5], which have been studied with both free and externally-driven dynamics, and for both classical and spin-S quantum cases. The aim has been to probe the quantum analogs of classically "chaotic" dynamics by taking advantage of some special features of certain spin systems -- the Hilbert space is compact so numerical accuracy is assured; and both the degrees of "quantumness" ($\hbar \sim S^{-1}$) and "integrability" can be controlled (including totally integrable limits). In this way new scaling behaviours (motivated by dynamical systems theory in classical problems) has been suggested for both energy level structures [5] and wave-functions [4] in the semi-classical regime $S \gg 1$.

(ii) 1-dimensional systems. We have examined both ferromagnetic and antiferromagnetic easy-plane spin chains with an in-plane magnetic field (modeling, e.g., $CsNiF_3$ and TMMC, respectively), and both classical dynamics and quantum thermodynamics. Concerning dynamics, we have focused most recently on the kink-solitons in the antiferromagnetic case and on their interactions [6]. For the Hamiltonian

$$H = \sum_n J \vec{S}_n \cdot \vec{S}_{n+1} + A(S_n^z)^2 - BS_n^x \tag{1}$$

($J, A > 0$, site index n), there are both "XY" and "YZ" solitons, corresponding to sublattice rotations by π principally in the XY and YZ plane, respectively -- at $B = B_c = (2/JS))(2A/J)^{\frac{1}{2}}$ (where the rest kink energies are equal [6]) there is a natural crossover. Using an analytic ansatz [6], we have found that both dynamic branches can be incorporated in a unified manner and that: (a) the XY and YZ branches always merge at a field-dependent kink velocity; and (b) the XY solutions have a "negative effective mass" branch for $B > B_c$, similar to the ferromagnetic case [7]. Kink-antikink collisions may be summarized as generally transmitting for YZ kinks (which are more sine-Gordon-like in character), whereas for XY they transmit for small B, bind (into "breathing" states) or annihilate for larger B ($\lesssim B_c$), and reflect for $B > B_c$ (i.e. for the negative mass branch): thus the XY kinks behave in these respects much as for the XY kink solitons in the corresponding ferromagnetic case [7].

Using quantum MC or quantum transfer matrix methods [8] has been instructive for the thermodynamics of materials such as CHAB or $CsNiF_3$ in magnetic fields, since precise experimental measurements of specific heat differences (i.e. with and without

a magnetic field) are now available. Thus an important test of proposed magnetic Hamiltonians is possible. Although agreements between the experiments and quantum simulations are very good, the systematic differences are important. They are outside the respective errors and of a percentage magnitude consistent with differences between various theoretical treatments (of semiclassical quantization, out-of-plane spin fluctuations, etc.). Thus, for these theoretical "disputes" to be meaningful requires that the original Hamiltonian be agreed to even greater accuracy -- including additional small terms (e.g. dipolar, in-plane symmetry-breaking).
(iii) 2-dimensional systems. Here again we have focused on easy-plane Heisenberg models and their classical thermodynamics [9], including crystalline symmetry-breaking terms in the easy-plane as appropriate (e.g. for magnetically-intercalated graphite or the layered magnet Rb_2CrCl_4 [10]). We have also made preliminary studies [11] of transverse instabilities and their nonlinearly saturated forms on moving domain walls. This is a topic of considerable general importance in the field of pattern formation and extended dynamical systems, and the magnetic examples deserve more careful reassessment and measurement in the light of current advances in this field. However, the remainder of this report will deal with recent attempts [1,2] to study classical spin dynamics above the Kosterlitz-Thouless transition in easy-plane Heisenberg models, emphasising possible signatures of unbound vortices. This is also a topic which has reached a stage of great opportunity because of improved and more varied quasi-2-d materials, and because of increased interest in making inelastic neutron scattering measurements at small frequency and wave-vector [12].

As in 1-d soliton systems, we adopt an "ideal gas phenomenology" as a first step, so as to explicitly recognize the vortex collective structures -- i.e. the spin field is decomposed into additive spin wave and vortex contributions[13]. Here we treat only the vortex contribution, which is itself considered as a noninteracting dilute gas of unbound vortex-antivortex pairs ($T \gtrsim T_c$) moving in a screening background of the remaining bound pairs.

We consider the anisotropic ferromagnetic Heisenberg Hamiltonian

$$H = -J \sum_{(m,n)} [S_x^m S_x^n + S_y^m S_y^n + \lambda S_z^m S_z^n] \tag{2}$$

with (m,n) near-neighbor pairs on a square lattice. In a continuum approximation, unit strength vortex configurations behave as $\phi = \tan^{-1}(y/x)$ and $\theta = \frac{1}{2}\pi(1 - \exp - (r/r_v))$ for $r \gg r_v$, $\theta \to 0$, as $r \to 0$, where $S_x = S \cos\phi \sin\theta$, $S_z = S \cos\theta$, $r^2 = x^2 + y^2$, and r_v is a vortex "core radius" a $[2(1-\lambda)]^{-\frac{1}{2}}$ (lattice constant a). It follows that S_z is only locally sensitive to the presence of vortices, whereas S_x and S_y are globally sensitive -- i.e. out-of-plane and in-plane spin correlations will reflect the vortex structure and mean separation (correlation length), respectively (c.f. solitons in 1-d easy-plane ferromagnets and antiferromagnets, respectively) and therefore the corresponding dynamic structure factors, $S(\vec{q},w)$, are quite distinct.

Calculations of $S_{xx}(\vec{q},w)$ and $S_{zz}(\vec{q},w)$, within the ideal vortex gas phenomenology, are given in Refs.[2]. The results are

$$S_{xx}^{vortex}(\vec{q},w) \simeq \frac{S^2}{2\pi^2} \frac{\gamma^3 \xi^2}{\{w^2 + \gamma^2[1 + (\ q)^2]\}^2} , \tag{3}$$

$$S_{zz}^{vortex}(\vec{q},w) \simeq \frac{S^2 \pi^{3/2}}{4} \cdot r_v^2 \cdot \frac{n_v r_v^2}{\bar{u}/r_v} \cdot \frac{e^{-w^2/q^2 \bar{u}^2}}{q r_v [1 + (q r_v)^2]^3} , \tag{4}$$

where $\gamma = \frac{1}{2}\pi^{\frac{1}{2}}\bar{u}/\xi$, (correlation length) $= \xi_0 \exp(b\tau^{-\frac{1}{2}}) = n_v^{-\frac{1}{2}}$, $\xi_0 = O(a)$, and $\bar{u}$ is the rms vortex speed. $\bar{u}$ and the forms (3,4) are calculated within an <u>assumed</u> Maxwellian vortex velocity distribution and fitted to the calculation of HUBER [13]: $u(\tau) = (\pi b)^{\frac{1}{2}}(Ja/\hbar)\tau^{-\frac{1}{4}}(\exp(-b\tau^{-\frac{1}{2}})$. Here we have assumed the equilibrium temperature ($\tau = T - T_c/T_c$) dependencies from Kosterlitz-Thouless thermodynamics. For the temperatures below, b is weakly τ-dependent in the range (0.3, 0.5) [14].

Note the (squared) Lorentzian and Gaussian forms of the predicted central peaks (3,4), which we may characterize with widths (Γ) and integrated intensities (I):

$$\Gamma_z \simeq uq \tag{5a}$$

$$\Gamma_x \simeq \begin{cases} \bar{u}/\xi, & q\xi \ll 1 \\ \bar{u}q, & q\xi \gg 1 \end{cases} \tag{5b}$$

$$I_z \simeq \tfrac{1}{4} S^2\pi^2 n_v r_v^4[1 + (qr_v)^2]^{-3} \tag{6a}$$

$$I_x \simeq (S^2/4\pi)n_v \xi^4[1 + (\xi q)^2]^{-3/2} \tag{6b}$$

We clearly observe the anticipated length scales r_v and ξ for out-of-plane and in-plane cases, respectively, yielding much larger intensities for in-plane scattering ($\sim(\xi/r_v)^4$ relative to out-of-plane). Recent MC-MD simulation [1,2,15] have indeed observed central peaks for $T > T_c$ and provided some striking agreements with the predicted forms (5) and (6). Examples are shown in Figs. 1(a) and 1(b) for the XY limit ($\lambda = 0$), where we have also indicated predictions assuming b = 0.5 and $\xi_0 = a$. Even better agreement may be obtained by <u>fitting</u> these parameters (since they are not exactly available from Kosterlitz-Thouless theory at these temperatures [14]). Remarkably, <u>independent</u> fitting for S_{xx} and S_{zz} gives agreement to within 20% for the parameters at several temperatures.

Theoretically, there are several important issues under continuing study, including: (i) a self-consistent calculation of the vortex velocity distribution function; (ii) vortex-vortex, vortex-spinwave, and multimagnon effects which appear to have significant effects on scattering intensity and q-dependence (especially for S_{zz}, c.f. 1-d easy-plane ferromagnets); (iii) extrinsic dissipation and impurity pinning effects; (iv) 3-d and Heisenberg crossovers; and (v) symmetry-breaking (e.g. crystalline) fields which produce important additional collective structures with slow dynamics, e.g. domains, interacting with the vortices. (These are important, e.g., in Rb_2CrCl_4 [12] and $CoCl_2$-intercalated graphite [10,16].) In addition, several quasi-2-d magnets are low-spin (e.g. K_2CuF_4 [17] and $BaCo_2(AsO_4)_2$ [12] are $S = \frac{1}{2}$) so that quantum effects should be significant. Thermodynamic studies [18] suggest that, as in 1-d [8], the main quantum effects are reductions of fluctuation intensities (e.g. of specific heat). However, adequate description of quantum <u>dynamics</u> remains a major challenge in all dimensions.

Regarding inelastic neutron scattering data on real materials, conclusions must be tentative at this stage because systematic information on central peaks as functions of q and T is only now being gathered. However, as discussed elsewhere [2], preliminary evidence for K_2CuF_4, Rb_2CrCl_4 and $BaCo_2(AsO_4)_2$ is quite consistent with our predictions at small q as far as central peak widths and some other trends (e.g. crossover from XY to Heisenberg behavior as q is increased at fixed $\lambda < 1$) are concerned. Much more data and less ad hoc form fitting is anticipated in the near future. It will be especially helpful if the (weak) S_{zz} correlations can be measured. A very attractive possibility is that quasi-2-d magnets may provide good candidates [12]

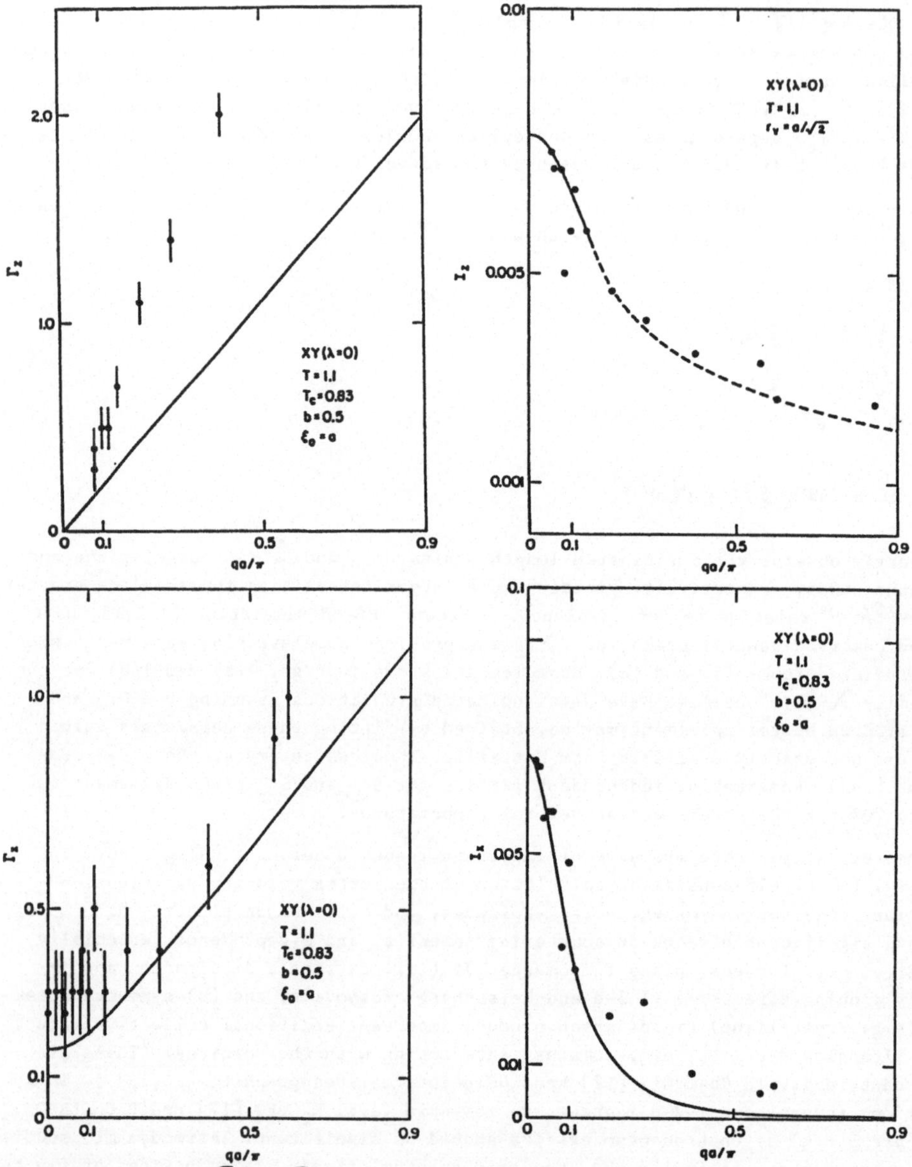

Figure 1. Widths (Γ_z and Γ_x) and intensities (I_z and I_x) of central peaks obtained in a numerical MC-MD simulation of the XY model ($\lambda = 0$) on a 100 x 100 lattice for (a) $S_{zz}(\vec{q},w)$, and (b) $S_{xx}(\vec{q},w)$ at a temperature $T/J = 1.1$ ($T_c/J \simeq 0.8$). Numerical data is shown as solid points. Solid lines result from eqns. (5) and (6) without fitting parameters. The dashed line is a guide to the eye only -- eqn.(6a) yields a significantly lower value at large q [2].

for studying frustration dynamics - another context where collective structures are expected to control global response time scales but where current theoretical understanding is extremely poor.

* Permanent address: Theoretical Division and Center for Nonlinear Studies, Los Alamos National Laboratory, Los Alamos, NM 87545, USA.

1. C. Kawabata, M.Takeuchi and A.R. Bishop, J. Mag. Mag. Mat.54-57, 871 (1986)
2. F.G. Mertens, A.R. Bishop, G.M. Wysin and C. Kawabata, preprints (1987)
3. J.M. Kosterlitz and D.J. Thouless, J. Phys. C6, 1181 (1973)
4. K. Nakamura, Y. Okazaki and A.R. Bishop, Phys. Rev. Lett. 57, 5 (1986)
5. K. Nakamura, Y. Okazaki and A.R. Bishop, Phys. Rev. B33, 1963 (1986), Phys. Lett. A117, 459 (1986)
6. G.M. Wysin and A.R. Bishop, J. Phys. C19, 221 (1986); and in press.
7. G.M. Wysin and A.R. Bishop, J. Phys. C17, 5975 (1984)
8. G.M. Wysin and A.R. Bishop, Phys. Rev. B34, 3377 (1986)
9. C. Kawabata and A.R. Bishop, Solid State Comm. 60, 169 (1986)
10. C. Kawabata and A.R. Bishop, Z. Physik B65, 225 (1986)
11. J.C. Ariyasn and A.R. Bishop, Phys. Rev. B (in press); J. Pouget, S.Aubry, A.R. Bishop, P.S. Lomdahl, preprint.
12. E.g. L.P. Regnault and J. Rosat-Mignod, in Magnetic Properties of Layered Transition Metal Compounds, eds. L.J. de Jongh and R.D. Willet; M.T. Hutchings et al., J. Mag. Mag. Mat. 54-57, 673 (1986)
13. See also earlier steps in this direction by D.L. Huber, e.g. Phys. Rev. B26, 3758 (1982)
14. S.W. Heinekamp and R.A. Pelcovits, Phys. Rev. B32, 4258 (1985)
15. A standard MC algorithm is used to initialize an equilibrium state for MD. Typically 5 averages on square lattices up to 100 x 100 have been used.
16. E.g. M. Elahy and G. Dresselhaus, Phys. Rev. B30, 7225 (1984)
17. E.g. K. Hirakawa et al., J. Phys. Soc. Jpn. 51, 2151 (1982)
18. E. Loh, Jr., D.J. Scalapino and P.M. Grant, Phys. Rev. B31, 4712 (1985)

Theory of Excitations in Incommensurably Modulated Magnets

P.-A. Lindgård

Risø National Laboratory, DK-4000 Roskilde, Denmark

1. Introduction

The theory of the excitation spectrum in a longitudinally modulated spin structure is intricate because the presence of an incommensurable modulation destroys the translational invariance, and consequently the crystal momentum **q** is not a good quantum number[1]. Recently progress has been made in this problem[2] by utilizing methods developed for describing the dynamics of disordered or quasi-periodic structures. The present problem is closely related to structurally modulated systems by charge density waves[3] or artificially constructed pseudorandom layered structures (by molecular beam epitaxy), to fractal structures as well as that of electrons moving in a quasi-periodic potential[4].

The interest in the magnetic systems derives from the fact that sinusoidally modulated structures are observed in several materials, Nd, Pr, $CeAl_2$, Er, Tm, Cr and that single crystal data on the spin dynamics is now beginning to be revealed, so far for Nd[5], $CeAl_2$[6] and Cr[7]. In order to demonstrate some of the interesting and quite unexpected consequences of the incommensurability let us study a very schematic and simplified model for these systems: the ANNNHE model[8] (anisotropic nearest and next nearest neighbour Heisenberg model). The even simpler Ising version the ANNNI model[9] is often used to discuss the static properties, showing that sinusoidal structures are only obtained close to the phase transition and that higher harmonics (squaring up) occurs at lower temperatures as well as chaotic structures.

The main description of these structures is that they consist of regions with large moments (like magnetic molecules) which are loosely coupled through walls with small moments. A crucial point is that the regions are similar, but not identical. This picture also applies to Nd, which was recently found[5] to be a two Q structure in a large temperature interval. Physically, one would expect the spectrum to consist of localised internal "molecular" modes with high frequencies and extended low frequency modes, considering the "molecules" as weakly coupled entities. It is interesting that these effects are already found in the simple single Q sinusoidal structure for incommensurate values of Q. This is surprising because when $Q = Q_c = nL/N$ is commensurate the structure can be divided into N sublattices (each with equal average moment) and the RPA dispersion relation found by an exact diagonalization of a $N \times N$ matrix. All commensurate modes are extended, have a dispersion and no width in q or ω, but are N times multivalued in the full Brillouin zone, and the density of states has in general of the order of N gaps. For a slightly incommensurate $Q \sim Q_c$ the modes become broad in q (consequently also effectively so in energy) and for a simple calculation of this a differential method was proposed[10]. However, considerably more structure of the incommensurate dynamics was revealed by considering the actual difference equations at hand[2]. It remains to be tested experimentally whether this survives the quantum and thermal fluctuations in the real magnetic systems.

2. Linear spin wave theories for an incommensurate sinusoidal structure

A sinusoidally modulated structure is defined by

$$\langle S^z_R \rangle = M\cos(\mathbf{Q}\cdot\mathbf{R}+\phi) \quad \text{or} \quad \langle S^z_q \rangle = M[\delta(\mathbf{q}+\mathbf{Q}) + \delta(\mathbf{q}-\mathbf{Q})]/2$$

where M is the amplitude, and we set $\phi = 0$. The longitudinal phason mode corresponding to this broken symmetry is not discussed in this paper. In the random phase approximation (RPA) this structure is stable for the following model Hamiltonian:

$$H = -\sum_{q} \left| J_q\left(S^x_q S^x_{-q} + S^y_q S^y_{-q}\right) + K_q S^z_q S^z_{-q} \right| \tag{1}$$

when $K_q > J_q$ for $q = Q$ at which K_q has a maximum and M is not too large. If we let $K_q = D + J_q$ it is an axial single ion anisotropy which keeps the spins in the z-direction, and prevents the system from ordering with a spiral structure with a constant amplitude on every site. There is no difference in principle if D is wave vector dependent, corresponding to exchange anisotropy, except that in this case the RPA treatment is more reliable. A more accurate treatment of a single ion anisotropy is desirable, in particular in the small moment walls. In doing so one is bringing in also the internal crystal field excitations, but the theory follows the same line as the simple RPA theory, which most clearly demonstrates the effect of the incommensurability. If we consider only the wave vector q along the ordering vector Q the problem is effectively one-dimensional.

Let us define the Green's function $G_n = G_{q+nQ}(\omega) = \langle\langle S^+_{q+nQ}; S^-_{-q}\rangle\rangle_\omega$ and an interaction $W_n = M(K_Q - J_{q+nQ})$. The RPA equation of motion can be formulated[10] exactly as

$$V_n(G_{n+1}+G_{n-1}-2G_n)+\Delta_n[(G_{n+1}-G_n)+(G_n-G_{n-1})]+(2V_n-\omega)G_n = -2\pi M, \tag{2}$$

where $V_n = \frac{1}{2}(W_{n+1}+W_{n-1})$ is the average interaction and $\Delta_n = \frac{1}{2}(W_{n+1}-W_{n-1})$ is the difference.

For simplicity consider first the almost ferromagnetic case $Q_c = 0$ and $Q = \delta Q$ small, we then expand in Q and write the differential equation

$$V_n Q^2(d^2G_n/dq^2)+2Q\Delta_n(dG_n/dq)+(2V_n-\omega)G_n = -2\pi M.$$

Let us assume that (1) contains nearest and next nearest neighbour interactions J_1 and J_2. We can eliminate J_2 by the RPA condition for the sinusoidal ground state: $J_1 = -4J_2\cos Q$ and define the RPA frequency for $Q_c = 0$:
$\omega_q = M[2D + J_1(1 - \cos q)^2]$ and expand V_n and Δ_n in Q. We then obtain the simple differential equation[10]

$$\tfrac{1}{2}Q^2(\omega_q G_q - 2\omega_q G_q + p_q G_q)+(\omega_q-\omega)G_q = -2\pi M, \tag{3}$$

where $p_q = 4J_1\cos q(1-\cos q)$ and $F_q = dF_q/dq$. The equation could be studied in more detail, which would be particularily interesting for $\omega \to 0$. However, for a simple evaluation of the q-broadening of the RPA solution we solve it to order Q^2 by writing G_q as a formal two-pole function

$$G_q = \frac{2\pi M}{\omega_q}\,\frac{\omega_q(\omega-\omega_q-2i\beta)+\alpha^2}{(\omega-\omega_q)^2-\alpha^2-2i\beta(\omega-\omega_q)}\,. \tag{4}$$

Inserting (4) into (3) for fixed ω, the equations for the real and imaginary parts determine α^2 and β^2 to be: $\alpha^2 = 2\beta^2 = Q^2\omega_q{}^2$ for $\omega = \omega_q$. The resulting spectrum $S(q,\omega) = \mathrm{Im}G_q(\omega)/\omega$ was calculated for $Q = 2\pi/50$ as an example. It spans the frequency band of the RPA spectrum with sharp cut-offs at the edges and broad peaks in the middle. The width is largest where the dispersion is steepest (ω_q large); note in this approximation there are no states for $\omega < 2MD$.

For $Q_c \neq 0$ we solve the $N\times N$ matrix exactly and find the q-broadening by expanding in $\delta Q = Q-Q_c$. There will be several gaps in the spectrum, which is now approximated by a 2N-pole function, where N could be taken very large. On physical grounds it was suspected, however, that details revealed for $N > 10$, say, would not be observable since the RPA basis is invalidated by fluctuations. It was recently demonstrated that one needs at least a five-pole G_q to describe several interesting aspects of the spectrum for $Q = 2\pi/7.7$. A broadening of the dispersion relation was also discussed by a continued fraction theory by Liu[12].

To gain more insight Ziman and Lindgård[2] returned to the difference equation (2), now written for the spin operators

$$i\frac{d}{dt}S^+_{q+nQ} = W_{n+1}S^+_{q+(n+1)Q} + W_{n-1}S^+_{q+(n-1)Q} \ . \tag{5}$$

Thus within the RPA theory, for Q incommensurable with the lattice, the normal modes and their frequencies are derived from an infinite tri-diagonal matrix, which can be made symmetric by defining the operators

$$a_n^+ = (W_n)^{\frac{1}{2}}S^+_{q+nQ} \tag{6}$$

Then the equations of motion for operators a_n^+ are those of a one-dimensional tight-binding model,

$$H = \sum_n t_n(a^+_{n+1}a_n + a^+_n a_{n+1}) \ , \tag{7}$$

with the quasi-periodic hopping terms $t_n = (W_n W_{n+1})^{\frac{1}{2}}$. Notice in this mapping the "sites" n correspond to the original q-space. The eigenmodes with frequency ω_k are $b_k^+ = \Sigma_n f_k(n) a_n^+$, where k is in the "reciprocal space" for the "sites" n. The scattering function, found by expanding S^+_{q+nQ} in these modes, is

$$S(q',\omega) \propto \frac{\pi M}{W_n}\sum_k |f_k(n)|^2\delta(\omega-\omega_k) \ . \tag{8}$$

For an infinite crystal the set $q' = \{q+nQ\}$, reduced to the first Brillouin zone, form a dense set for incommensurate Q.

It is instructive to recover the results for a simple commensurable magnetic phase: the two-sublattice antiferromagnet with $Q = \pi$. In this case for any n, $t_n = t(q) = M((K_Q-J_q)(K_Q-J_{q-Q}))^{\frac{1}{2}}$. By translational invariance the eigenoperators and eigenvalues are

$$b_k^+ = \sum_n e^{ikn}a_n^+ \ , \quad \omega_k = 2t(q)\cos k. \tag{9}$$

In this case, however, out of the infinite set of frequencies ω_k, only two are physically relevant, since for all n the operator a_n^+ is identical to a_{n+2}^+. Thus, except when k is 0 or π the factors e^{ikn} interfere destructively. For $Q = \pi$ the result is a pair of frequencies $\omega_k = \pm 2t(q)$ in agreement with well known results.

The Hamiltonian (7) differs from Harper's equation[4], which has been analyzed by Aubry and others[13], in that it has no diagonal term. In our case it is the off-diagonal term that is modulated. The same methods used, essentially adapted from those applied to random systems, can be applied here. In particular, we can define[14] a complex "wavevector" k(E) corresponding to (9). The imaginary part $\lambda(E)$ is the Lyapunov exponent. If $\lambda > 0$ the wave function decreases exponentially. It is thus localized, but for $\lambda = 0$ it is extended. The real part I(E) gives the number of nodes and thus the integrated density of states below the energy E[14].

The difference equation for eigenvectors of (7),

$$Ea_n = t_n a_{n+1} + t_{n-1}a_{n-1}, \tag{10}$$

may be iterated numerically starting from arbitrary boundary conditions (a_0,a_1). The Lyapunov exponent is defined by

$$\lambda(E) = \lim_{N\to\infty}\left[\frac{\ln|a_N|}{N}\right], \tag{11}$$

and the integral of the density of states

$$I(E) = \int_{-\infty}^{E}\rho(E')dE' = \lim_{N\to\infty}\frac{\Sigma(E,N)}{N} \tag{12}$$

by counting the number of sign changes $\Sigma(E,N)$ of the sequence of real numbers $(a_0,a_1,\ldots,a_N)$. Note that this procedure does not involve diagonalization of the matrix, but gives only spectral information, i.e., it determines $S(q,\omega)$ averaged over q.

By an exact numerical calculation of I(E) for a chain of 10000 spins we find a number of gaps in the spectrum with finite $\lambda(E)$, but $\lambda(E)=0$ for all E in the bands so all states are extended in q-space.

3. Physical Picture of the Energy Gaps

To understand the physics of this result let us discuss (7) semiclassically following Harper[4)] and Wilkinson[13)]. In the continuum approximation we replace $q+nQ$ by the continuous variable x and expand a_{n+1} in (7) as

$$a(x+Q) = e^{i\hat{k}}\, a(x)$$

where $\hat{k} = -iQd/dx$. The eigenvalue equation (10) can then be obtained by quantizing the classical Hamiltonian

$$H(x,k)_{classical} = 2t(x)\cos k,$$

$$Ea(x) = \hat{H}(x,\hat{k})a(x) = \{t(x)e^{i\hat{k}} + e^{-i\hat{k}}t(x)\}a(x), \tag{13}$$

where $t(x) = t_n$ with $q+nQ = x$. The semiclassical approximation is valid for $Q\to 0$, however it appears to explain the numerical results in a quite substantial range of Q. Note that in (13) we have expanded to infinite order in Q, whereas in (3) we expanded only to δQ^2, but from a commensurate value Q_c.

To construct a semiclassical wave function we first consider energy contours of the function H(x,k). In Fig. 1 we draw these for k in the range $[0,2\pi]$ and x in the range $[0,4\pi]$ for parameters corresponding to a simplified model of Nd with $J_1/D = -0.386$ and $Q = 2\pi/7.7$, which is relatively small (but even closer to an eight sublattice model). The classical equations of motion have orbits that are either closed, or open in the x direction.

The closed orbits occur around maxima A,A' of the surface H, or minima. The contour B separates the closed from the extended orbits whose energies are smaller in absolute value. Eigenfunctions corresponding to a finite set of closed orbits occur at discrete energy levels given by Bohr-Sommerfeld quantization with approximately harmonic energies $E = E_0+(p+\frac{1}{2})\Delta$. Because of tunneling these levels are broadened into bands, the width of which depends on overlap of wave functions localized close to equivalent maxima, around A and A', for example. Consequently, we might expect that all eigenstates are extended when tunneling is taken into account. This was verified by calculation of the Lyapunov exponent numerically where we found $\lambda(E) = 0$ for all E. Open orbits of the classical picture become, with the inclusion of quantum-mechanical interference, a relatively broad continuum at lower energies, which develops gaps from the periodicity in the x direction.

Another feature suggested by Fig. 1 is the special nature of the zero-energy contour, an axis of symmetry. In fact, one can explicitly determine the two degenerate zero-energy eigenstates and there is a response at low energies[2)]:

$$\lim_{\omega\to 0} \mathrm{Im}G_q(\omega)/\omega \propto W_n^{-2} = |K_Q - J_q|^{-2}.$$

Thus there is a "quasi-elastic" band which has maxima at Q and 2π-Q and which is relatively broad in energy, since it appears in the region of open orbits. These excitations have the greatest amplitudes in the low moment walls. This was again substantiated by the numerical calculation. This was of course not found by the perturbation approach[10,11)]. It is not connected with the phason mode, which occurs in the longitudinal response function.

4. Discussion

It is only in the frequency dependence that sharp structure is displayed, the q dependence is smooth. This is a consequence of the fact that the states are extended in q space. To evaluate the actual q dependence of the response the spectral weight of eigenmodes was calculated[2)] by the procedure outlined above with of the order of 10000 modes; we then determine the momentum dependence of $G_q(\omega)$ in each frequency range by

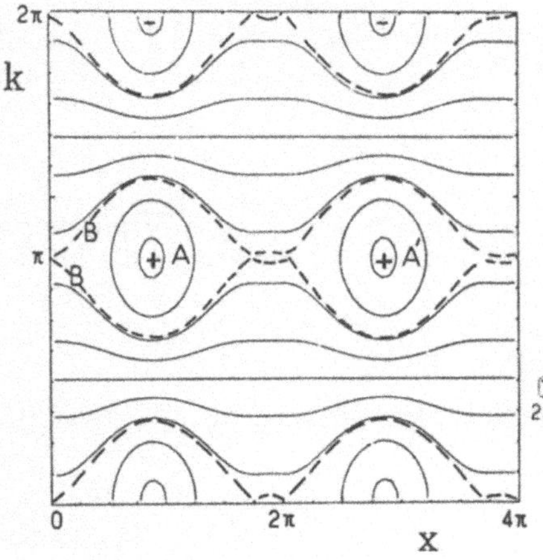

Fig. 1. Energy contours of H(x,k) (13) for $J_1/D=-0.386$ and $Q=2\pi/7.7$. A and A' are equivalent maxima; B is the separatrix separating closed and open orbits. The zero-energy lines are horizontal at $k=\pi/2$ and $3\pi/2$.

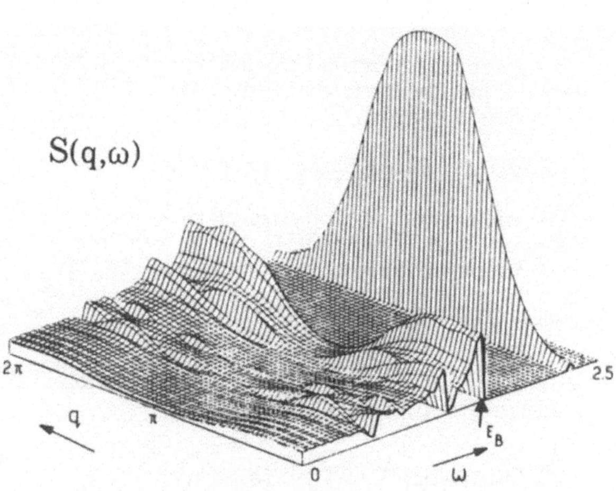

Fig. 2. Scattering function $S(q,\omega) = \mathrm{Im}G_q(\omega)/\omega$ for the same parameters (exact numerical solution of RPA for 10^4 spins). Note the sharp high frequency band ($2.2411 < E < 2.2547$), the gap down to the first open orbit with energy $E_B = 1.481$, the gaps at lower frequency and finally the broad low frequency response.

a diagonalization of a tridiagonal matrix corresponding to a few hundred modes and fold the results together.

In Fig. 2 we show the dynamical susceptibility function $S(q,\omega) = \mathrm{Im}[G_q(\omega)/\omega]$ calculated for the case of modulation with the relatively small $Q = 2\pi/7.7$. It is verified that in addition to the smearing out of spin-wave excitations of the ferromagnetic state $Q = 0$, as predicted by the perturbative approach[10,11], there are two qualitatively new features.

(i) In the high-frequency region at least one band of response is extremely sharply defined in energy near the band-edge magnons in the commensurable case. For $Q = 2\pi/50$ we find at least seven such narrow "harmonic oscillator" bands, corresponding to the fact that H(x,k) in Fig. 1 has deeper minima and higher maxima A,A' for smaller Q. The q-dependence of the dynamical susceptibility corresponds to a particle in a bound state of a potential giving the closed orbits in Fig. 1. As the energy ω decreases, the bands broaden in energy and have more maxima as a function of q, corresponding to the structure of less tightly bound states.

(ii) A quasi-elastic, i.e. low-energy, feature that has a broad maximum at the incommensurable vectors $\pm Q$. The total weight vanishes with Q and the energy width is not exponentially small. As we have stressed, this response is associated with the regions of low average moment in which the linearization of the equations of motion is least likely to be accurate.

Mathematically the quasi-periodic RPA theory (2) and (5) is very interesting. The spectrum may consist[13] of a Cantor set of gaps on all scales (the number of sublattices $N \to \infty$). Physically, however, the validity of the RPA equations is limited by both quantum and thermal fluctuations, which will smear out the excitations also in energy. It will nevertheless be very interesting to have more experimental search in magnetic systems for some of the exact features of the RPA theory we have described here. Scans of $S(q,\omega)$ at constant q or constant ω should be quite different.

5. References

1. B.R. Cooper, R.J. Elliott, S.J. Nettel, and H. Suhl: Phys. Rev. 127, 57 (1962).
2. T.M. Ziman and P. A. Lindgård: Phys. Rev. B33, 1976 (1986).
3. A recent review is by R. Currat, in "Multicritical Phenomena", Ed. R. Pynn and A. J. Skjeltrop (Plenum, New York 1984) p. 178.

4. P.G. Harper: Proc. Phys. Soc. London A68, 874 (1955).
5. K.A. McEwen and W.G. Stirling: J. Magn. Magn. Mat. 30, 99 (1982), K.A. McEwen, E.M. Forgan, M.B. Stanley, J. Bouillot and D. Fort, Physica 130B, 360, 1985.
6. R. Osborn, M. Loewenhaupt, B.D. Rainford and W.G. Stirling(preprint).
7. S.K. Burke, W.G. Stirling, K.R.A. Ziebeck and J.G. Booth: Phys. Rev. Lett. 51, 494 (1983).
8. R.J. Elliott: Phys. Rev. 124, 346 (1961).
9. P. Bak and J. von Boehm: Phys. Rev. B21, 5297 (1980), and M.E. Fisher and W. Selke: Phys. Rev. Lett. 44, 1502 (1982).
10. P.A. Lindgård: J. Magn. Magn. Mat. 31-34, 603 (1983), Risø Report 441, 18 (1980).
11. W. Lovesey (preprint).
12. S.H. Liu: J. Magn. Magn. Mat. 22, 93 (1980).
13. S. Aubry and G. André, in Proceedings of the Israel Physical Society, edited by C.G. Kuper (Hilger, Bristol, 1979), Vol. 3, p. 133; S. Ostlund and R. Pandit, Phys. Rev. B 29, 1394 (1984); M. Wilkinson, Proc. R. Soc. London, Ser. A 391, 305 (1984); recent mathematical work is reviewed by B. Simon, Adv. Appl. Math. 3, 463 (1982).
14. D.J. Thouless, J. Phys C 5, 77 (1972).

Linear Spin Dynamics of Incommensurably Modulated Magnets: Numerical and Analytic Results

S.W. Lovesey[1] *and A.P. Megann*[2]

[1]Rutherford Appleton Laboratory, Chilton, Oxfordshire, OX11 0QX, United Kingdom
[2]Department of Physics, The University, Southampton, SO9 5NH, United Kingdom

1 Introduction

A potentially useful model of spin dynamics in incommensurably modulated magnets is constructed by linearizing equations of motion derived from a Heisenberg Hamiltonian. If the favoured spin axis is in the z-direction, the equation of motion for the transverse spin operator $S^+ = S^x + iS^y$ contains the product S^+S^z and linearization is achieved by replacing S^z by its molecular field value. A longitudinally modulated spin structure described by a single wave vector Q is realized at elevated temperatures in a Heisenberg model with competing exchange interactions and single-site anisotropy [5]. The molecular field value of the spatial Fourier transform of S^z is

$$\langle S^z_q \rangle = (S/2)\,\{\delta_{q,Q} + \delta_{q,-Q}\} \;, \tag{1}$$

for which the corresponding equation of motion reads

$$i\,\partial_t S^+_n = W_{n+1}\, S^+_{n+1} + W_{n-1}\, S^+_{n-1} \;, \tag{2}$$

where S^+_n is the spatial Fourier transform at the wave vector $q' = q + nQ$, n is an integer and the coefficient W_n is defined by

$$W_n = 1 + \alpha \cos\{q + nQ\} + \beta \cos\{2(q + nQ)\} \;, \tag{3}$$

with $(\alpha/\beta) = -4\cos Q$. Here we have chosen q (like Q) to be parallel with the z-axis. The parameters α, β are proportional to the nearest and next-nearest exchange interactions along the z-axis, and the transverse exchange cancels out in the equation of motion. The unit of energy contains the single site anisotropy parameter and Q.
we find

$$\begin{aligned} \mathrm{Im.}\,\chi(q,\omega) &= (\omega/2)\ \chi(q)\ |B_{N-1}|/\sqrt{L};\ L > 0 \\ &= 0\ ;\ L < 0 \;. \end{aligned} \tag{9}$$

The solutions ω of $L = 0$ for a given q determine the band edges. Formulas (7)-(9) apply for periodic systems in which $Q = (2\pi\, M/N)$ and can be implemented numerically, by computing A_{N-1}, B_{N-1} and B_N for given values of ω, q, or analytically to provide closed expressions for the dynamic susceptibility.

From these results we find ($Q = \pi/2$, $\omega_b = 2t_0$)

$$\mathrm{Im.}\,\chi(q',\omega) = (S/2W_n)\ \{\omega/\sqrt{(\omega_b^2 - \omega^2)}\}\ , \tag{10}$$

which agrees with the result obtained in [4] by an alternative argument.

3 General Numerical Method

The numerical results presented here were obtained using the method described in reference [1]. The q-dependence is derived by diagonalizing a matrix corresponding to 400 modes while the density of states is evaluated for a system of 2×10^4 sites.

We have obtained results for the parameters quoted in [1] which reproduce their plot of the dynamic susceptibility to a good approximation; in Figure 1 we show the numerical result for the case $Q = \pi/2$, which compares well with our exact result (9), and provides confidence in our numerical method. However, a shortcoming of the method is that it misrepresents the q-dependence of the dynamic susceptibility although the analytic results show that this is a minimal effect.

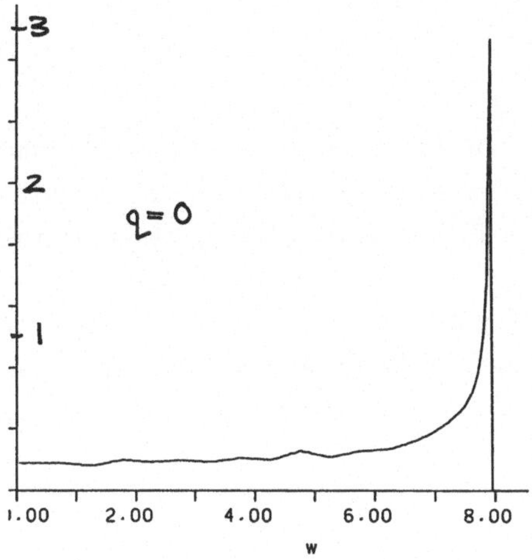

Figure 1

The quantity $\mathrm{Im.}\chi(q,\omega)/\omega$ is plotted as a function of ω for $Q = \pi/2$ and $q = 0$. The ω scale is a factor 4 larger than units used in the text.

4 Results and Conclusions

In Figure 2 we show the values of the susceptibility for antiferromagnetic-like parameters, which complements the ferromagnetic-like results provided in [1]. To be precise, in [1] the parameters are $Q = 0.260\pi$, $\alpha = -0.386$ and $\beta = 0.141$, whereas in Figure 2 the parameters α, β are chosen to give $Q = 0.740\pi$.

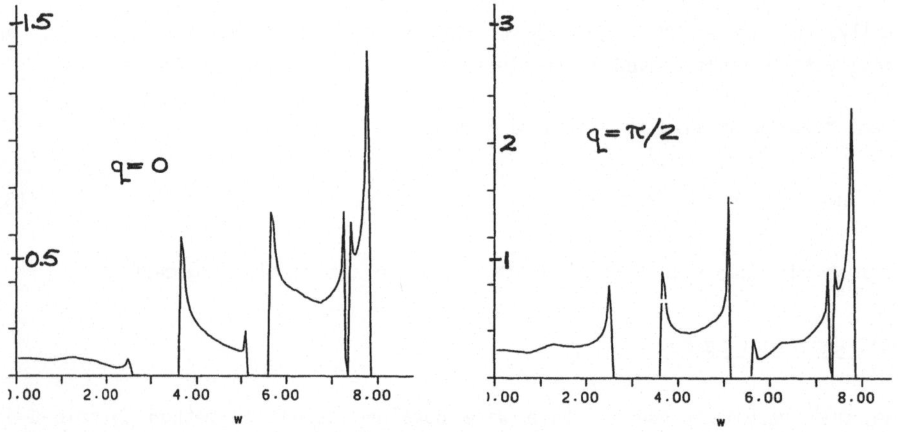

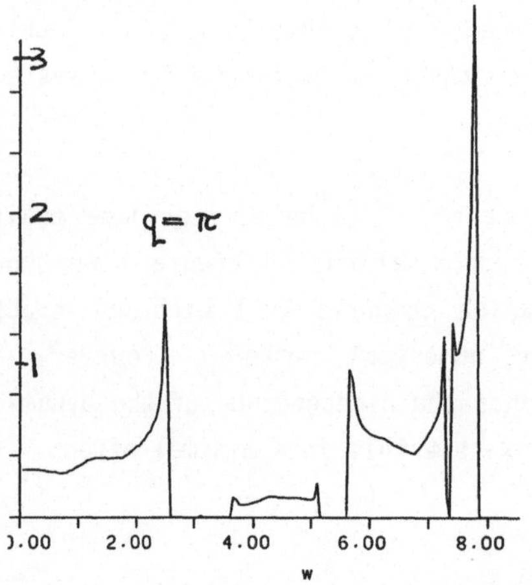

Figure 2

The susceptibility shown in Figure 1 is displayed for $Q = 0.74\pi$ and $q = 0, \pi/2$ and π.

Several authors have considered the properties of the transverse spin fluctuation spectrum generated by (2) [1,2,3,4]. In this paper we present numerical results which taken together with the work reported in [1] provide a full picture of the dynamics as a function of Q. We also report an analytic expression for the dynamic susceptibility that is valid for $Q = (2\pi\, M/N)$ where M and N are integers. The expression can be exploited to give closed analytic expressions, or as an efficient numerical scheme which seems to be the most practical route by which to obtain results for large N.

2 Analysis

The equation of motion (2) can be diagonalized in terms of eigenfrequencies ω_α and eigenvectors $f_\alpha(n)$ that satisfy the difference equation

$$\omega_\alpha f_\alpha(n) = t_n f_\alpha(n+1) + t_{n-1} f_\alpha(n-1) \quad , \tag{4}$$

with

$$t_n^2 = W_n W_{n+1} \quad . \tag{5}$$

The imaginary, or dissipative, part of the transverse dynamic susceptibility measured in neutron scattering is

$$\text{Im.}\,\chi(q',\omega) = (\pi S\omega/W_n) \sum_{\alpha} |\, f_\alpha(n)\, |^2\, \delta(\omega + \omega_\alpha) \tag{6}$$

and the corresponding static susceptibility $\chi(q) = (2S/W_0)$ [4]. A method of computing the susceptibility (6) is described in the following section.

For the moment we observe that for special values of the parameters an exact analytic result is readily recovered. When $Q = (2\pi\, M/N)$ where M and N are integers, and we take $M < (N/2)$, $t_n^2 = t_{n+N}^2$ so the difference equation is periodic. In this instance the dynamic susceptibility can be expressed in terms of quantities A_m, B_m which are computed from the fundamental recurrence formula ($Z_m = A_m, B_m$),

$$Z_{m+1} = \omega Z_m - t_m^2 Z_{m-1} \; ; \; m = -1, 0, 1, \ldots \tag{7}$$

starting with $A_{-1} = 1$, $A_0 = 0$ and $B_{-1} = 0$, $B_0 = 1$. Defining the function

$$L = L(q,\omega) = (2t_0 \ldots t_{N-1})^2 - (A_{N-1} + B_N)^2 \tag{8}$$

We draw attention to the marked difference in the ferro- and antiferromagnetic-like spectra. For the former there is a substantial quasi-elastic response and a high frequency component that is both well separated and highly q-dependent in amplitude. Referring to Figure 2 we note that the distinct high frequency component is absent; there are essentially three broad components to the spectrum centred on $\omega = 0$, 4 and 7, and gaps occur between the components. This structure is tolerably reproduced by the simple analytic approximation given in reference [4], as can be seen from the results presented in Figure 3.

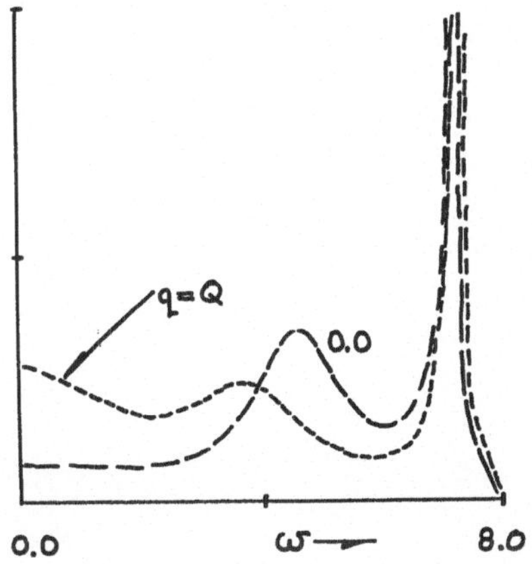

Figure 3

The susceptibility shown in Figures 1 and 2 calculated from the approximation derived in [4] is shown for $Q = 0.74\pi$ and $q = 0, Q$.

We have demonstrated through a numerical evaluation of the susceptibility (6) that the inelastic neutron scattering spectrum varies strongly with the exchange parameters that determine Q. The numerical results for $Q = 0.26\pi$ [1] and $Q = 0.74\pi$, taken together with the analytic result (7) for $Q = \pi/2$, provide a fairly complete picture of the behaviour of the spectrum as a function of Q. However, an analytic scheme is provided for periodic structures $Q = (2\pi\ M/N)$ which in principle, of course, is the general solution for the linear spin dynamics of a single-Q, longitudinally modulated magnet.

Acknowledgement

One of us (SWL) is grateful to Professor Mario Rasetti for discussions on the properties of periodic systems.

References

1. T Ziman and P-A Lindgard B**33** 1976 (1986)
2. S H Liu J Magn Magn Mater **22** 93 (1980)
3. P-A Lindgard J Magn Magn Mater **31-34** 603 (1983)
4. S W Lovesey Z Phys B **67** 525 (1987)
5. P J Jensen, K A Penson and K H Bennemann Phys Rev **B35** 7306 (1987)

On the Interpretation of Electronic and Phonon Anomalies in Mixed Valence Compounds

A. Stasch

Max-Planck-Institut für Festkörperforschung, Heisenbergstr. 1, D-7000 Stuttgart 80, Fed. Rep. of Germany

I. Introduction

Neutron and Raman scattering have revealed the existence of very interesting phenomena in systems with 4f and 5f electronic levels. Those most investigated include: 1) mixed valence phase transitions, 2) the occurrence of heavy fermions, 4) the presence of the rare earth elements in the high T_c superconductors.

In the present paper we would like to demonstrate ways of interpreting existing Raman data in the most intensively studied mixed valence system with 4f electrons, SmS, and the interesting 5f superconducting crystal, α-U, on the basis of a model including d-f hybridization. In Sect. II we propose the use of the electronic Raman effect from the crystal field states of Sm^{3+} as a tool to investigate the electronic properties of SmS before the mixed valence transition as well as during processes in which the external pressure forces the system towards the mixed valence state [1].

In the case of α-U, in Sect. III, we present a qualitative interpretation of the strong enhancement of the transition temperature to the superconducting state under very modest pressure of 11 kbar on the basis of d-f hybridization using the semiquantitative theory applied to 3d superconductors in [2].

II. Electronic Raman scattering and d-f hybridization in SmS

In order to simplify calculations of the electronic scattering tensors in rare earth systems, it is convenient to introduce irreducible spherical tensors with components α_Q^K [3], where Q = -K, -K + 1, ..., +K and

$$(\alpha_Q^K)_{ki}=F(K,\nu)\sum_{\gamma,S,L,J,J_z,\gamma'S,L',J',J_z'} a^*(i;\gamma',S,L',J',J_z')\, a(k;\gamma',S,L,J,J_z)\cdot\langle\gamma',S,L',J',J_z'|U_Q^K|\gamma,S,L,J,J_z\rangle \quad (1)$$

Here $|i\rangle$ and $|k\rangle$ are initial and final states of the electronic system and the coefficients a define the wave functions of the crystal field states in terms of the Russel-Saunders states. The matrix element of the unit tensor U between these states is then defined in terms of 3j symbols by

$$\langle\gamma'SL'J'J_z'|U_Q^K|\gamma SLJJ_z\rangle=(-1)^{J'-J_z'}\begin{Bmatrix} J' & K & J \\ -J_z' & Q & J_z\end{Bmatrix}\langle\gamma'SL'J'\|u^*\|\gamma SLJ\rangle , \quad (2)$$

where

$$\langle\gamma SL'J'\|u^*\|\gamma SLJ\rangle=(-1)^{S+L'+J'+K}[(2J+1)(2J'+1)]^{1/2}\begin{Bmatrix} L' & J' & S \\ J & L & K\end{Bmatrix} \times\langle\gamma'SL'\|u^*\|\gamma SL\rangle \quad (3)$$

and the numerical values of the reduced matrix elements have been tabulated by Nielsen and Koster [4]. The function $F(K,\nu)$ describes the dependence of the Raman tensor on the energy of the initial and intermediate states and on the frequency of the incident laser radiation [3]. The unperturbed electronic wave functions ψ_i are obtained from the diagonalization of the crystal field Hamiltonian in the form of the linear combinations of the free ion wave functions:

$$\psi_i = \sum_{S,L,J,J_z} a(i;\Upsilon,S,L.J.J_z)|\Upsilon,S,L,J_z> . \qquad (4)$$

Finally, the hybridization of these crystal states for f and d shells is described phenomenologically by introducing the perturbation term

$$H' = \sum_{i,t_{2g}} V_{n,d}\, c^{\dagger}_{i}, c_{t_{2g}} + c.c. \quad , \qquad (5)$$

where $c^{\dagger}_{i}$ and $c^{\dagger}_{t_{2g}}$ denote the electron creation operators in the f (i) and d (t_{2g}) type crystal field states, respectively.

In the case of SmS, which has Fm3m $[O_h^5]$ symmetry, the crystal field Hamiltonian can also be written in the explicit form

$$H_c = A\,\beta_J\,[O_4^0 + 5\,O_4^4] + C\,\Upsilon_J\,[O_6^0 - 21\,O_6^4] \quad , \qquad (6)$$

where A and C are crystal field parameters and O_4^0, O_4^4, O_6^0 and O_6^4 denote the Stevens Equivalent Operators. The J-dependent parameters β_J and Υ_J have been calculated using the method elaborated by Racah, Judd, Ofelt and Stevens. As the next step we determine the symmetry of the ^7F_i, i=1,2,...,6, manifolds using the standard group analysis technique. We thus obtain the following pictures of the crystal field splitting:

$$^7F_0 - A_{1g}, ; \quad ^7F_1 - F_{1g}, ; \quad ^7F_2 - E_g + F_{2g}, ; \quad ^7F_3 - A_{2g} + F_{1g} + F_{2g}, ;$$

$$^7F_4 - A_{1g} + E_g + F_{1g} + F_{2g}, ; \quad ^7F_5 - E_g + 2F_{1g} + F_{2g}, ;$$

$$^7F_6 - A_{1g} + A_{2g} + E_g + F_{1g} + 2F_{2g}, . \qquad (8)$$

They are in good agreement with the Raman measurements of Güntherodt and coworkers [6], who found the electronic Raman transitions with the antisymmetric scattering tensor at 173 cm^{-1}, which corresponds to the $^7F_0 \rightarrow {}^7F_1$ transition from Table I. From the energy of this transition we then obtain the spin orbit constant equal to 86.5 cm^{-1}. In a similar way the symmetry properties of the Raman tensors and the Raman energies enable us to find the crystal field parameters A and C and finally to examine the validity of the crystal field model in the description of the electronic states in SmS and other rare earth systems.

The six most recently observed electronic transitions in SmS [1] show that the simple crystal field model used for the derivation of the electronic Raman tensors for the $^7F_0 \rightarrow {}^7F_i$ transition describes the SmS behaviour only in the absence of external pressure. According to the data of Elmiger [1], the Raman processes disappear successively with increasing external pressure starting with the peak at the highest energy, i.e. the $^7F_0 \rightarrow {}^7F_6$ transition. We would like to suggest that this behaviour is a consequence of the resonant interaction between long-lived states of ^7F_i levels and short-lived electrons of the t_{2g}, d band. If such hybridization occurs we have to introduce a linewidth of

Table I. Raman tensors for $^7F_0 \rightarrow {}^7F_1$ and $^7F_0 \rightarrow {}^7F_0$ transitions.

Multiplet	Crystal field state	Spherical scattering tensor	Cartesian Raman tensor
$^7F_1\,(F_{1g})$	$\|1, 0\rangle$ $\|1, 1\rangle$ $\|1, -1\rangle$	$-(1/21)^{1/2}\, F(1,v)\, (\alpha^1_0 + (1/2)^{1/2}\, (\alpha^1_1 - \alpha^1_{-1}))$	$0,\ a_1,\ 0$ $-a_1,\ 0,\ b_1$ $0,\ -b_1,\ 0$
$^7F_2\,(F_{2g})$	$\|2, 1\rangle$ $\|2, -1\rangle$ $(1/2)^{1/2}\, (\|2, 2\rangle + \|2, -2\rangle)$	$(1/70)^{1/2}\, F(2,v)\, [(\alpha^2_1 + \alpha^2_{-1}) + (\alpha^2_2 + \alpha^2_{-2})]$	$a_2,\ 0,\ 0$ $0,\ -a_2,\ b_2$ $0,\ b_2,\ 0$
$^7F_2\,(F_{2g})$	$\|2, 0\rangle$ $(1/2)^{1/2}\, (\|2, 2\rangle - \|2, -2\rangle)$	$-(1/35)^{1/2}\, F(2,v)\, (\alpha^2_0 + (1/2)^{1/2}\, (\alpha^2_1 + \alpha^2_{-1}))$	$a_3,\ b_3,\ 0$ $b_3,\ a_3,\ 0$ $0,\ 0,\ c_3$

$$\frac{1}{\pi}\,\frac{h/\tau_i}{(h\omega_1 - h\omega_2 - E_{F_i} + E_{F_0})^2 + (h/\tau_i)^2}\,, \tag{8}$$

and instead of the lifetime of the ^7F_i state τ_i, the lifetimes of the d-f hybrids, τ_I, τ_{II}, which are given by

$$\frac{1}{\tau_I} = \frac{1}{\tau_i}\,|a_+|^2 + \frac{1}{\tau_{t_{2g}}}\,|v_+|^2 \;;\quad \frac{1}{\tau_{II}} = \frac{1}{\tau_i}\,|a_-|^2 - \frac{1}{\tau_{2g}}\,|v_-|^2 \tag{9}$$

with

$$a_\pm = \frac{2V_{t_{2g},i}}{\{4|V_{t_{2g},i}|^2 + [E_{t_{2g}} - E_i \pm ((E_{t_{2g}} - E_i)^2 + 4|V_{t_{2g},i}|^2)^{1/2}]\}^{1/2}}\,,$$

$$v_\pm = \frac{E_{2g} - E_i \pm [(E_{t_{2g}} - _i)^2 + 4|V_{t_{eg}}\,{}_{7_i}|^2]^{1/2}}{\{4|V_{t_{2g}i}|^2 + [E_{t_{2g}} - E_i \pm ((E_{t_{2g}} - E_i)^2 + 4|V_{t_{2g},i}|^2)^{1/2}]\}^{1/2}}\,. \tag{10}$$

$V_{t_{2g},i}$ is the d-f interaction parameter given in (5). In the case of exact hybridization, i.e. $E_{t_{2g}} = E_i$, we obtain

$$\frac{1}{\tau_{I,II}} = \frac{1}{2}\left(\frac{1}{\tau_{t_{2g}}} + \frac{1}{\tau_i}\right) \cong \frac{1}{2\tau_{t_{2g}}} \ll \frac{1}{\tau_i}\,, \tag{11}$$

which causes broadening of the Raman line by some orders of magnitude, so making practical observation impossible.

III. Pressure-enhanced superconductivity, d-f hybridization and phonon softening in α-U

Neutron scattering studies [5] demonstrate the existence of strong phonon softening in the superconducting 5f metal phase of α-U, suggesting a correlation between the lattice instabilities and the strong sensitivity of the transition temperature T_c to the external pressure.

We start our interpretation of this phenomenon with the McMillan formula, which describes the superconducting transition temperature T_c [2]:

$$T_c = \frac{\theta}{1.45} \exp\left(- \frac{1.04(1+\gamma)}{\lambda-\mu(1+0.62\lambda)}\right) , \qquad (12)$$

where θ is the Debye temperature, and λ and μ stand for the electron-phonon and electron-electron coupling constants, respectively. The electron-phonon interaction parameter λ can be written explicitly as

$$\lambda = \int_0^{\omega_D} \alpha^2(\omega)\, F(\omega)\, \omega^{-1} d\omega \quad , \qquad (13)$$

where $F(\omega)$ is the phonon density of states and α^2 is an average electron-phonon interaction, which is given by [2]

$$a^2F(\omega) = \frac{1}{Mn(E_F)N} \sum_{ll',tt',\vec{k},\vec{q},j} \exp(i\vec{k}\cdot\vec{R}_l)\exp[i(\vec{k}+\vec{q}).\vec{R}_{l'}](1-VN)^{-1}{}_{lt}$$
$$\times [F^t(\vec{q})\cdot\vec{e}(\vec{q},j)][F^{t'}(\vec{q})\cdot\vec{e}(\vec{q},j)](1-VN)^{-1}_{t'l'}1\delta(E(k)-E_F)\delta(E(\vec{k}+\vec{q})-E_F)$$
$$\times \delta(\omega-\omega(\vec{q},j)). \qquad (14)$$

Here S is the screening matrix, M the mass of the α-U ion, and $\vec{q} = \vec{k}-\vec{k}'$. $\omega(\vec{q},j)$ and $E(\vec{q})$ denote phonon and electron energies. $\vec{e}(\vec{q},j)$ is the polarization vector of the phonon with wave vector $\vec{q}$. If we now follow McMillan and factorize λ by assuming $\alpha^2(\omega)$ = const., we obtain [7]

$$\lambda = \frac{N(E_F)\langle g^2\rangle}{M\langle\omega^2\rangle} \quad ,$$

where g describes the electron-phonon coupling. It shows that the value of λ increases monotonically with density of states at the Fermi surface and is inversely proportional to the second moment of the phonon frequency. The third important factor giving the large values of λ is the resonance-like d-f enhancement of the d-f screening as a result of the d-f hybridization described by the matrix 1-VN in (14). In summary, we have also found three important factors, which can explain a sharp increase of T_c under pressure as an effect of the d-f hybridization. In addition our results suggest that this hybridization, together with the screening resonance, may also be the decisive mechanism for the phonon softening.

Acknowledgements

The author wishes to express his gratitude to Professor H. Bilz for his interest, encouragement and support of this work. He would also like to thank Dr. W. Kress for very useful discussions about the phonon behaviour in the considered system.

References

1. M. Elmiger, P. Wachter: private communication
2. W. Hanke, J. Hafner and H. Bilz: Phys. Rev. Lett. 37, 1560 (1976)
3. A. Stasch, W. Piechocki and J.A. Konigstein: Chem. Phys. 105, 317 (1986)
4. C.W. Nielsen and G.F. Koster: Spectroscopic coefficient for p^n, d^n and f^n configurations (MIT Press, 1963)
5. W. S. Crummett, H.G. Smith, R.M. Nicklow and N. Wakabayashi: Phys. Rev. B19, 6028 (1979)
6. G. Güntherodt, R. Merlin, A. Frey and M. Cardona: Solid State Commun. 27, 551 (1978)
7. W. Weber: Phys. Rev. B8, 5093 (1973)

Bloch Lines as Magnetic Excitations in Domain Wall Dynamics with a Large Sensitivity to Initial Conditions +

J.J. Zebrowski and A. Sukiennicki

Institute of Physics, Warsaw Technical University,
ul. Koszykowa 75, PL-00-662 Warszawa, Poland

During the motion of a Bloch wall in a magnetic material with a large uniaxial anisotropy two modes of the precession of magnetic moments necessary to obtain a displacement of the wall are possible [1]: for $H_Z \leq H_W$ the precession is bounded (stationary motion of the wall) and for $H_Z > H_W$ the precession is unbounded (running-oscillatory motion [2]) with the Walker critical field $H_W = \alpha 2\pi M$, while α is the Gilbert phenomenological damping parameter and $4\pi M$ is the saturation magnetization. In terms of nonlinear dynamics the two modes of motion of the wall are equivalent - in appropriately chosen coordinates - to a fixed point attractor (stationary motion) and a limit cycle (running-oscillatory motion) [2,3].

The purpose of this paper is to study numerically the nonlinear dynamics of the Bloch wall in bulk magnetic materials with a large uniaxial anisotropy. We treat the wall as a thin membrane of magnetic moments (the Bloch surface) and show that solitary wave-like excitations propagate along it. When the drive $H_Z > H_W$, the wall exhibits a large sensitivity to initial conditions, a number of topological solitary wave-like excitations is generated and a strange attractor is found in phase space [3]. For low drive fields both topological and nontopological solitary wave-like excitations are possible.

For the case of a large uniaxial anisotropy material the equations of motion of the Bloch wall are [3,4]

$$\dot{q} = \gamma\Delta\left[2\pi M \sin(2\varphi) - \frac{2A}{M}\frac{\partial^2\varphi}{\partial z^2}\right] + \alpha\Delta\dot{\varphi},$$

$$\dot{\varphi} = \gamma\left[H_Z + \frac{2A}{M}\frac{\partial^2 q}{\partial z^2}\right] - \alpha\frac{\dot{q}}{\Delta},$$

where A is the exchange constant, γ is the gyrotropic ratio, Δ is the Bloch wall width parameter, and q(z,t) is the position of the Bloch wall surface, while $\varphi(z,t)$ is the azimuthal angle of the magnetization. z is the coördinate along the direction of the applied field H_Z and the dot over a symbol denotes differentiation over the time. The above equations of motion were solved numerically using a vector version of the DuFort-Frankel explicit finite difference scheme [3]. Periodic boundary conditions were used to simu-

+Supported by the University of Lodz under project CPBP 01.08.B1.1

late a bulk material with a 20 μm period over which 200 grid points were evenly spaced. The time step was 0.05 ns and the results were found to be numerically stable up to at least 2000 ns. For initial conditions $\varphi(z,0)=0$ and $q(z,0)=0$ were used (flat initial conditions). In certain cases a rectangular pulse in $\varphi(z,0)$ of the proper width and height was superimposed on the otherwise flat initial conditions.

For all drive fields smaller than the Walker field H_W the solutions of the equations of motion of the Bloch wall for the flat initial conditions always reproduced the results of the Walker model [1]. The solutions then always converged onto a fixed point attractor [3]. When the drive field was larger than the Walker field, the solutions also reproduced the results of the Walker model [1] giving — in the $\dot{\varphi}(\varphi-\langle\varphi\rangle)$ space with $\langle\varphi\rangle$ the least squares fit average — the simple limit cycle state of the wall (running oscillatory motion).

We simulated the natural small fluctuations of material parameters which are bound to occur in a real physical situation by requiring that during the calculation at t=20 ns at a single grid point the drive field value be decreased by 8 % for the very short time of only 0.1 ns (collision of the Bloch wall with a point defect). The result of such a numerical procedure for $H_Z=12$ Oe > $H_W=10.92$ Oe is given in fig.1 where $\varphi(z,t)$ is given in part a, $q(z,t)$ - in part b and a phase portrait $q(z,t)-\overline{q(t)}$ vs $\varphi(z,t)-\overline{\varphi(t)}$ is depicted in part c. The bar over the symbol denotes a spatial average over the period of the periodic boundary conditions. The curves in part a and b of fig.1 are depicted every 20 ns. It can be seen in fig.1a that some 60 ns after the collision a structural transition occurs in the wall: a number of pairs of solitary wave like excitations are generated. The kinks belonging to adjacent pairs move towards each other, collide and next pass through each other. The wall surface (which was initially flat) is highly distorted (fig.1b) due to the motion of the kinks in fig.1a. The system is now in a chaotic state with a *high sensitivity to initial conditions*: the trajectory shown in fig.1c is a two-dimensional projection of a four-dimensional strange attractor [3]. A similar result as the one shown in fig.1 may be obtained if random fluctuations on the level of 10^{-12} are added at each grid point to the numerical solutions at each time step [3]. The transition to chaos then occurs after about 200 ns of the motion of the wall but the character of the final state is indistinguishable from the character of the one shown in fig.1.

The use of the above-described perturbation techniques for $H_Z<H_W$ does not give any change in the fixed point state of the wall (stationary motion)[3]. Below, a different approach was used. The domain wall in stationary motion was perturbed by applying an (unphysical) initial condition in the form of a narrow pulse in the azimuthal angle $\varphi(z,0)$. Depending on the height of this pulse as well as on its width different solitary wave-like structures were

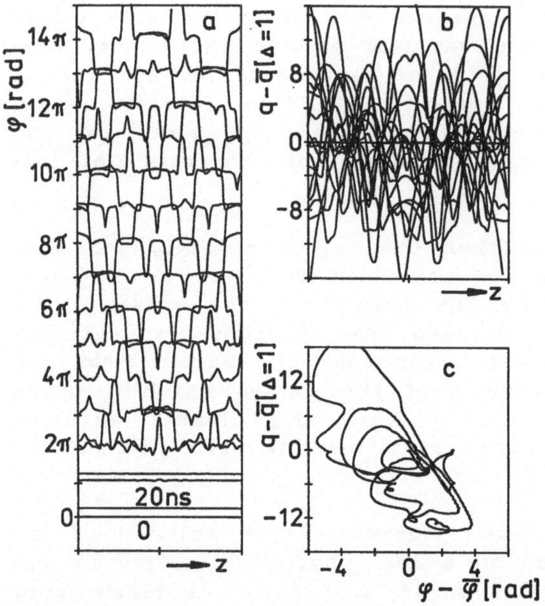

Fig.1 Deterministic chaos during Bloch wall motion.

generated within the structure of the domain wall. Note that the domain wall was moving under the influence of the static drive field at the time these structures were generated. The largest drive fields, for which the domain wall still does not exhibit a large sensitivity to initial conditions [3] i.e. it remains in the stationary motion state, were used in this part of the study. Only the results for the $\pi/2$ high initial condition pulse are quoted here.

If the width of the pulse in the initial condition $\varphi(z,0)$ exceeded 0.5 μm, only two topological solitary waves (a pair of Bloch lines [3]) in the wall structure were generated. This can be seen in fig.2a for a 0.6 μm wide initial pulse in a Bloch wall moving in a static drive of 10.9 Oe. The solitary wave is topologically stable because, in the Bloch wall, both $\varphi = n\pi$ and $\varphi = (n+1)\pi$, $n=0,\pm1,\pm2,\pm3...$, are different minimum energy orientations of the magnetic moment. A small "forerunner" pulse precedes each of the kinks in fig.2a. Note that the drive field rotates the magnetic moments outside the solitary wave by some moderate angle away from the exact static minimum of energy position.

The part of the wall where the topological solitary waves are created lags behind the rest of the wall so that a "well" in the shape of the wall surface (fig.2b) forms simultaneously with the pair of kinks in the wall structure in fig.2a.

As the motion of the wall continues the kinks move in opposite directions. Because of the periodic boundary conditions used in the calculations, the left kink of the pair wraps around and reappears on the right of fig.2. This is as if, in an infinitely long wall,

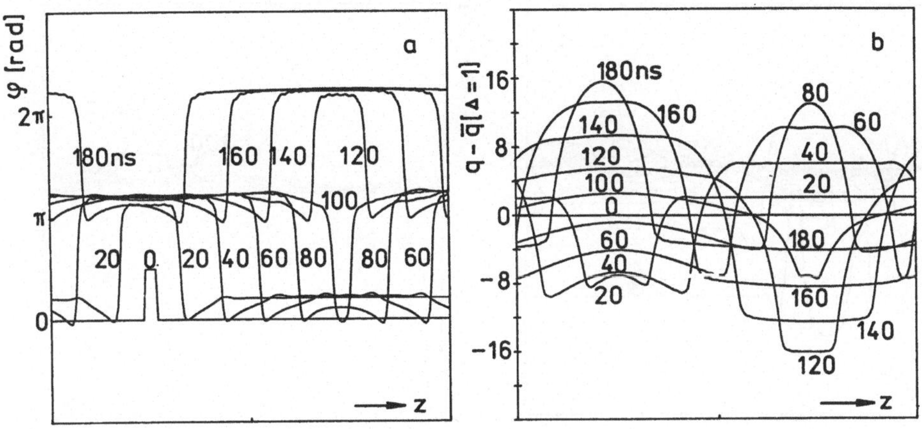

Fig.2 Topological excitations and the related wall shape.

there were several points at which pairs of Bloch lines were generated so that, after a certain time of motion of the wall, the kinks belonging to adjacent Bloch line pairs approached each other. Next, the kinks collide and pass through each other conserving their identity. To see that the collision of the solitary waves results in their passing through each other rather than reflecting off each other, note that in fig.2b the curvature of the wall shape changes sign during the collision process. With periodic boundary conditions in effect the wrap-around-collision-wrap-around sequence of events will continue indefinitely always with only the single pair of kinks taking part. This is in marked contrast to what occurs at drive fields larger than the Walker field: as discussed above a state of deterministic chaos is found with several Bloch lines generated and propagating along the wall structure. The wall shape then is also very complicated and dynamically changing.

When the non-topological solitary waves obtained from a 0.4 μm wide initial condition pulse collide, they pass through each other and the curvature of the wall changes sign i.e. the solitary waves conserve their identity (fig.3a - wall structure up to the moment of collision, part b - after the collision). However,due to dissipation, the solitary waves which emerge from the collision are very small (note the twice larger scale in fig.3b). All the same they decay very slowly so that in fig.3 they may be seen to traverse a certain distance (about 10 μm) within the time limit of the calculation (260 ns) and, in the periodic world of this study, another collision is imminent in fig.3.

In some circumstances the nontopological solitary wave excitations may become unstable. In that case a transition to the topological wave occurs. Typically this transition occurs when the initial condition pulse is wider than 0.4 μm but less than 0.5 μm. The transition takes place some distance from the point of generation but before the excitations have travelled far enough to meet in colli-

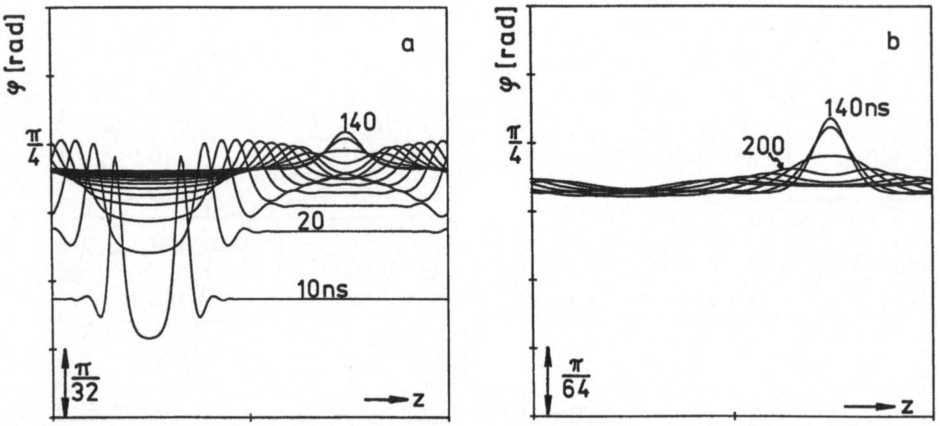

Fig.3 Collisions of nontopological solitary waves.

sion. During such a transition the solitary waves are stopped. Such a transition may also occur when the period of the boundary conditions is reduced (e.g. to 12 μm). Then instead of passing through each other the two colliding nontopological waves are transformed into a pair of topological solitary waves. After the transition the rest of the evolution very much resembles the one shown in fig.2 just after the generation of the Bloch line pair. The transition to topological waves looks as if the nontopological solitary wave was a bound state of a pair kink-antikink.

In summary, the Bloch wall in materials with a large uniaxial anisotropy, when moving in a static drive field which is larger than the Walker critical field, exhibits an extreme sensitivity to initial conditions. Very small perturbations may then induce a state of deterministic chaos in the wall. In such a state topologically stable solitary wave-like excitations are present in the wall structure and collide with each other conserving their identity even though this nonlinear system is dissipative and driven. By strong perturbations, topological and nontopological solitary waves may be generated in a Bloch wall moving in a field smaller than the Walker critical field. These solitary waves conserve their identity during collisions but, though nontopological waves seem to decay due to the dissipation acting in the system, they are able to traverse macroscopic distances of the order of tens of micrometers. In some cases the nontopological waves undergo a transition and become pairs of topological solitary waves.

1. N.L.Schryer, R.L.Walker, J.Appl.Phys.45, 5406 (1974).
2. J.Hołyst, A.Sukiennicki, Żebrowski, Phys. Rev.B33, 3492 (1986).
3. J.J.Żebrowski,A.Sukiennicki,to be published in Acta Phys. Pol. (Proc.of 3rd Int. Conf. Physics of Magnetic Materials,Szczyrk 1986).
4. J.C.Slonczewski, J.Appl.Phys.44, 1759 (1973).

Dynamics of Magnetic Domain Walls and Barkhausen Noise in Metallic Ferromagnetic Systems

G. Bertotti

Istituto Elettrotecnico Nazionale Galileo Ferraris and GNSM-CISM, Corso Massimo d'Azeglio 42, I-10125 Torino, Italy

1. INTRODUCTION

The aim of the present paper is to show how a stochastic differential equation similar to that proposed by Langevin for the description of Brownian motion can be fruitfully applied to the interpretation of Bloch wall dynamics in a ferromagnetic system magnetized by the action of an external magnetic field.

In order to be able to state our main concepts in clear form and also to give an intuitive feeling of the proposed approach, we shall concentrate on a particularly simple case, that of a ferromagnetic slab of rectangular cross section S, in which the variations of magnetization are produced by an individual plane 180° Bloch wall, extending longitudinally along the slab and vertically across S, and moving horizontally in S. If the internal degrees of freedom related to the wall flexibility are neglected, the wall motion can be described by a single degree of freedom. This may be the horizontal wall coordinate x in S, or, equivalently, the magnetic flux $\Phi = 2I_s dx$, where I_s represents the saturation magnetization and d the slab thickness.

This is essentially the situation considered by WILLIAMS et al. [1] in their classical paper on wall dynamics, as well as by NEEL [2] in his interpretation of hysteresis effects at low magnetic fields, in the so-called Rayleigh region. The fundamental result obtained in [1] - and subsequently confirmed by several authors - is that, in metallic materials, the wall motion is governed by the simple equation (the dot means time derivative)

$$\sigma G \dot{\Phi} = H - H_c = H_{exc} \quad . \tag{1}$$

This equation states that, due to the damping effect of induced electric currents, the flux rate $\dot{\Phi}$ (proportional to the wall velocity) depends linearly on the difference H_{exc} between the magnetic field H acting on the wall and the coercive field H_c experienced by the wall. The proportionality constant is given by the product of the electrical conductivity σ and of a dimensionless coefficient which, in the limit of a wide slab ($S/d \gg d$) takes the value $G = (4/\pi^3) \Sigma 1/(2k+1)^3 = 0.1356$.

Néel's model, on the other hand, does not consider any dynamic effect, but shows that ferromagnetic hysteresis at low fields can be explained by assuming that the coercive field H_c is a random function of Φ, and by investigating the statistical distribution of possible wall positions resulting from this random character of H_c. Our contribution in this paper is just to show that, by bringing together (1) and the idea that H_c shows random fluctuations, one obtains a stochastic differential equation which provides a proper theoretical basis for the description of Bloch wall dynamics in a perturbed medium and, in particular, for the interpretation of the Barkhausen effect commonly observed in ferromagnetic materials.[3]

2. THE MODEL

2.1 Derivation of the Stochastic Differential Equation

The mentioned stochastic differential equation governing wall motion is explicitly worked out by a careful analysis of the physical meaning of the various terms appearing in (1). The main points to be considered are the following.

- The connection between $\dot{\Phi}$ and H_{exc} expressed by (1) is meaningful only if $H_{exc} > 0$. The very fact that H_c represents a coercive field implies that, when $H < H_c$, the wall will be locked in its position, i.e. $\dot{\Phi} = 0$. We shall take account of this fact by writing, instead of (1),

$$\sigma G \dot{\Phi} = H_{exc}\, \theta(H_{exc}), \tag{2a}$$

$$H_{exc} = H - H_c\,, \tag{2b}$$

where $\theta(x)$ represents the Heaviside step function.

- The fluctuations of H_c are expected to originate either from the action of the wall internal degrees of freedom, related to its flexibility, or by the interaction of the moving wall with the surrounding medium, as considered by Néel. In both cases, H_c will fluctuate only when the wall is moving, i.e. $H_{exc} > 0$, and we are led to express the rate of change of the instantaneous coercive field value in the form $(dH_c/dt)\,\theta(H_{exc})$, where dH_c/dt is a suitable stochastics process, to be better specified in the following.

- The magnetic field H is expected to be made up of two basic contributions, the applied field H_c and a demagnetizing counterfield H_{dem}. If we consider a portion of the hysteresis loop characterized by a constant differential permeability μ, we expect $\dot{H}_{dem} = -\dot{\Phi}/S\mu$ and $\dot{H}_a = \langle\dot{\Phi}\rangle / S\mu$, so that the rate of change of H will be

$$\dot{H} = (\langle\dot{\Phi}\rangle - \dot{\Phi})/S\mu = (\sigma G \langle\dot{\Phi}\rangle - H_{exc}\,\theta(H_{exc}))/\tau\,, \tag{3}$$

where the time constant τ is equal to

$$\tau = \sigma G S \mu \quad . \tag{4}$$

By taking the time derivative of (2b) and introducing the dimensionless variables $v = H_{exc}/H^{(w)}$, $w = -H_c/H^{(w)}$, with $H^{(w)} = \sigma G \langle\dot{\Phi}\rangle$, we obtain the stochastic differential equation

$$\frac{dv}{dt} + \frac{v\theta(v) - 1}{\tau} = \frac{dw}{dt}\,\theta(v) \quad . \tag{5}$$

The information on the microscopic interactions of the Bloch wall with the surrounding medium is contained in the stochastic process dw/dt. In the spirit of the Langevin approach, and also on the basis of experimental results about the statistical properties of H_c in ferromagnetic systems containing only one moving domain wall [4,5], we shall assume that w is essentially a Wiener process. Two main classes of models can then be considered, which will be now discussed separately.

2.2 Time Wiener Process

We shall first consider the case where w is a Wiener process in time, with

$$\langle |dw|^2 \rangle = a\, dt \quad . \tag{6}$$

With this assumption, (5) is but a Langevin equation for (v-1) in the region v>0, with, however, a highly non-trivial complication at v=0, where the character of the equation changes completely as a consequence of the presence of the mentioned coercive field threshold. This situation can be properly dealt with by considering the Fokker-Planck equation for the probability density P(v,t) associated with (5) in the region v>0 and by carefully investigating, then, the various terms which contribute to the probability flow across the boundary v=0. We have not the space to discuss here the details of this analysis, and we simply present the obtained results. We shall make use of the new variables $u = t/\tau$ and $z = z_0(v-1)$, with $z_0 = \sqrt{2/a\tau}$. Notice that the point v=0 will now correspond to $z = -z_0$. The main results characterizing the statistical properties of the process are the following.

- Any permitted probability density P(z,u) is made up of a regular part f(z,u), defined in the region $z > -z_0$ and of a Dirac singularity at $z = -z_0$, according to the following expression,

$$P(z,u) = f(z,u) \left(1 + (1/z_0)\delta(z+z_0) \right) \quad . \tag{7}$$

The Dirac singularity determines the finite probability that, at a generic instant of time, the wall is not moving but locked in its position by the coercive field.

- f(z,u) obeys the Fokker-Planck equation

$$\frac{\partial f}{\partial u} - \frac{\partial}{\partial z} \left(zf + \frac{\partial f}{\partial z} \right) = 0 \tag{8}$$

with the boundary condition

$$\frac{\partial f}{\partial u} - z_0 \left(zf + \frac{\partial f}{\partial z} \right) = 0 \qquad \text{at } z = -z_0 \quad . \tag{9}$$

Equation (9) is obtained by imposing, in (7), the conservation of total probability.

- Any solution of (8) and (9) decaying exponentially when $z \to \infty$ can be expressed as a series of eigenfunctions of the form

$$f_q(z,u) = \exp(-qu)\exp(-z^2/4)\, D_q(z), \tag{10}$$

where $D_q(z)$ represents the parabolic cylinder function of index q, and the set of permitted values of q is given by the solutions of the equation

$$q\,(q-1)\, D_{q-2}(-z_0) = 0 \quad . \tag{11}$$

- The eigenfunctions defined by (10) and (11) provide a proper orthogonal basis for the representation of the probability densities of the form (7), since any $P_q(z,u)$ obtained by substituting (10) into (7) turns out to be orthogonal to $f_r(z,u)$ if $q \neq r$.

- The autocorrelation function of the process is a simple exponential of time constant τ (see (4)).

- The stationary probability density of the process (corresponding to q=o) is given, as can be seen from (10) and (7), by a truncated gaussian, supplemented by the mentioned singularity at $z = -z_0$ (i.e. v = 0). As previously mentioned, the presence of this singularity implies that there is a finite probability that the wall is, at a given instant, not moving, which means that the wall motion has an intrinsic intermittent character, a fact well confirmed by Barkhausen noise experiments. When the average wall velocity (i.e. $\langle\dot{\Phi}\rangle$) increases, $z_0 \to \infty$ and the Dirac singularity tends to vanish. In this limit, the wall motion becomes continuous and the fluctuations of v around its average value are described by a simple Ornstein-Uhlenbeck process.

2.3 Space Wiener Process

Let us now consider the case where w is a Wiener process with respect to Φ ,

$$<|dw|^2> = b_0 \, d\Phi = bv\theta(v) \, dt \quad . \tag{12}$$

This assumption is directly suggested by Neel's model - where H_c is just considered to be a random function of Φ - and also supported by the results reported in[5,6]. The Fokker-Planck equation associated with (5) and (12) in the region $v>0$ is easily written and the boundary $v = 0$ turns out to behave like an impenetrable barrier. In terms of the variables $u = t/\tau$ and $z = cv$, with $c = 2/b\tau$, we obtain

$$\frac{\partial P}{\partial u} - \frac{\partial}{\partial z}\left((z-c+1)P + \frac{\partial P}{\partial z} \right) = 0 \quad . \tag{13}$$

The eigenfunctions of the problem are of the form

$$P_n(z,u) = \exp(-nu) \, z^{c-1} \exp(-z) \, L_n^{c-1}(z) \quad ; \quad n=0,1, \ldots \tag{14}$$

where $L_n^{c-1}(z)$ is the Laguerre polynomial of order n. Even in this case, the antocorrelation function of the process is an exponential of time constant τ . The stationary probability density is, however, different from that obtained in section 2.2. Instead of a truncated gaussian with a Dirac singularity at v=0, we obtain from (14)

$$P_{st}(v) = K \, v^{c-1} \exp(-cv) \quad . \tag{15}$$

Since the parameter $c \propto <\dot{\Phi}>$, P_{st} will exhibit a power law divergence at $v \to 0$ when $<\dot{\Phi}>$ is so low that $c<1$, this divergence disappearing then when c becomes >1.

According to (15), therefore, the intermittent character of wall motion predicted by the model discussed in section 2.2 is now replaced by a more complex behaviour, controlled by a power law. Due the the very presence of this power law, however, the wall motion at low values of v turns out to exhibit an interesting fractal structure. In fact, let us choose a velocity threshold v_0 so small that $cv_0 \ll 1$, and let us consider the probability $Pr[v < v_0]$ that $v < v_0$. According to (15)

$$Pr[v < v_0] \sim (v_0)^c \qquad \text{when } cv_0 \ll 1 \quad . \tag{16}$$

This result can be visualized by considering a specific time behaviour of v(t), and by marking, on the unit time interval, all the sub-intervals in which $v < v_0$. The total length of these sub-intervals will of course represent $Pr[v < v_0]$ or, in other words, the probability that, with a velocity resolution v_0, the wall is not moving. When v_0 is decreased, each of these sub-intervals will progressively split into several disjoint pieces and the corresponding total length will decrease according to (16). This means that the considered set of sub-intervals is actually a fractal, whose fractal dimension is, as shown by (16), $D = 1-c < 1$. A related fractal structure is also expected to be found in the fine behaviour of magnetization vs. applied field along the hysteresis loop, for it will essentially reproduce the behavior of $\int v dt$ vs. t.

3. CONCLUDING REMARKS

Although the dynamic system considered in this paper is a very simple one, involving only one moving Bloch wall, we believe that the results here derived should provide a proper theoretical background also for the interpretation of the magnetization process in ferromagnetic systems characterized, as is usually the case, by a complex, intricate magnetic domain structure. Actually, several considerations [7] suggest that even in these cases (1) should be applicable to the description of the collective behaviour of Bloch walls inside a characteristic correlation region of size determined by the microstructural properties of a given material.

On the other hand, we point out that the model discussed in section 2.3 may be improved and made more realistic by taking into account that, from a physical viewpoint, the correlation in the fluctuations of H_c at two different wall positions is expected to decay to zero when the distance between these positions increases. This aspect of the problem is completely neglected by describing the fluctuations of H_c in terms of a pure Wiener process. Actually, we expect that this Wiener process should represent an approximation, valid for small wall displacements, of an Ornstein-Uhlenbeck process, describing the real character of the fluctuations of H_c. Preliminary investigations of the consequences of this assumption seem to indicate that a good deal of the complex phenomenology of the Barkhausen effect in ferromagnetic materials can be interpreted as a direct, natural consequence of this description of the stochastic behaviour of H_c.

4. REFERENCES

1. H.J. Williams, W. Shockley and C. Kittel, Phys. Rev. 80, 1090 (1950).
2. L. Néel, Cahiers de Phys. 12, 1 (1942).
3. H. Bittel, IEEE Trans. Magn. MAG-5, 359 (1968).
4. W. Grosse-Nobis, J. Magn. Magn. Mat. 4, 247 (1977).
5. R. Vergne, J.C. Cotillard and J.L. Porteseil, Revue Phys. Appl. 16, 449 (1981).
6. R. Vergne, J.L. Porteseil and J.C. Cotillard, J. Magn. Magn. Mat. 15-18, 1470 (1980).
7. G. Bertotti, in Proc. 3rd Int. Conf. on Physics of Magn. Mat., Szczyrk (Poland) 1986, World Scientific Singapore.

Part IV

Spin Excitations: Experiments and Computer Simulations

Magnetic Fluctuations and Excitations in the S=1 Antiferromagnetic Chains of NENP

L.P. Regnault[1], *W.A.C. Erkelens*[1*], *J. Rossat-Mignod*[1], *J.P. Renard*[2], *M. Verdaguer*[3], *W.G. Stirling*[4], *and C. Vettier*[4]

[1]Centre d'Etudes Nucléaires, DRF-G/SPh-MDN, 85 X, F-38041 Grenoble Cedex, France
[2]Institut d'Electronique Fondamentale, Université Paris IX, F-91405 Orsay Cedex, France
[3]CNRS (UA 420), Université Paris-Sud, F-91405 Orsay Cedex, France
[4]Institut Laue-Langevin, 156 X, F-38402 Grenoble Cedex, France

1. Introduction

In this paper we report experimental results concerning the magnetic properties of the S = 1 antiferromagnetic chain, as observed in the new system $N_i(en)_2NO_2ClO_4$ (alias NENP). Conventional ideas on the ground state properties of antiferromagnetic quantum chains have been recently challenged by HALDANE /1/. According to his theory, the S = 1 Heisenberg antiferromagnetic chain (1D-HAF) would exhibit an ***energy gap*** between a non-magnetic ***singlet*** ground state and the first excited states, in sharp contrast with the S = 1/2 1D-HAF which exhibits a continuum of excited states /2/. Such a difference between integer and half integer spins has been more clearly verified by numerical calculations /3,4,5/. The best evaluations agree with a rather large value of the energy gap $E_G \sim 0.4|J|$ for the case S = 1 /5/. However, the existence of the Haldane gap would affect largely the magnetic properties. The most characteristic features would be an exponential decrease of all the susceptibilities when $T \lesssim E_G/k$, and the absence of long-range ordering (LRO) down to very low temperature, due to the finite size of the correlation length /1,3/. Unfortunately, the existence of such properties is not a unique signature of the Haldane conjecture. Indeed, similar effects are predicted and observed in the S = 1/2 alternating chain /6,7/ as in the S = 1/2 spin-Peierls system /8,9/. This makes more difficult to obtain an unambiguous description of the experimental results.

2. Magnetic Properties of NENP

NENP crystallizes in an orthorhombic structure of space group Pnma at all temperatures. In this compound, the N_i^{2+} ions are strongly linked by NO_2 groups along the chain axis $\vec{b}$ ($\parallel \hat{z}$), whereas the chains are well isolated from each other by ClO_4groups, yielding a rather good 1d-character /10/. The high-temperature susceptibilities are characteristic of a S = 1 antiferromagnetic chain described by the Hamiltonian /10,11/

$$\mathcal{H} = |J| \sum_i \vec{S}_i\vec{S}_{i+1} + D \sum_i \left(S_i^z\right)^2 + \mu_B \sum_i \bar{\bar{g}}\, \vec{S}_i . \vec{H} \qquad (1)$$

with the parameters $g_{\parallel} \simeq 2.15$, $g_{\perp} \simeq 2.22$, $|J| \simeq 50$ K and $D \simeq 1\text{-}1.5$ K corresponding to a small planar anisotropy with $D/|J| \sim 2$ % /10/.

[*]On leave from Kamerlingh Onnes Laboratory, University of Leiden, The Netherlands

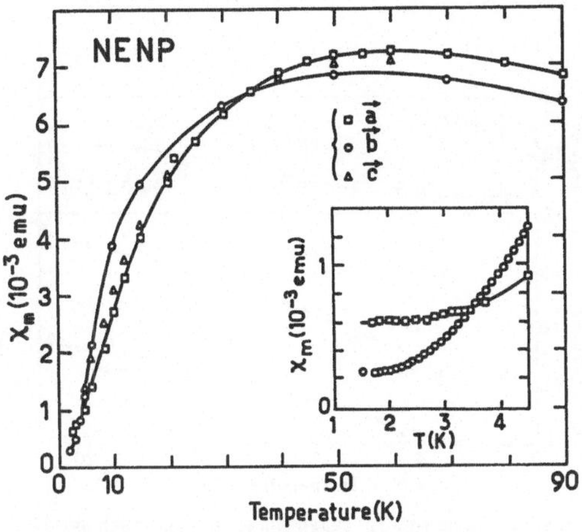

Fig. 1 Molar susceptibilities of NENP as a function of temperature.

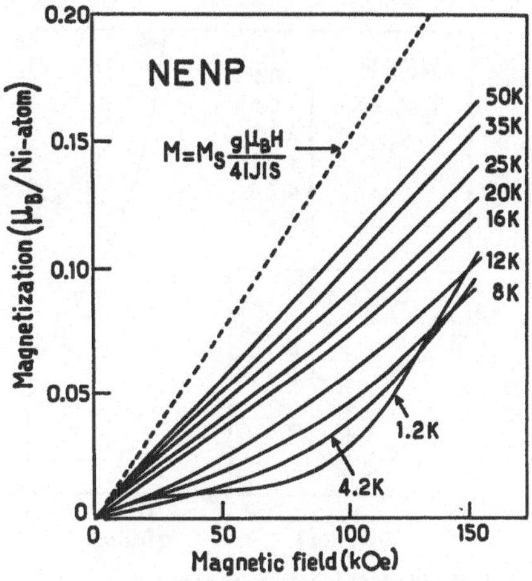

Fig. 2 Magnetization curves of NENP as a function of field and temperature.

At low temperatures, the susceptibilities deviate from the prediction of the S = 1 HAF chain and fall down abruptly as the temperature is lowered below ~ 15-20 K (Fig. 1). This behaviour can be roughly fitted to the exponential relation /11/

$$\chi^{\alpha}(T) \simeq \chi^{\alpha}(0) + C^{\alpha} \exp(-E^{\alpha}_G/kT) \qquad (\alpha = a, b, c)$$

in which $\chi^{\alpha}(0)$ represents a weak residual susceptibility, typically $\chi^{\alpha}(0)/\chi^{\alpha}_{max} \sim 5$ %, and with $E_G \sim 14$ K, which clearly indicates the existence of a ***gap*** between a ***non-magnetic*** ground state and the excited states.

The low-temperature magnetization reflects also this behaviour as shown in Fig. 2. At low field, the small value of the induced magnetic moment is confirmed, despite the existence of a small parasitic moment ($\sim 0.01\ \mu_B$) due to the presence of intrinsic impurities in the polycrystalline sample used for these measurements. At higher fields ($H \gtrsim 100$ kOe), the non-magnetic state is destroyed and the magnetization increases approximatively linearly, with a slope very close to the classical value, in agreement with theoretical predictions of PARKINSON *et al.* for the S = 1 1d-HAF model /4/. The critical field is estimated to be $H_c \simeq 90$ kOe at T = 1.2 K, in agreement with the value deduced from the energy gap and based on the relation $H_c \sim E_G/g\mu_B$. It is argued that such a behaviour cannot be understood easily in terms of single ion anisotropies and, for example, is completely different from that expected for the spin-flop transition. This point is enhanced by the absence of any 3d-LRO down to 1.2 K, as shown from specific heat, NMR and elastic neutron scattering experiments indicating either a rather small interchain coupling J' or an anomalously limited correlation length on basis of the approximate relation $kT_N \sim z'|J'|\xi(T_N)$.

Nevertheless, the best way to have a more accurate proof of the existence of this gap consists in measuring directly the excitation spectrum by inelastic neutron scattering (INS). In a conventional INS experiment on NENP, two gaps associated with magnetic fluctuations along (∥) and perpendicular (⊥) to the chain axis have been observed, with energies at $q \simeq 1$ $E^{\parallel}_G \simeq 2.6$ meV ($\simeq 29$ K) and $E^{\perp}_G \simeq 1.2$ meV ($\simeq 14$ K), respectively. The dispersion curves of these two modes are given in Fig. 3, in which the solid lines represent fits to a classical spin wave relation which, around $q \simeq 1$, can be written as

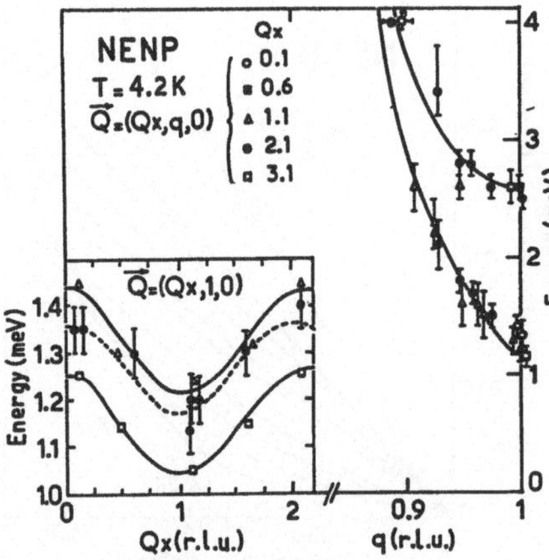

Fig. 3 Dispersion curves along and perpendicular to the chain. The lines are fits to the spin wave theory.

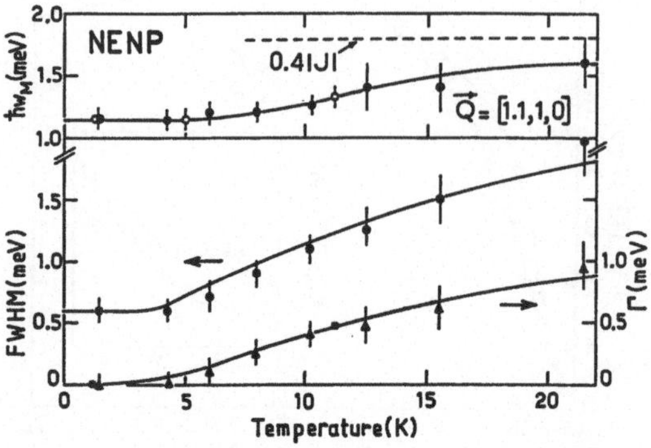

Fig. 4 Temperature dependences of the peak position (ω_M), experimental width (FWHM) and damping parameter (Γ) of the low-energy mode.

$$\omega_q^{\parallel,\perp} \simeq \left(E_q^{\parallel,\perp} + (2\pi|J|Sq^*)^2\right)^{1/2} \tag{2}$$

with $q^* = 1 - q$ and $|J| \simeq 55$ K, in quantitative agreement with the previous determination /11/. A small wave vector dependence of the lowest energy mode (~ 0.2 meV) is observed along the [100] direction, indicating a weak interchain coupling J'. Actually, high-resolution INS experiments indicate the existence of *two* modes rather than one at low energy as shown in Fig. 3, which have similar dispersion curves shifted by only 0.2 meV. Although the origin of this splitting remains not clear, the observed dispersion can be accounted for by the classical spin wave approximation, which yields a ratio $|J'/J| \sim 4 \times 10^{-4}$, emphasizing the good 1d-character of NENP.

The effect of the temperature on the low-energy mode has been investigated. Two interesting features have been observed (Fig. 4). Firstly, the energy is found to increase slightly with increasing temperature, i.e. as soon as the Heisenberg character becomes more pronounced. At the highest temperature, a value $E_G^{\perp}/|J| \simeq 0.35$ has been observed, a value very close to the theoretical prediction for the Haldane gap. Together with this shift a progressive damping of the excitation is also observed which is characterized by a damping parameter Γ following approximatively the exponential relation : $\Gamma \simeq \Gamma_0 \exp\left(- E_G^{\perp}/kT\right)$ with values $E_G^{\perp} \simeq 14$ K and $\Gamma_0 \simeq 20$ K. The excitation becomes rapidly overdamped above 20 K and disappears around 30 K, whereas a quasi elastic contribution centered around $\omega = 0$ is growing up.

From the magnetization measurements, we have seen that the non magnetic state is destroyed above a critical field $H_c \simeq 90$ kOe. An interesting point to be discussed now is the effect of a magnetic field on the energy gap. Figure 5 summarizes the main characteristics of the field dependences. For the low-energy mode, a non conventional behaviour is observed, characterized by a very weak field-dependence of the peak position and by a progressive decrease of the intensity up to a factor two at the critical field H_c, after which the intensity remains constant. Such a behaviour is quite surprising within the "classical" model based on an energy gap induced by a small Ising-type anisotropy. The mode around 2.6 meV displays a more classical behaviour, at least at low field. Nevertheless, the situation appears to be more complex when considering the polarization of these modes. The low-energy mode of XY-type at low field (H < 25 kOe), becomes of Y-type at about and above H_c, the magnetic field being applied along the $\vec{c}$ axis ($\hat{y} \parallel \vec{c}$). The higher energy mode, which is of z-type at low field becomes mostly of X- or/and Y-type at higher fields, with a field dependence close to the two-magnon banch at $2g_{\perp}\mu_B H$. At about H_c the situation is in fact very similar to that observed in TMMC /12/, in which there is an anticrossing between the out-of-plane fluctuations and the two-magnon modes due to the canting of the spins at high field from their spin-flop positions /13/.

3. Discussion

The susceptibilities, magnetization and INS experiments described above give evidence for energy gaps with temperature and field dependences which cannot be explained easily within the framework of the classical spin wave theory involving OP and IP anisotropies. At least three arguments can justify this point of view :
i/ The order of magnitude of the OP anisotropy determined at ***low temperature*** ($D \simeq 5$ K) is not consistent with the high-temperature determination by a factor of about 4.
ii/ The order of magnitude of the IP anisotropy (~ 1 K) needed for explaining the smallest energy gap is in contradiction with the absence of LRO down to 1.2 K and with the weak values of the susceptibilities $\chi^{\alpha}(0)$.
iii/ The observed field dependence is not consistent with such an in-plane Ising-type anisotropy. In that case, a field dependence such as $\sim \left(E_G^2 + (g\mu_B H)^2\right)^{1/2}$ is expected, which is far from being verified experimentally (Fig. 5). Moreover, the field dependence of the intensity shows that the system is in a rather good ***planar*** state in zero field.

Two alternative explanations for this energy gap can be developed in the light of existing theories. Either NENP verifies the Haldane conjecture or it is an example for an alternating chain. In the last case, the alternation can be due either to a permanent distortion of the chain yielding to non equivalent exchange paths or, less probably, to displacements of the Ni^{2+} ions at low temperature as for the spin-Peierls transition. Unfortunately, the lack of ***accurate*** theoretical

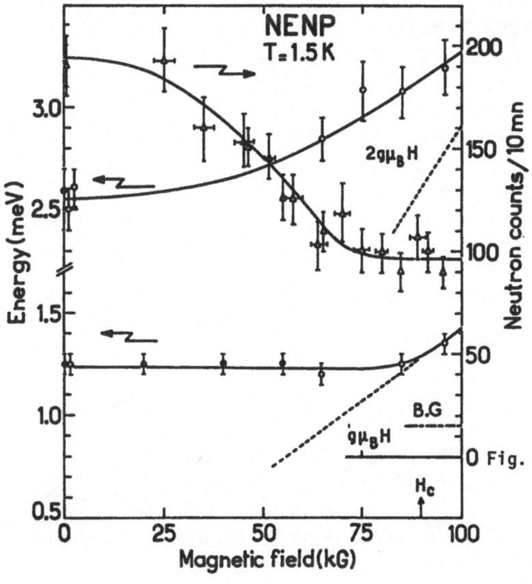

Fig. 5 Field dependences of the magnetic excitations around q=1 : energies (●,o) and intensity at maximum for $\vec{Q}$=(0.5,1,0). The full lines are guides to the eye. Dashed lines : classical one-and two-magnon modes.

results for a $S = 1$ AF chain in both cases prevents from deciding *unambiguously* between the two hypotheses.

• Haldane conjecture : the situation in NENP is a little more complex than in the ideal system. The planar anisotropy splits the gap into two components $E_G^{\parallel}$ and $E_G^{\perp}$. But quite interestingly, the average of the experimental values $E_G^0 = \left(E_G^{\parallel} + E_G^{\perp}\right)/2 \simeq 0.42|J|$ (reflecting more or less the limit of the pure HAF model, since it realizes a cancellation, at first order, of the anisotropy effect), agrees *quite well* with the numerical calculations of NIGHTINGALE *et al.* /5/ giving $E_G^0 \simeq 0.41|J|$. Concerning the dependence of the energy gap with the anisotropy, a rather strong effect is found experimentally : $E_G^{\parallel,\perp}/|J| \simeq E_G^0/|J| \pm KD/|J|$ with $K \simeq 7$. This anisotropy dependence is much less marked in the numerical calculations which give only a value $K \simeq 2$, but with $E_G^0/|J| \sim 0.26$ /3/. To our opinion, this discrepency appears minor, since neither the experimental value of D at low temperature, nor the theoretical value of K are known with enough accuracy. More accurate theoretical developments are strongly needed to clarify this important question.

• Alternating chain : only the case $S = 1/2$ has been studied both experimentally and theoretically. In particular the size of the energy gap has been related to the alternation parameter /6,7/. For $CuCl_2$ (γ-picoline) /14/ or PHCC /15/ alternation parameters $\alpha \sim 0.6$ have been measured yielding an energy gap $E_G/|J| \sim 0.5$, while the chains appear to be structurally uniform. But in these compounds the exchange interactions involve Cu-Cl-Cu bonds which are known to be very sensitive to both the distance and bond angles. For NENP, the situation is drastically different since the exchange involves flat Ni-NO_2-Ni bonds which are known to be very covalent and then weakly sensitive to the distance /10/. A more quantitative analysis of our results with this last hypothesis would need further theoretical investigation of the $S = 1$ alternating chain which is not available at this time. Nevertheless, the second explanation seems to be less probable than the first one from crystallographic and chemical bonding arguments. We are left therefore with only the Haldane conjecture for a quantum energy gap for explaining consistently all our experimental data.

1. F.D.M. Haldane: Phys. Rev. Lett. 50, 1153 (1983).
2. J. Des Cloizeaux, J.J. Pearson: Phys. Rev. 128, 2131 (1962).
3. B. Botet, R. Jullien, M. Kolb: Phys. Rev. B28, 3914 (1983).
4. J.B. Parkinson, J.C. Bonner: Phys. Rev. B32, 4703 (1985).
5. M.P. Nightingale, H.W.J. Blöte: Phys. Rev. B33, 659 (1986).
6. M. Duffy, K.P. Barr: Phys. Rev. 165, 647 (1968).
7. L.N. Bulaevskii: Sov. Phys. JETP 17, 684 (1963).
8. J.W. Bray, L.V. Interrante, I.S. Jacobs, J.C. Bonner: in Extended Linear Chain Materials, ed. by J.S. Miller, vol. 3 (Plenum, New York, 1982) p. 353.
9. E. Pytte: Phys. Rev. B10, 4637 (1974).
10. A. Meyer, A. Gleizes, J.J. Gired, M. Verdaguer, O. Kahn: Inorg. Chem. 21, 1729 (1982).
11. J.P. Renard, M. Verdaguer, L.P. Regnault, W.A.C. Erkelens, J. Rossat-Mignod, W.G. Stirling: Europhys. Lett. 3, 945 (1987).
12. I.U. Heilmann, J.K. Kjems, Y. Endoh, G.F. Reiter, G. Shirane: Phys. Rev. B24, 3939 (1981).
13. K. Osano, H. Shiba, Y. Endoh: Prog. Theor. Phys. 67, 995 (1982).
14. H.J.M. De Groot, L.J. De Jongh, R.D. Willet, J. Reedijk: J. Appl. Phys. 53, 8038 (1982).
15. A. Daoud, A. Ben Salah, C. Chappert, J.P. Renard, A. Cheikhrouhou, Tranqui Duc, M. Verdaguer: Phys. Rev. B33, 6253 (1986).

Study of Longitudinal Fluctuations in $CsNiF_3$ by Means of Inelastic Polarized Neutron Scattering

U. Balucani[1], *B. Dorner*[2], *K. Kakurai*[3], *R. Pynn*[4], *M. Steiner*[3], *V. Tognetti*[5], *and R. Vaia*[1]

[1]Istituto di Elettronica Quantistica CNR, I-50127 Firenze, Italy
[2]Institut Laue-Langevin, 156 X, F-38042 Grenoble, France
[3]Hahn-Meitner-Institut, D-1000 Berlin 39, Germany
[4]LANSCE, Los Alamos National Laboratory, Los Alamos, NM 87545, USA
[5]Dipartimento di Fisica dell'Università degli Studi and CISM-GNSM, I-50125 Firenze, Italy

1. INTRODUCTION

The polarized neutron scattering technique has been long applied in the study of magnetic systems mainly for elastic scattering, e.g. to separate unambiguously the magnetic from the nuclear scattering. Although the 3-axis polarization analysis (PA) technique was introduced by Moon et al./1/, it became generally available only very recently when strong neutron sources and powerful polarizing devices had been developed /2/. The inelastic PA technique can now be applied for a detailed analysis of the spin dynamics by separating the transverse and longitudinal response. As a result of this new opportunity, many theoretical studies can be tested on a much sounder basis. Since the conventional neutron spectra are dominated by the transverse components, up to now the test for the longitudinal part has been done for classical spin chains, using dynamical simulation /3/. In this case the thermally activated spin-energy coupling was shown /4/ to be an essential mechanism to explain the features of the magnetization fluctuation spectra of isotropic magnetic chains in an applied field H, at intermediate temperatures T and wavevectors k. Here we present polarized neutron scattering experiments in the 1-D ferromagnet $CsNiF_3$, showing the relevance of the spin-energy coupling in this regime (28 kG $< H < 70$ kG, $T = 25$ K). This experimental situation differs from the soliton-bearing regime (continuum limit, $H < 10$ kG, $T < 12$ K) and from the hydrodynamic regime ($k \longrightarrow 0$, possible second magnon mode). The classical approximation for the spin model is no longer valid for this real experiment. Therefore a quantum version of the previous theory /4/, based on the competition between the ordering effect of the field and the thermal disorder, induced by spin-energy coupling, is presented.

2. THE REAL COMPOUND AND THE MODEL

$CsNiF_3$ is a quasi 1-D ferromagnet /5/, which shows 3D antiferromagnetic ordering below $T_N = 2.6$ K only. The spins of magnitude $S = 1$, associated with the Ni^{2+} ions, are arranged in chains with spacing $c = 5.22$ Å and their behavior in an applied field H can be described by the Hamiltonian

$$\mathcal{H} = -J\sum_i \vec{S}_i \cdot \vec{S}_{i+1} + D\sum_i (S_i^z)^2 - g\mu_B H \sum_i S_i^y, \tag{1}$$

where $g = 2.4$, the easy plane anisotropy $D = 9.0$ K and the exchange coupling $J = 23.6$ K. In the isotropic case (D=0) the spin-wave frequencies turn out to be

$$\omega_k = g\mu_B H + 2JS(1-\cos k) \tag{2}$$

(lattice units are assumed). In the presence of anisotropy the spin wave frequency becomes

$$\omega_k^D = \omega_k\left[1 + (2\tilde{D}S/\omega_k)\right]^{1/2} , \qquad (3)$$

where $\tilde{D} = D(1-1/2S)$.

Under the experimental conditions the anisotropy is sufficiently small with respect to temperature and magnetic field so that, as a suitable approximation, the isotropic model can be used. However, the classical approximation turns out to be wrong, because T = 25 K = 2.15 meV is very low with respect to the zone-boundary magnon energies ($\omega_{ZB} \sim 9$ meV). Furthermore the Hartree-Fock renormalization factor α turns out to be in good agreement with the experimental data, deduced by the transverse part of the experimental spectra (Fig.3), only if calculated by the quantum approach ($\alpha \sim 0.9$). The classical counterpart of the calculation gives $\alpha \sim 0.6$ in a perturbative approach and cannot even be evaluated in the self-consistent scheme.

3. THE EXPERIMENT

The polarization analysis (PA) experiment was carried out on the polarized beam 3-axis instrument IN20 at ILL, Grenoble. All measurements were made with fixed final neutron energy at 14.7 meV. Second order neutrons were suppressed by a pyrolytic graphite filter. For measurements with unpolarized neutrons, vertically-curved pyrolytic graphite monochromator and analyser were used. PA was achieved with focussing Heusler alloy monochromator and analyser. The change of the monochromators is controlled by the computer, so that a change from the unpolarized to polarized neutron set up or vice versa is obtained within a few hours. The collimations used were 30'-30'-40'. A direct current flipper placed after the monochromator allowed the spins of the incident neutrons to be chosen up or down. In PA mode the observed flipping ratio was 14 measured in the (100) Bragg peak of $CsNiF_3$ at H = 70 kG. The $CsNiF_3$ sample was mounted in an asymmetric, split-pair cryomagnet providing a vertical field. The crystal was mounted with its (100) and (001) directions in the scattering plane. (001) is the magnetic chain direction. Measurements were made at $\vec{Q} = (0.7\pi,0,k)$ with k in the range $0 \leq k \leq 0.35\pi$.

Fig.1 shows data obtained using unpolarized neutrons at $\vec{Q} = (0.7,0,0.25)\pi$ in a field H = 70 kG for different temperatures. In addition to the spin-wave peak at energy transfer ω = 2.4 meV, there is temperature-dependent intensity at lower energies. The temperature variation of this intensity at ω = 1.5 meV is shown in Fig.2. Also shown in Fig.2 is the temperature dependence of the intensity at ω = 1.05 meV in H = 28 kG. The field dependence of the intensity clearly shows the magnetic origin of the scattering. In order to establish the nature of the spin fluctuations responsible for this low-energy scattering, PA measurements were performed. Since all scattering in this experiment is magnetic, non-spin-flip (n.s.f.) and spin-flip (s.f.) scattering are respectivelv related to the longitudinal and transverse components of the spectrum, $S_{\parallel}(k,\omega)$ and $S_{\perp}(k,\omega)$. The s.f. and n.s.f. spectra at $k = 0.25\pi$ obtained with H = 70 kG at T = 1.7 K, apart from the spin-wave peak in the s.f. scattering, appear essentially featureless and represent the experimental background. When an average value of this background is subtracted from the data obtained at 25 K, the s.f. and n.s.f. difference signals shown in Figs.3 and 4 are obtained. In both figures appropriate

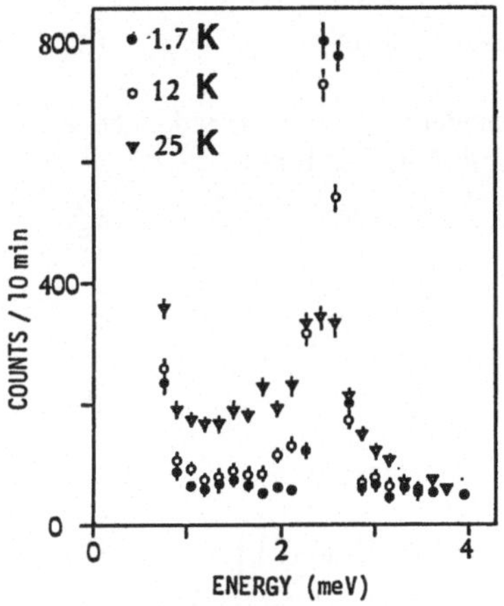

Fig.1 - Spectra obtained using unpolarized neutrons at different temperatures, for H = 70 kG and $k = 0.25\pi$.

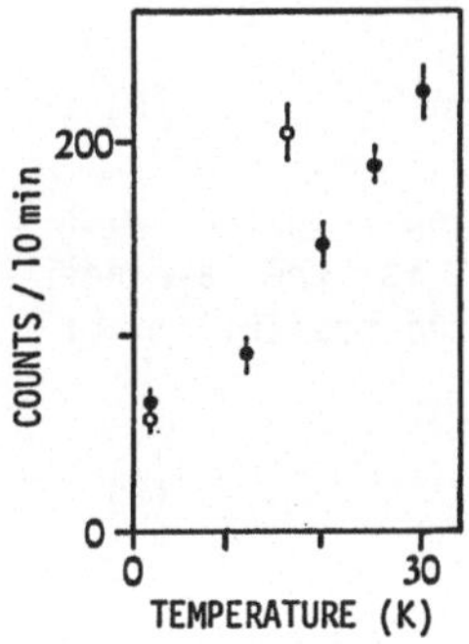

Fig.2 - Temperature dependence of the unpolarized neutron intensity in H = 70 kG (•) and 28 kG (∘), at the energy transfer of 1.5 meV and 1.05 meV, respectively.

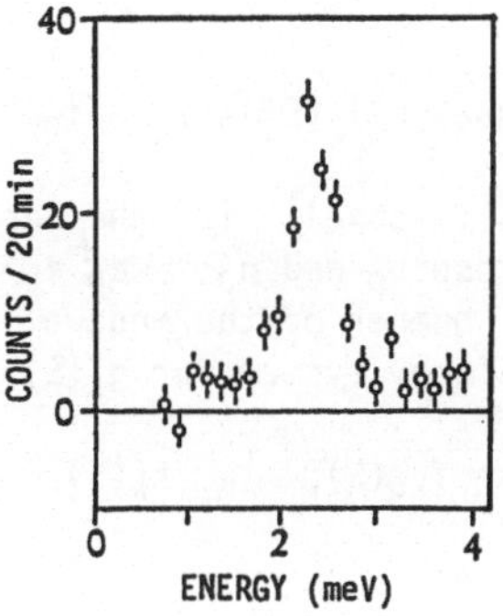

Fig.3 - Background corrected spin flip scattering at T = 25 K, for H = 70 kG and $k = 0.25\pi$.

corrections have been made for the experimental flipping ratio. The s.f. scattering (Fig.3) shows a broadened, but underdamped, magnon peak. The n.s.f. scattering (Fig.4) displays a broad, finite energy peak with an upper energy-cutoff slightly above the spin-wave energy. Comparison of Figs.3 and 4 indicates that most of the scattering shown in Fig.1 at 1.5 meV is due to longitudinal spin fluctuations. Thus Fig.2 is indeed a reasonable representation of the temperature dependence of this scattering. This temperature dependence explains why the previous search /6/ for finite energy, longitudinal fluctuations in $CsNiF_3$ was unsuccessful: the experiment was performed for $T \leq 8$ K, where $S_{//}(k,\omega)$ is too small to be detected.

4. THE THEORY

We refer to the classical calculations previously performed within the Green-function nonperturbative approach in Ref.4. Since normal ordering problems do not arise, it is easy to translate the theory into the quantum framework. The spectrum of the magnetization fluctuations can be expressed by means of a spectral identity as

$$S_{//}(k,\omega) = (2\pi)^{-1}\int_{-\infty}^{\infty}dt\, e^{-i\omega t}\langle S^y_{-k}S^y_k(t)\rangle = -2\mathrm{Im}\, g_k(\omega+i0^+)\, N(\omega), \tag{4}$$

where $N(\omega) = [1-e^{-\beta\omega}]^{-1}$, $\beta = 1/T$, and we have introduced the retarded Zubarev-Green-function $g_k(E)$ associated with the magnetization fluctuations:

$$g_k(E) = \langle\langle S^y_k ; S^y_{-k}\rangle\rangle_E = N^{-1}\sum_q f_{kq}(E), \tag{5}$$

where $f_{kq}(E)$ is expressed by Bose operators as

$$f_{kq}(E) = \sum_p \langle\langle a^\dagger_{k+q}\, a_q ; a^\dagger_{p-k}\, a_p\rangle\rangle_E . \tag{6}$$

In the ladder approximation the following equation is recovered:

$$f_{kq}(E) = f^{(0)}_{kq}(E)\left\{1+(4\, J/N)\sum_p[\cos p+\cos(k+q)-\cos k-\cos(p-q)]\, f_{kp}(E)\right\} \tag{7}$$

where

$$f^{(0)}_{kq}(E) = (2\pi)^{-1}(n_{k+q}-n_q)\,(E+\Omega_{k+q}-\Omega_q)^{-1} \tag{8}$$

and $\Omega_q = JS[h + 2\alpha(1-\cos q)]$ is the Hartree-Fock renormalized frequency with $\alpha = 1-(NS)^{-1}\sum_p(1-\cos p)n_p$ and $n_p = \langle a^\dagger_p a_p\rangle = [\exp(\beta\Omega_p)-1]^{-1}$ is the average quantum occupation number of the mode p. Equation (7) gives the possibility of obtaining a closed linear system for $g_k(E)$ together with

$$h_k(E) = N^{-1}\sum_q \cos q\, f_{kq}(E), \qquad t_k(E) = N^{-1}\sum_q \sin q\, f_{kq}(E), \tag{9}$$

which gives as solution

$$g_k(E) = \frac{g^{(0)}_k(E) + 4\pi J\left\{[g^{(0)}_k(E)]^2 - [h^{(0)}_k(E)]^2 - [t^{(0)}_k(E)]^2\right\}}{1 + 4\pi J\left\{(1+\cos k)\,[g^{(0)}_k(E) - h^{(0)}_k(E)] + \sin k\, t^{(0)}_k(E)\right\}}, \tag{10}$$

where the superscript (0) means replacing f_{kq} with $f^{(0)}_{kq}$ in the respective definitions. In the free magnon approximation one finds $g_k(E) = g^{(0)}_k(E)$ (for consistency, in this approximation one must take $\alpha = 1$). The curly-bracketed term in the denominator of eq.(10) is proportional to the spin-energy coupling $\langle\langle E_k ; S^y_{-k}\rangle\rangle^{(0)}_E$, where E_k is the Fourier transform of the exchange energy density. The full first order calculations have been performed numerically, because of the presence of the complicated boson factors, which prevent from proceeding analytically as in the classical case. In Fig.4 the results of this quantum mechanical theory are compared with the experimental data and with the free magnon spectrum. The latter shows a square-root divergence $(\omega_c-\omega)^{-1/2}$ at the cutoff frequency $\omega_c = 4JS\sin(k/2)$. This singularity occurs in 1-D, also in the classical case, as a consequence of the divergence in the two-magnon density of states. The spin-energy coupling strongly modifies the non-interacting spectrum: the singularity is washed out and an inelastic peak clearly appears: the ordering effect of the field tends to become less and less significant. Although these qualitative features were also present in the classical case /4/, it is essential to note that the good agreement shown in Fig.4 is crucially dependent on the use of the quantum framework: the aforementioned breakdown of the classical renormalization theory would at best lead to a severe shrinkage of the frequency band, and the agreement with the data would be completely spoiled.

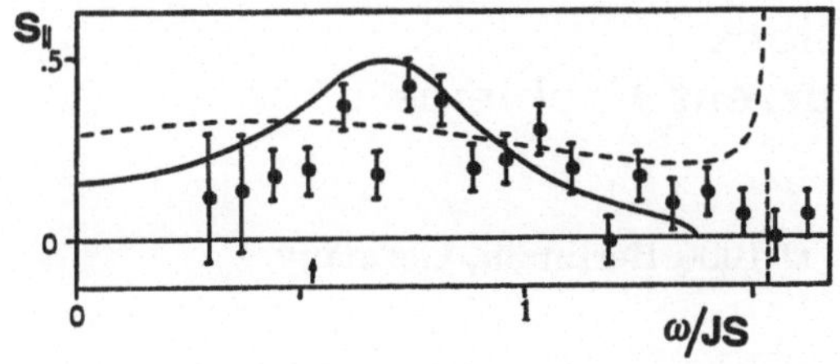

Fig.4 - Background corrected non-spin-flip scattering at T = 25 K, for H = 70 kG and k = 0.25 π. Dashed line: theoretical free-magnon spectrum. Solid line: spin-energy coupling theory. The arrow marks the position of the spin-wave renormalized frequency in the zero-field case.

The experimental results have been reported after a scaling of the neutron intensity with respect to the associated transverse intensity, so that no fitting parameters are present. The interpretation of these data in the framework of the present theory shows that the nonperturbative spin-energy coupling approach appears to be appropriate in this range of wavevectors, fields and temperatures.

REFERENCES

1. R.M.Moon, T.Riste, W.C.Koehler: Phys. Rev. 181, 920 (1984)
2. R.Pynn: Rev. Sci. Instr. 55, 837 (1984)
3. V. Tognetti, A.Rettori, M.G.Pini, J.M.Loveluck, U.Balucani, E.Balcar: J. Phys. C16, 5641 (1983)
4. U.Balucani, M.G.Pini, V.Tognetti, A.Rettori: Phys. Rev. B26, 4974 (1982)
5. M.Steiner, K.Kakurai, J.K.Kjems: Z. Phys. B53, 117 (1983)
6. M.Steiner, K.Hirakawa, G.F.Reiter, G.Shirane: Solid. State Comm. 40, 65 (1981)

Dynamics of Critical Fluctuations in Isotropic Ferromagnets: Current Problems

F. Mezei

Hahn-Meitner-Institut, Postfach 390128, D-1000 Berlin 39, Germany

1. Introduction

In recent years new developments in high-resolution neutron spectroscopy, primarily the appearance of Neutron Spin Echo (NSE) spectroscopy, have opened up the way for a really probing study of critical dynamics in ferromagnets. A most important new possibility is the investigation of the fluctuations at wave vectors ($q \simeq 0.1 - 0.5$ Å^{-1}) much smaller indeed than the Brillouin zone, i.e. we can better approach the limit of large distances, which limit is the very sense of criticality. As a result, today we possess an accurate picture of the critical dynamics of various typical isotropic ferromagnets. Ironically, the new experimental findings literally raised more questions than they answered. Theory and experiment are at flagrant variance in two main respects: the role played by dipolar interactions and the dynamic (energy) line shape of the fluctuations at T_c. Even the only marked agreement between theory and experiment, namely the nice $q^{5/2}$ power law predicted by dynamical scaling for the relaxation rate at T_c, is more puzzling than expected, since it should not be valid for much of the q range where dipolar interactions are important.

2. Fundamentals

It is a bad omen that in the field of critical dynamics the most fundamental quantity itself, the dynamic exponent z, is a subject of misunderstanding. The original [1] and only correct definition of z is the dynamic scaling equation

$$\omega_c(q,T) = Aq^z f(\kappa/q), \tag{1}$$

where $\omega_c(q, T)$ is the energy (frequency) characterizing the fluctuations at wavenumber q, A is a constant, f(x) is the scaling function and $\kappa = \xi^{-1}$ is the inverse correlation length. A direct determination of z can be obtained by measuring ω_c (which simply means the linewidth Γ if a Lorentzian line is assumed) as a function of q at $T = T_c$, i.e. $\kappa = 0$,

$$\omega_c(q,T=T_c) = Aq^z, \tag{2}$$

where we take by definition $f(0) = 1$. This experiment can only be performed by neutrons. For the purpose of bulk measurements another equation has been deduced from (1),

$$\omega_c(q,T) = Aq^z f(\kappa/q) = A\kappa^z g(\kappa/q), \tag{3}$$

where the new scaling function $g(\kappa/q) = (q/\kappa)^z f(\kappa/q)$. This formally leads to a definition of z from bulk measurements (which we denote by z^*):

$$\omega_c(q=0,T) = A\kappa^{z^*} g(\infty) . \tag{4}$$

In the spin non-conserved universality class $g(\infty)$ has a well-defined value and thus (2) and (4) define the same quantity $z = z^*$[1]. Unfortunately it is often overlooked that the spin conserving character of the Heisenberg universality class leads to $g(\infty) = 0$, thus (4) leaves z^* undefined. Of course, spin conservation cannot be perfect and therefore there will still be a finite $q=0$ relaxation rate Γ_0 due to some parasitic interactions. Therefore z^* as formally deduced from (4) has a priori nothing to do with z as defined by (2). Actually, if conventional Onsager theory $\Gamma_0 \propto 1/\chi$ holds, the divergence of the susceptibility χ governed by static scaling results in a critical slowing down in Γ_0 with $z^* = \gamma/\nu \simeq 2$.

3. Statics: The Demagnetization (Dipolar) Anisotropy

With smaller q values now accessible, we started to explore critical fluctuations in domains where the intrinsic susceptibility of the sample becomes very high, actually comparable to unity. Therefore, as it is well known from classical magnetostatics, demagnetization effects become very important. In our case this means that the internal magnetic fields created by the critical fluctuations themselves actually tend to reduce them. Thus the usual Ornstein-Zernicke susceptibility $\chi_q^{(0)}$ has to be replaced by

$$\chi_q = \frac{\chi_q^{(0)}}{1 + 4\pi m \chi_q^{(0)}} = \frac{C}{q^2 + \kappa^2 + m q_d^2} , \tag{5}$$

where C is a constant and $q_d^2 = 4\pi C$ is the definition of q_d. In usual small-angle scattering experiments the neutrons only see the $\chi_q^\perp$ magnetic fluctuations transverse with respect to **q**, i.e. we probe plane wave fluctuations with the magnetization lying in the infinite planes. In this geometry m is obviously 0, i.e. $\chi_q^\perp = \chi_q^{(0)}$. On the other hand, the magnetization fluctuation longitudinal to q (i.e. perpendicular to the wave planes) experience demagnetization with m = 1, i.e. the longitudinal susceptibility $\chi_{q=0}^{\parallel}$ does not diverge at T_c, but saturates at a value $C/q_d^2 = 1/4\pi$. This phenomenon has recently been experimentally observed for the first time in a polarized neutron scattering study of the critical fluctuations around Bragg reflections in EuS and EuO single crystals [3]. Note that the demagnetization factor can generally be written down for the α component of the magnetization as $m = (q_\alpha/q)^2$ [4]. The more common appellation for these demagnetization effects is dipolar interaction, and q_d is called the dipolar wave vector.

Table I shows the values of q_d and the corresponding temperature T_d, defined by the relation $\kappa(T_c+T_d) = q_d$. Thus below T_c+T_d the intrinsic Ornstein-Zernicke susceptibility (m = 0 in (5)) exceeds $1/4\pi$. The values in the table have been determined from data on the volume susceptibilities and κ [5, 6]. It is obvious that the dipolar effects are substantial over most of the critical q and T range in EuS and EuO and in a good part of it in Fe. Surprisingly, as we will see, the only experimentally established manifestation of these dipolar effects is that of the dipolar anisotropy of χ_q as given by (5).

Table 1. The dipolar temperature T_d and wavevector q_d characterizing the extent of the dipolar regime around the Curie point T_c in various ferromagnets.

	T_c[K]	T_d[K]	q_d[Å^{-1}]
Fe	1040	8.6	0.045
Ni	631	2.4	0.013
EuO	69.1	8.2	0.147
EuS	16.6	4.8	0.24

4. Dynamics: a Catalogue of Contradictions

To be more precise, in what follows we are concerned with the dynamics $\chi_q^{\perp}$ only, which is observed in standard neutron scattering experiments. The study of the dynamics of $\chi_q^{\parallel}$ could be possible in the vicinity of a Bragg reflection in a single crystal and is being planned [7], but it is less than straightforward.

The most spectacular feature of the new results in both Fe [8] and EuO [5, 9] is the confirmation of the power law (2) over an impressive energy range of 4 orders of magnitude in energy with $z = 2.5$ as predicted by dynamic scaling for the Heisenberg universality class (neglecting an insignificant correction involving the Fisher exponent η). The value of the coefficient A is also rather well predicted (within 20 %) by an approximate expression proposed by Riedel [10] which relates A to T_c and q_d [5]. What is very surprising about the nice power law is that it continues to be valid for $q < q_d$, where the dipolar effects should dominate, and a cross-over has been predicted to a power law with $z = 2$ [11] or $z = 1$ [12].

Another casualty of the recent more precise and more extensive studies was the only previously accepted experimental confirmation [13] of the (1) scaling law, actually obtained in Fe. The breakdown newly observed in this material [8] roughly corresponds to adding a non-scaling term $\Gamma_0 = B\kappa$ to (1) which implies $z^* = 1$ in contrast to $z = 2.5$. This term has a strong influence at small q values. Another aspect of scaling, the universality, could not be confirmed either: there is little resemblence between the behaviour of Fe and that of EuO [14].

The most puzzling contradiction, however, is that between the low q neutron data and the q=0 relaxation rate Γ_0 determined from the bulk high-frequency response in EuO [15]. These results are in quantitative agreement with dipolar theories both at $T > T_c+T_d$ [16] (non-critical regime) and $T_c < T < T_c+T_d$ [11] (dipolar critical regime with $z^* = 2$). In view of the above comments on the difference between z and z^*, the fact that different values have been observed for z and z^* is a priori not a contradiction. The real problem is that at various temperatures $T > T_c$ the observed Γ_0 values are higher than Γ_q observed at finite q's, as illustrated in Fig. 1. It is hard to imagine how to reconcile these findings.

The final contradiction in this list concerns the lineshape. All theoretical approaches predict a lineshape virtually identical with the one illustrated in Fig. 2 by the dashed line [17, 18] where the linewidth was treated as a free parameter. A comparison to recent most detailed coupled-mode calculations [19] (dotted line) also shows that the theoretically deduced linewidths are at variance with

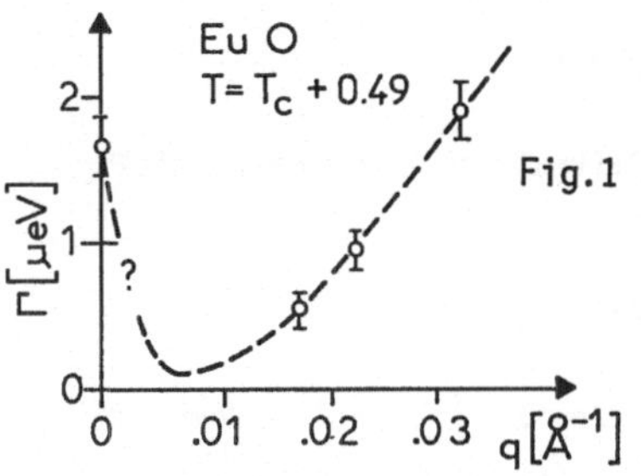

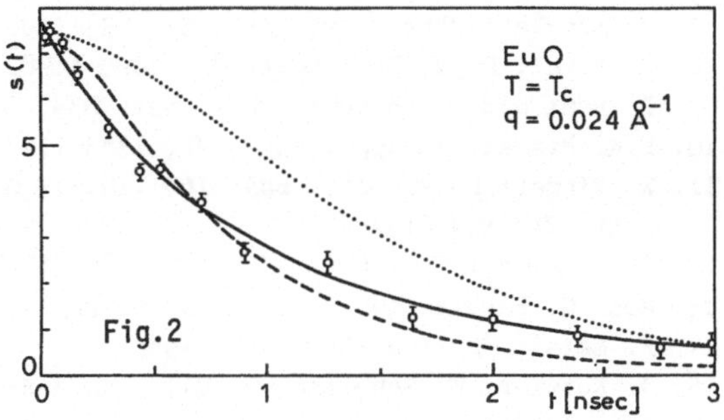

Fig.1. Contradiction between the q=0 [15] and the finite q (present work) relaxation in EuO at $T=T_c+0.5K$.

Fig.2. Comparison between the lineshape (as observed in the variable time domain by NSE spectroscopy) and various theoretical predictions in EuO. Dashed line: best fit to first neighbour mode-coupling [17] and renormalization group [18] theories with the linewidth taken as a free parameter. Dotted line: first and second neighbour mode coupling theory with the linewidth as predicted [19]. Continuous line: Lorentzian lineshape with the linewidth calculated assuming $A=8.3$ meVÅ$^{5/2}$. The lines were numerically evaluated with the finite q resolution of the experiment taken into account.

those observed. On the other hand, the NSE results shown are fully compatible with a Lorentzian line the width of which has been calculated from best experimental value of A in (2) fitted over the whole q range of 0.015 - 0.3 Å [9]. Note that at higher q values, $q > q_d$ a good agreement has been recently found in EuO between experimental and theoretically predicted lineshapes [20], but obviously the discrepancy between the linewidths predicted by [19] and those observed persists.

With the exception of this last problem of the linewidth, it is worth noting that all anomalies listed above occur for q values less than q_d, i.e. in a domain where the dipolar interaction should be important. Unfortunately all existing predictions on dipolar effects in the dynamics near T_c $(4\pi\chi_q^{(0)}>1)$ were proved to be utterly wrong by the experiments. Therefore it may be hoped that a correct dipolar theory yet to come will eventually resolve all or most of these problems.

Sure enough, I am unable to provide clues to such a theory, but I can certainly see the vast theoretical difficulties in treating a problem which concerns time and length scales (order of 10^{-9} s and 100 Å, respectively) between the microscopic and the macroscopic ones.

References

1. B.I. Halperin and P. C. Hohenberg: Phys. Rev. 177, 417 (1969)
2. P. Resibois and C. Piette: Phys. Rev. Lett. 24, 514 (1970)
3. J. Kötzler, F. Mezei, D. Görlitz and B. Farago: Europhys. Lett. 1, 675 (1984)
4. M.E. Fisher and A. Aharony: Phys. Rev. Lett. 30, 559 (1973)
5. F. Mezei: J. Magn. Magn. Mat. 45, 67 (1984)
6. J. Kötzler and M. Muschke: Phys. Rev. B34, 3543 (1986)

7. J. Kötzler: Phys. Rev. Lett. **51**, 833 (1983)
8. F. Mezei: Phys. Rev. Lett. **49**, 1096 (1982)
9. P. Böni and G. Shirane: Phys. Rev. **B33**, 3012, (1986)
10. E.K. Riedel: J. Appl. Phys. **42**, 1383 (1971)
11. W. Finger: Phys. Lett. **60A**, 165 (1975); G. B. Teitelbaum: Sov. Phys. JETP Lett. **21**, 154 (1975)
12. S.V. Maleev: Sov. Phys. JETP **39**, 889 (1974)
13. see. G. Parette and R. Kahn, J. de Phys., **32**, 447 (1971)
14. F. Mezei: Physica **136B**, 417 (1986)
15. J. Kötzler, W. Scheithe, R. Blickhan und E. Kaldis: Solid. State. Comm. **26**, 641 (1978)
16. R. Raghavan and D.L. Huber: Phys. Rev. **B14**, 1185 (1976)
17. F. Wegner: Z. Phys. **216**, 433 (1968); J. Hubbard: J. Phys. C. **4**, 53 (1971)
18. R. Folk and H. Iro: Phys. Rev. **B32**, 1880 (1985)
19. S.W. Lovesey and R.D. Williams: J. Phys. C: Solid State Phys. **19**, L253 (1986)
20. P. Böni, M.E. Chen and G. Shirane: preprint

Magnetism of MnSi and FeSi

G. Shirane

Physics Department, Brookhaven National Laboratory,
Upton, NY 11973, USA

A review is presented of neutron scattering studies of isostructural MnSi and FeSi. MnSi exhibits typical characteristics of an itinerant electron ferromagnet. On the other hand, FeSi is a unique example of temperature induced magnetism through a narrow band gap.

1. Introduction

Transition-metal monosilicides MnSi and FeSi show a variety of interesting magnetic and transport properties. They both crystallize in a cubic structure (space group $P2_13$) in which magnetic atoms are located at displaced face centered positions. MnSi is a spiral ferromagnet with the Curie temperature T_c at 29K. This fascinating crystal has been extensively investigated by Ishikawa and his colleagues [1] and it is now considered a prototype of an itinerant electron ferromagnet [2]. FeSi, on the other hand, does not order magnetically. It shows, however, a very unique magnetic behavior which, until very recently, remains essentially unexplained.

The different characteristics of MnSi and FeSi may stem from their electronic band structures. Nakanishi et al. [3] calculated the density of states for these two crystals. As shown in Fig. 1, the Fermi energy E_F of MnSi falls at a location of high density of state. On the other hand, E_F of FeSi coincides with a narrow gap in the density of states curve. This calculation supports the gap model originally proposed by Jaccarino et al. [4] and later refined by Takahashi and Moriya [5].

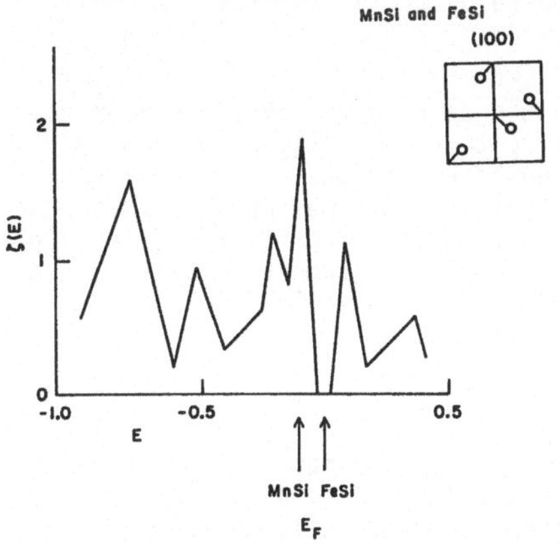

Fig. 1 Band calculation for MnSi at FeSi by Nakanishi et al. [3]

The electrons in FeSi are excited over the gap at elevated temperature and induce the local moments. However, the direct support for this gap model by neutron scattering has not been provided for a long period. On the contrary, a recent X-ray photoelectron experiment [6] was interpreted as an evidence against the temperature induced magnetism in FeSi. Later we will present the current Brookhaven neutron scattering experiment on FeSi [7], which gives a concrete support for the gap model.

2. MnSi

Magnetic properties of this crystal have been extensively studied [1,2] and details of magnetic scattering are well established. The most important feature is the persistence of magnetic correlation well above the Curie temperature, 29K. As shown in Fig. 2, the equi-intensity contours indicate very strong correlations even at 300K, which corresponds to 10 T_C. Detail study of line width Γ and their temperature dependence [1] also supports the theoretical model of itinerant electron magnetism [2].

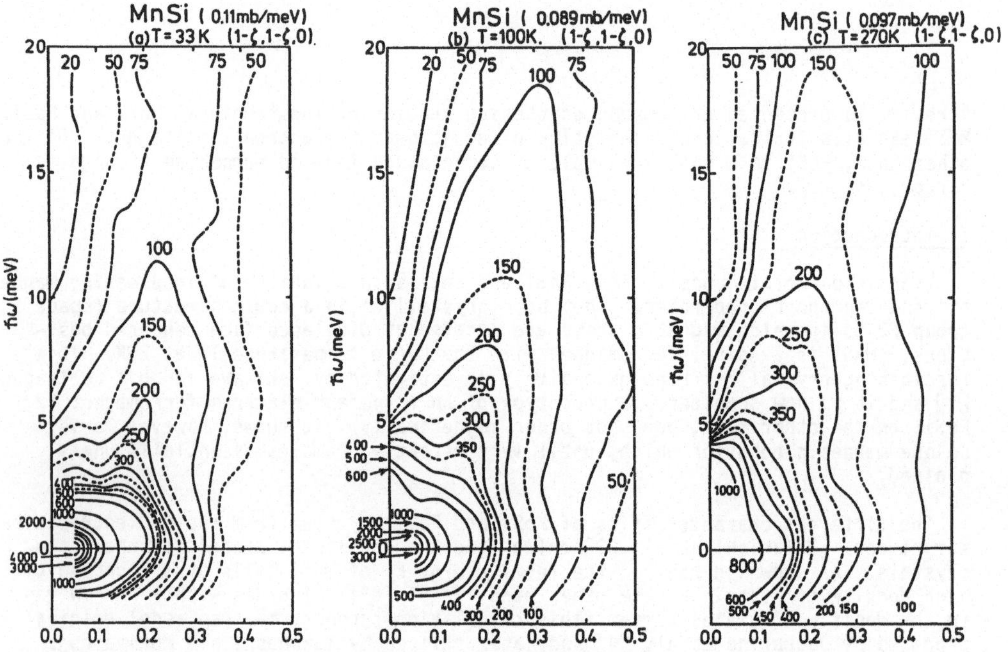

Fig. 2 Magnetic intensity contours along the [011] direction in MnSi. After Ishikawa et al. [1]

Polarized neutron measurement on MnSi powder was carried out by Ziebeck et al. [8] and this study covers the temperature range up to 20 T_C. Figure 3 shows M(Q), which is derived from the measured magnetic intensities divided by magnetic form factor squared. Intentionally poor instrumental resolution (43 meV FWHM) permits the effective integration over ω. One can see that the correlation remains strong up to 20 T_C; moreover, the moment derived by the observed neutron magnetic scattering at high temperature corresponds to 1.4 μ_B, compared with 0.4 μ_B below T_C.

3. FeSi

FeSi has been the subject of continuous investigation for many years because of the dramatic temperature dependence of its susceptibility (Fig. 4). In 1967 Jaccarino et al. [4] proposed that FeSi is a nearly ferromagnetic small-gap semiconductor as an alternative to models involving strong exchange coupled isolated spins or an antiferromagnetic phase below a (rather high) ordering temperature. As already mentioned in the Introduction, the gap model was supported by a band calculation [3], while a spin fluctuation model by Takahashi and Moriya [5] gave a satisfactory account for $\chi(T)$ and the specific heat. Evangelou and Edwards [9] then emphasized the itinerant nature of FeSi and pointed out the likelihood of ferromagnetic correlation.

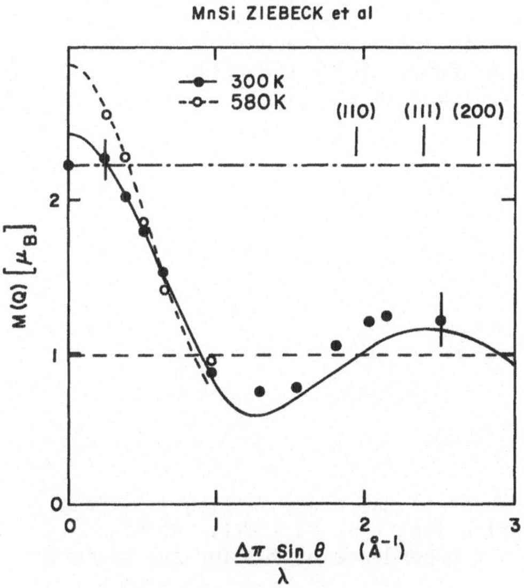

Fig. 3 Paramagnetic scattering in MnSi powder at 300K (10 T_C) at 580K (20 T_C). The chain line corresponds to the d.c. susceptibility and the broken line to the low temperature saturation moment. After Ziebeck et al. [8]

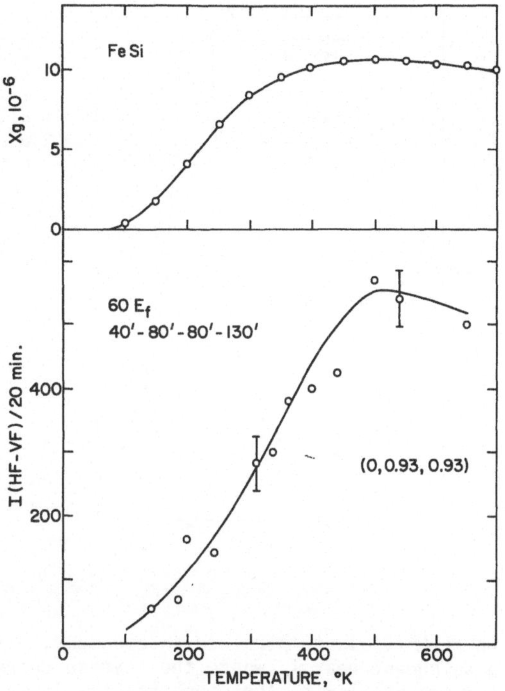

Fig. 4 Comparison of the magnetic susceptibility (Ref 4) and the magnetic scattering of FeSi. The neutron data taken near (011) with energy resolution FWHM of 15 meV. After Shirane et al. [7]

Neutron experiments [10,11,12] prior to 1983 failed to detect any magnetic scattering. More recently Ziebeck et al. [13] observed a magnetic signal from FeSi powder using polarization analysis; their 500K data are reproduced as the insert of Fig. 5. The increase in magnetic scattering at small q was interpreted as a signature of ferromagnetic correlations, while it was claimed that the coupling changes to antiferromagnetic at low temperature.

The current polarized neutron experiment at Brookhaven on a large FeSi single crystal [7] shows that $S(q,\omega)$ indeed exhibits a strong peak at (011). The data shown in Fig. 5 are taken at 500K as normalized to the form factor for comparison with reference 13. Our low-q data suggest ferromagnetic correlations, consistent with Ziebeck et al. [13], which are much more clearly evident around the (011) Bragg point. The magnetic intensity at (0, 0.93, 0.93), near (011), is shown in Fig. 4 as a function of temperature. The resemblance of this intensity to the susceptibility is striking, indicating that the gap model is basically correct.

The intensity profile in Fig. 5 around (011) obviously indicates ferromagnetic correlation, as observed in EuO and Fe. In these, paramagnetic scattering function $S(Q,\omega)$ can be approximated by a simple double Lorentzian

$$S(q,\omega) = M^2(0) \frac{\kappa_1^2}{\kappa_1^2 + q^2} \frac{\Gamma}{\Gamma^2 + \omega^2}, \qquad (1)$$

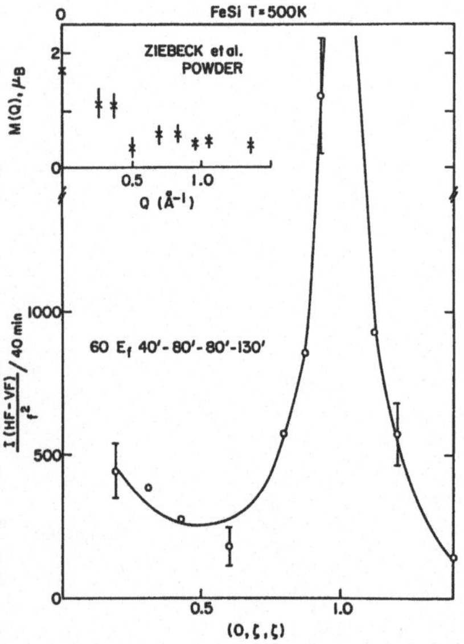

Fig. 5 The Q dependence of magnetic cross section along the [011] direction, normalized by f^2. Powder data by Ziebeck et al. [13] is shown in the insert. $d^* = 1.98\ \AA^{-1}$ at 300K. Both data are obtained by polarization analysis: the difference of horizontal field (HF) and vertical field (VF) intensities. After Shirane et al. [7]

where $M^2(0)$ is temperature independent. The shape of the cross section is strongly q dependent near T_c where the inverse correlation length κ_1 is small. At high-temperature limit, the cross section loses its q dependence except for the magnetic form factor f.

The magnetic scattering from FeSi behaves very differently from those of regular ferromagnets even though the steep rise toward (110), shown in Fig. 5, suggests familiar ferromagnetic correlation. First of all, this shape is temperature <u>independent</u> between 150 and 600K. On the other hand, the prefactor $M^2(0)$ increases strongly with temperature as shown in Fig. 4. This is the signature of temperature-induced magnetism in FeSi. When we integrate the magnetic intensity I_{mag} over energy ω

$$\int I_{mag} d\omega = M^2(q) = M^2(0)\,\frac{\kappa_1^2}{\kappa_1^2 + q^2} \qquad (2)$$

$M^2(q)$ is nearly q independent! The strong q dependence observed in Fig. 5 entirely comes from the strong q dependence of the line width $\Gamma(q)$. The details of the line shapes and their temperature dependence are now under study.

Thus FeSi represents a truly novel type of magnetic scattering. Since $M^2(q)$ is q independent, one should not use "ferromagnetic" correlations to describe FeSi. On the other hand q in $\Gamma(q)$ starts at the ferromagnetic Brillouin center and not the antiferromagnetic one. So far we have failed to observe the energy gap directly through the energy dependence of magnetic scattering. More definitive experimental observations, as well as theoretical calculations, are needed to clarify this important point.

I would like to thank P. Böni, Y. Endoh, J. E. Fischer, T. Moriya and K. Tajima for many informative discussions. Magnetic scattering studies of MnSi as FeSi at Brookhaven were initiated by the late Y. Ishikawa of Tohoku University and carried out as part of the U.S. Japan Cooperative Program on Neutron Scattering. Work at Brookhaven supported by the Division of Materials Sciences, U. S. Department of Energy, under contract DE-AC02-76CH00016.

4. References

1. Y. Ishikawa, Y. Noda, Y. J. Uemura, C. F. Majkrzak, and G. Shirane, Phys. Rev. B31, 5884 (1985) and references therein.
2. T. Moriya, Spin Fluctuations in Itinerant Electron Magnetism. Springer Verlag, 1985.
3. O. Nakanishi, A. Yanase and A. Hasegawa, J. Magn. Magn. Mat. 15-18, 879 (1980).
4. V. Jaccarino, G. K. Wertheim, J. H. Wernick, L. R. Walker and Sigurds Arajs. Phys. Rev. 160, 476 (1967).
5. Y. Takahashi and T. Moriya, J. Phys. Soc. Jpn. 46, 1451 (1979).
6. S. J. Oh, J. W. Allen and J. M. Lawrence, Phys. Rev. B35, 2267 (1987).
7. G. Shirane, J. E. Fischer, Y. Endoh and K. Tajima, Phys. Rev. Lett. (submitted).
8. K. R. A. Ziebeck, H. Capellmann, P. J. Brown and J. G. Booth, Z. Phys. B48, 241 (1982).
9. S. N. Evangelou and D. M. Edwards, J. Phys. C16, 2121 (1983).
10. H. Watanabe, H. Yamamoto, and K. Ito, J. Phys. Soc. Jpn. 18, 995 (1963).
11. K. Motoya, M. Nishi and Y. Ito, J. Phys. Soc. Jpn. 49, 1931 (1980).
12. M. Kohgi and Y. Ishikawa, Solid State Commun. 37, 833 (1981).
13. K. R. A. Ziebeck, H. Capellmann, P. J. Brown, and P. J. Webster, J. Magn. Magn. Mat. 36, 160 (1983).

Neutron Scattering from Heavy Fermion Systems

C. Broholm[1], *J.K. Kjems*[1], *G. Aeppli*[2], *E. Bucher*[2,3], *and W.J.L. Buyers*[4]

[1]Risø National Laboratory, DK-4000 Roskilde, Denmark
[2]AT&T Bell Laboratories, Murray Hill, NJ 07974, USA
[3]University of Konstanz, D-7750 Konstanz, Fed. Rep. of Germany
[4]Atomic Energy of Canada Limited, Chalk River, Ontario, Canada K0J 1J0

We review results of neutron scattering experiments on the heavy fermion systems URu_2Si_2, U_2Zn_{17} and UPt_3. In the antiferromagnetic superconductor URu_2Si_2 the excitations in the ordered phases are propagating, longitudinal and similar to those in rare earth singlet ground state systems. However, the ordered moment of 0.03 μ_B is anomalously small and the transverse response is overdamped. In the antiferromagnet U_2Zn_{17}, overdamped antiferromagnetic fluctuations dominate the excitation spectrum above and below T_N. In the superconductor UPt_3 we have established the existence of spin fluctuations with characteristic energies of 0.5 meV. We discuss a simple model of exchange coupled overdamped responses which gives a consistent description of the diffusive magnetic fluctuations in several heavy fermion systems.

1. Introduction

Single site properties in ordinary rare earth magnets are well acounted for by treating the local spin operator as an approximately conserved quantity. Dipolar excitations thus have a discrete spectrum which can be dispersed or broadened by conduction electron mediated inter site interactions.

In the recently discovered class of heavy fermion systems [1] the distinction between conduction electrons and localized f states is less obvious, since the 5f states, as a consequence of the strong hybridization with conduction electrons, are not approximate eigenstates of the problem. This fact is, as we shall see, directly observed in neutron scattering experiments probing the dipolar response function of the heavy fermion quasiparticles. In general the heavy fermion systems are characterized by continuous metallic excitation spectra in the energy region dominated by crystal field excitations in the ordinary rare earth magnets. In spite of this important distinction, the collective magnetic properties arising from RKKY interactions can be described in a RPA formalism analogous to the one used in ordinary rare earth magnetism.

In this paper we discuss, with emphasis on the aspects mentioned above, the magnetic excitation spectra of the three heavy fermion systems URu_2Si_2, U_2Zn_{17} and UPt_3. Finally we discuss a simple model which has been successful in describing the collective magnetic properties of several heavy fermion systems.

2. Experimental

The present neutron-scattering experiments were performed using the cold-neutron source of the DR3 reactor of Risø National Laboratory, and the thermal-neutron source at the NRU reactor of the Chalk River Nuclear Laboratories. We have employed standard triple axis techniques.

2.1. URu_2Si_2

URu_2Si_2 is a heavy fermion system (γ = 180 mJ/mole-U-K^2) with two low temperature phase transitions of which the lower at T_c = 1 K is to a strongly anisotropic superconducting state [2]. Neutron scattering has shown that an antiferromagnetically ordered moment of 0.03 μ_B, two orders of magnitude smaller than the fluctuating moment deduced from high temperature susceptibility [2], develops at T_N = 17.5 K and coexists with superconductivity below T_c = 1 K [3].

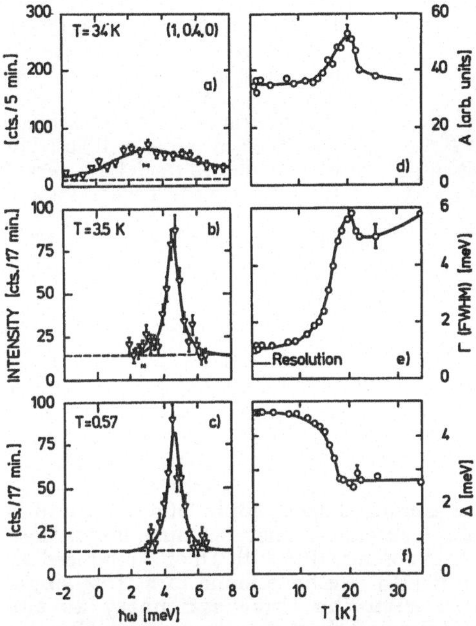

Fig. 1. Constant-q scans in URu_2Si_2 at q = (1,0.4,0) in (a) the paramagnetic, (b) the antiferromagnetic, and in (c) the superconducting phase. The data in (a) were taken with a different configuration than (b) and (c), but have been normalized so equivalent cross-sections correspond to the same height over the background. The solid lines in (a)-(c) are fits as described in the text. (d)-(f) show the temperature dependence of the three parameters obtained from fits to data similar to (a). The lines through these points are guides to the eye.

Figures 1 (a)-(c) show typical constant q = (1,0.4,0) scans in the paramagnetic, antiferromagnetic and superconducting phase. Figures 1 (d)-(f) summarize the temperature dependence of the excitation by showing the amplitude, A, damping, Γ (FWHM), and first moment, Δ, of a resolution and background corrected Lorentzian response which describes our data well.

In the paramagnetic phase the magnetic excitations are heavily damped with an energy width Γ (FWHM) $\simeq$ 6 meV at 20 K nearly 3 times larger than the thermal energy. Up to temperatures higher than 3 T_N, there are strong antiferromagnetic correlations in the basal plane of the body centered tetragonal structure. The spin dynamics change qualitatively at T_N = 17.5 K although only 1% of the paramagnetic moment actually orders. The overdamped paramagnetic fluctuations develop into propagating dispersive longitudinal crystal field-like excitations below T_N. The phase transition cannot, however, be explained by conventional crystal field theory. In particular the ratio of the ordered moment to the exciton transition matrix element is too small compared to the ratio of k_BT_N to the splitting of the two lowest lying levels. We believe that the phase transition at 17.5 K marks a more fundamental change in the properties of the magnetic quasiparticles and that the small ordered moment is only a consequence of this.

2.2. U_2Zn_{17}

U_2Zn_{17} is a heavy fermion system (γ = 550 mJ/mole-U-K^2) which undergoes an antiferromagnetic phase transition at T_N = 9.7 K with a concomitant reduction in the linear specific heat by a factor of three [5]. The ordered moment is 0.8 μ_B [6], well below the paramagnetic moment of about 3 μ_B deduced from high temperature susceptibility measurements.

In contrast to URu_2Si_2 the qualitative features of the magnetic excitation spectrum are unchanged at T_N [7]. Figure 2 shows a set of constant energy scans in the antiferromagnetically ordered phase at T = 2 K ~ T_N/5. Despite the antiferromagnetic order no spin waves are observed, only a ridge of quasi-elastic scattering at the antiferromagnetic zone center. The magnetic excitations can thus be described as overdamped, diffusive modes, which are broad in energy but well correlated over 5-10 unit cells.

2.3. UPt_3

At high temperatures UPt_3 has a paramagnetic Curie Weiss susceptibility with an effective moment of 3 μ_B, but it becomes superconducting below T_c = 0.5 K. Inelastic

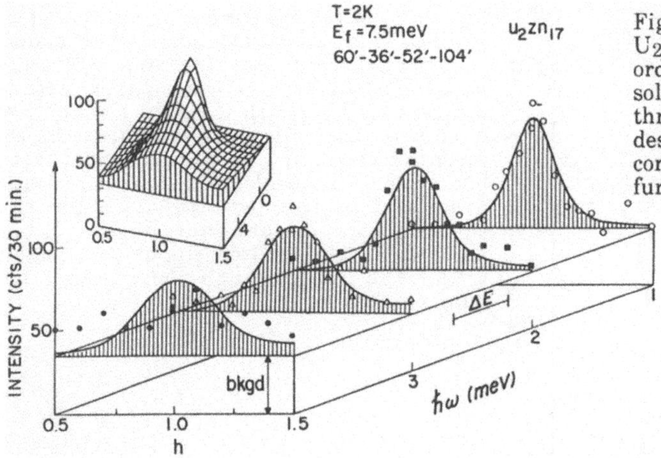

Fig. 2. Constant-energy scans in U_2Zn_{17} along (h,0,3-h) in the ordered phase at T=2 K. The solid lines are the result of a three-parameter fit to the model described in the text. Inset: The corresponding model scattering function in a perspective view.

neutron scattering has shown that the magnetic excitation spectrum in this compound is quasi-elastic and at low temperatures, as in U_2Zn_{17}, extends to energies much larger than the thermal energy [8,9]. Recently it was shown that doping UPt_3 with Th or Pd induced an antiferromagnetic phase transition at T_N = 6.5 K [10]. We have found in a UPt_3 single crystal a low energy quasielastic response with the same symmetry as the antiferromagnetic order in the doped samples [11]. Figure 3 shows low temperature (T = 1.9 K) constant q scans at ($\frac{1}{2}$01), which is the antiferromagnetic zone centre, and (001) which is where the high energy (~6 meV) quasi-elastic scattering is maximal [9]. The energy scale of the quasi-elastic response at ($\frac{1}{2}$01) is below our incoherent energy resolution of ~0.2 meV FWHM. Indeed preliminary neutron scattering experiments show that an antiferromagnetically ordered moment of 0.02 μ_B develops at 5 K in pure UPt_3 and coexists with superconductivity down to 100 mK [11].

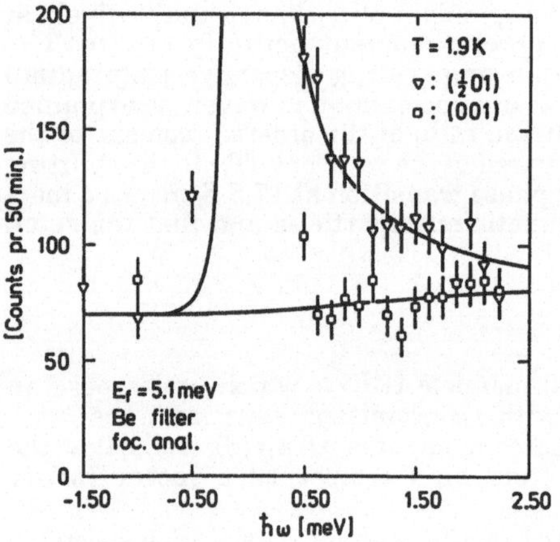

Fig. 3. Constant q scans in UPt_3. The solid lines are the result of fits to the model described in the text.

3. Discussion

The characteristic feature of the magnetic excitations of the heavy fermionsystems is the existence at low temperatures of a continuous spectrum extending to energies much higher than the available thermal energy. This shows that the states of the 5f J multiplet are not approximate eigenstates in these systems. In URu_2Si_2 a transition at T_N = 17.5 K which is currently not well understood renders the system much like crystal field dominated rare earth systems, although the transverse response remains overdamped [4]. However, in both U_2Zn_{17}, UPt_3 and indeed most other heavy fermion systems which have been investigated by inelastic neutron scattering, the magnetic excitations remain overdamped to the lowest temperatures, even in magnetically ordered phases. We believe that the large linear term in the specific heat of the heavy fermion systems is a direct consequence of this [11]. The entropy associated with the high temperature paramagnetic susceptibility is not removed in Schottky anomalies as for crystal-field-dominated rare earth magnets.

A simple phenomenological model incorporates the experimentally observed aspects of the overdamped spin dynamics [12-14,7]. In analogy to the single site susceptibility of rare earth crystal field systems we assume a non interacting susceptibility of the form

$$\chi_0(\omega) = \frac{\chi_0 \Gamma}{\Gamma - i\omega},$$

which accounts for the observed spectral properties. $\chi_0(\omega)$ is also of the form relevant for the single site Kondo problem. Introducing RKKY interactions characterized by the Fourier transformed exchange constant J(q), we obtain in the RPA the generalized susceptibility of the interacting system,

$$\chi_q^{-1}(\omega) = \chi_0^{-1}(\omega) - J_q = \frac{\Gamma - i\omega}{\chi_0 \Gamma} - J_q = \frac{\Gamma_q - i\omega}{\chi_0 \Gamma},$$

where $\Gamma_q = \Gamma(1-\chi_0 J_q)$ is a q dependent relaxation frequency. This simple model gives a good description of the neutron scattering cross-sections of U_2Zn_{17} [7] (see the solid lines in fig. 2) and UPt_3 [11]. Work continues to determine to what extent the model can account quantitatively for the bulk properties of heavy fermion systems.

4. Acknowledgements

The authors are grateful to Z. Fisk, A.Goldman, P. Matthews, A.A. Menovsky, J.A. Mydosh, H.R. Ott, T.T.M. Palstra, S.M. Shapiro, G. Shirane and J.L. Smith for the pleasent collaboration leading to the present results. They also thank P.-A. Lindgård and C.M. Varma for useful discussions.

5. References

1. G.R. Stewart, Rev. Mod. Phys. 56, 755 (1984).
2. T.T.M. Palstra, A.A. Menovsky, J. van den Berg, A.J. Dirkmaat, P.H. Kes, G.J. Nieuwenhuys, J.A. Mydosh, Phys. Rev. Lett. 55, 2727 (1985).
3. C. Broholm, J.K. Kjems, W.J.L. Buyers, P. Matthews, T.T.M. Palstra, A.A. Menovsky and J.A. Mydosh, Phys. Rev. Lett. 58, 1467 (1987).
4. P. Matthews, W.J.L. Buyers, J.K. Kjems, C. Broholm, T.T.M. Palstra, A.A. Menovsky and J.A. Mydosh, to be published 1987.
5. H.R. Ott, H. Rudigier, P. Delsing and Z. Fisk, Phys. Rev. Lett. 52, 1551 (1984).
6. D.E. Cox, G. Shirane, S.M. Shapiro, G. Aeppli, Z. Fisk, J.L. Smith, J.K. Kjems, H.R. Ott, Phys. Rev. B33, 3614 (1986).
7. C. Broholm, J.K. Kjems, G. Aeppli, Z. Fisk, J.L. Smith, S.M. Shapiro, G. Shirane, H.R. Ott, Phys. Rev. Lett. 58, 917 (1987).
8. G. Aeppli, E. Bucher, G. Shirane, Phys. Rev. B32, 7579 (1985).
9. G. Aeppli, A. Goldman, G. Shirane, E. Bucher, M.-Ch. Lux-Steiner, Phys. Rev. Lett. 58, 808 (1987).
10. A.P. Ramirez, B. Batlogg, A.S. Cooper, and E. Bucher, Phys. Rev. Lett. 57, 1072 (1986); G.R. Stewart, A.L. Giorgi, J.O. Willis, and J.O. Rourke, Phys. Rev. B34, 4629 (1986); A. de Visser, S.C.P. Klasse, M. van Sprang, J.J.M. Franse, A. Menovsky, and T.T.M. Palstra, J. Magn. and Magn. Mat. 54-57, 375 (1986); A. I. Goldman, G. Shirane, G. Aeppli, B. Batlogg, and E. Bucher, Phys. Rev. B34, 6564 (1986); P. Frings, B. Renker, and C. Vettier, preprint (1986).
11. G. Aeppli, J.K. Kjems, C. Broholm, E. Bucher, A. Goldman, G. Shirane to be published (1987).
12. G. Aeppli, H. Yoshizawa, Y. Endoh, E. Bucher, J. Hufnagl, Y. Onuki, and T. Komatsubara, Phys. Rev. Lett. 57, 122 (1986).
13. E. Abrahams, J. Magn. Mang. Mat. 63-64, 234 (1987); C.M. Varma in Proc. of Eighth Tanaguchi Symposium. Eds. T. Kasuya and T. Saso, 277 (Springer, Berlin, 1985). pr
14. C.J. Pethick, D. Pines, Nordita preprint 86/27 S (1986).

Linear and Nonlinear Excitations in the $S=1/2$ Ferromagnetic Chain System $[C_6H_{11}NH_3]CuBr_3$ (CHAB)

K. Kopinga and W.J.M. de Jonge

Department of Physics, Eindhoven University of Technology, P.O. Box 513, NL-5600 MB Eindhoven, The Netherlands

1 Introduction

$[C_6H_{11}NH_3]CuBr_3$ (CHAB) is a system built up from ferromagnetic $S = \frac{1}{2}$ chains with a nearest-neighbor interaction of about 55 K, which contains 5 % easy-plane anisotropy. The interchain interactions are smaller by three orders of magnitude [1], and induce a three-dimensional magnetic ordering below $T_c = 1.50$ K. If at temperatures above T_c a symmetry-breaking field is applied within the easy (XY) plane, the equation of motion of the spins in this compound can be mapped to the sine-Gordon (sG) equation, at least under certain approximations [2]. Apart from linear excitations (magnons), this equation has nonlinear solutions called kink solitons. We have analyzed the influence of the linear and nonlinear excitations on various experimentally observable properties of CHAB. As an extension of earlier investigations on the magnetic heat capacity of this compound [3], we present in this paper an analysis of the nuclear spin-lattice relaxation time T_1 of the hydrogen nuclei and the magnetization, both measured in the presence of an external field along the c axis.

The crystallographic structure of CHAB is orthorhombic, space group $P2_12_12_1$. The magnetic properties of the individual chains can be described by the Hamiltonian [1,4]

$$\mathcal{H} = -2\sum_i \left[J^{xx}S_i^xS_{i+1}^x + J^{yy}S_i^yS_{i+1}^y + J^{zz}S_i^zS_{i+1}^z\right], \tag{1}$$

with $J^{xx}/k = 55 \pm 5$ K, $J^{zz}/J^{xx} = 0.95$, and $(J^{xx} - J^{yy})/J^{xx} = 5 \times 10^{-4}$. The y axis coincides with the crystallographic c axis, whereas the x axis lies within the ab plane at an angle φ from the b axis. Two symmetry-related types of chains are present, with $\varphi = -25^o$ and $\varphi = 25^o$, respectively. We will confine ourselves to measurements collected with the external field $B \parallel c$, which is located in the XY plane for both types of chains.

2 Nuclear Spin Lattice Relaxation

The spin-lattice relaxation rate T_1^{-1} of the hydrogen nuclei was measured for $1.2 \text{ K} < T < 6.5$ K and external fields $0 < B < 70$ kG by means of a spin-echo technique. Since the relaxation of the nuclear spin system towards equilibrium is induced by fluctuations in the electron spin system, it can be expressed in terms of the various elementary magnetic excitations. We have interpreted the experimental data by assuming that the dominant excitations for B // c are magnons and solitons. To enable a quantitative comparison of the effect of these excitations, we have calculated their respective contributions by exploiting the relation between T_1^{-1} and the dynamic structure factors $\mathcal{S}^{\alpha\beta}(q,\omega)$ [5]. This approach has the advantage that the contribution of the solitons can be evaluated without approximating their actual shape by, for instance, a rectangular profile [6]. In our case this relation can be written as

$$T_1^{-1}(j) = \sum_{\alpha}\sum_{\beta}\sum_{m=0}^{\infty} G^{\alpha\beta}(j,m)\,\frac{2\pi}{N}\sum_{q}\mathcal{S}^{\alpha\beta}[q,\omega_N(j)]\cos(mqa) \ . \tag{2}$$

In this equation j refers to a set of nuclear spins, α and β run over the coordinates x,y,z, and $G^{\alpha\beta}(j,m)$ are the so-called geometrical factors describing the interaction between the set of nuclear spins and electron spin system. $\omega_N(j)$ represents the Larmor frequency of the nuclear spins j, which depends on both the external and the internal (dipolar) field.

Following the usual approach [6] we plotted the data as $\log(T\,T_1^{-1})$ against $\sqrt{B}/T$ in Fig. 1. In the figure we have included the calculated contribution of the two-

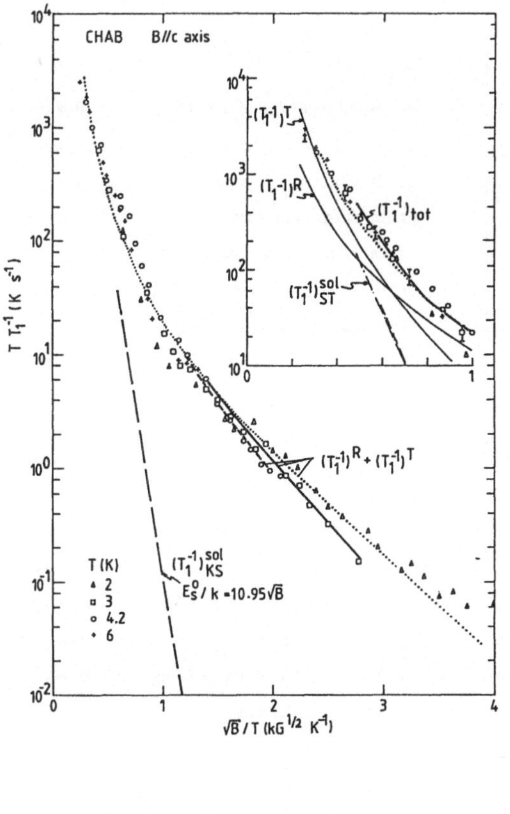

Fig. 1
Spin-lattice relaxation of the hydrogen nuclei in CHAB for ***B // c*** *plotted in reduced form. The meaning of the various theoretical predictions is explained in the text*

magnon (Raman) process $(T_1^{-1})^R$, the three-spinwave process $(T_1^{-1})^T$, and that arising from solitonlike excitations $(T_1^{-1})^{sol}$. It appears that for $\sqrt{B}/T > 1$ the field and temperature dependence of the relaxation rate can fully be explained by the two-magnon process. At lower values of $\sqrt{B}/T$, the data can be described fairly well by the sum of two- and three-magnon processes (dotted curve). All these processes are calculated from standard linear spin-wave theory, based on Eq. (1) and the parameters appropriate to CHAB. The inclusion of soliton processes seems to improve the description of the data below $\sqrt{B}/T = 0.6$, (see inset), but their effect on the relaxation rate is very small.

3 Magnetization

The magnetization (M) was measured with a commercial (PAR) vibrating sample magnetometer for $0 < B < 50$ kG and $1.4 < T < 10$ K. Since for several magnetic model systems, including the sG model, the reduced magnetization M/M_S is a universal function of $T/\sqrt{B}$, we plotted our experimental results in such a way in Fig. 2.

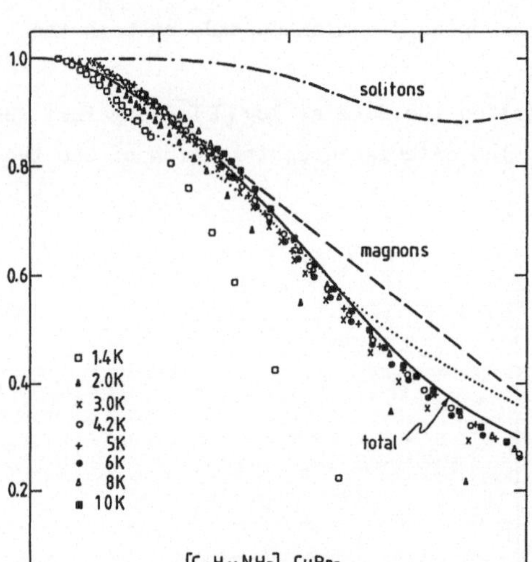

Fig. 2
Magnetization M of CHAB for B // c plotted in reduced form together with the decrease of M resulting from magnons and that resulting from kink solitons. The dotted curve reflects numerical results for the sG model [8]

This figure shows that the data collected above 3 K almost perfectly collapse onto a single curve, suggesting universal behavior. At lower T systematic deviations occur, which are due to the small interchain interactions. In the figure we have included several theoretical predictions. The dashed curve represents the decrease of M from its saturation value M_S, calculated from linear spin-wave theory using Eq. (1) without any adjustable parameters. It is obvious that for $T/\sqrt{B} > 1$ systematic deviations between this prediction and the data occur, suggesting the presence of other excitations, which are not included in the theory.

Within the framework of the classical sG model, the free energy can be written [7] as $F = F_{sol} + F_m$, where F_{sol} is proportional to the soliton density n_{sol} and F_m reflects the contribution of linear excitations (magnons). The decrease of M resulting from solitons, calculated by differentiation of F_{sol} with respect to B, is denoted by the dashed-dotted curve. If we add this decrease to that calculated from linear spin-wave theory, thus replacing F_m in first order by its quantummechanical counterpart, we obtain the result reflected by the solid curve, which describes the experimental data very well. The dotted curve represents (exact) numerical calculations on the sG model [8]. These calculations, in which all elementary excitations are implicitly included, systematically deviate from the data, also at low values of $T/\sqrt{B}$, where the contribution of solitons is insignificant. Possibly, this is caused by the presence of spin components out of the XY plane, which are not taken into account in this model. However, attempts to describe the observed magnetization by other classical models, i.e., discrete systems of classical spins having, alternatively, two or three nonzero components, were also unsuccessful. One might therefore conclude that for a correct description of the present compound for $B /\!/ c$ a quantum treatment of the linear excitations is necessary. Such a conclusion is supported by measurements of the heat capacity C [3], which revealed that, in contrast to C itself, the excess heat capacity $\Delta C = C(B) - C(0)$, which is dominated by nonlinear excitations [7,8], can be described fairly well by the classical sG model.

4 Discussion

In concluding we would like to remark that a consistent description of various magnetic properties of CHAB for $B /\!/ c$ seems possible if the linear excitations are described by conventional spin-wave theory and the nonlinear excitations by a classical model. In the presence of such an external field the sG model appears to be appropriate, as can be inferred from the magnetization data presented above as well as from measurements of the excess heat capacity reported before [3]. To investigate whether such a description is also appropriate to $CsNiF_3$, we measured the magnetization of this compound for $1.2 \lesssim T \lesssim 20$ K and fields up to 50 kG in the XY plane. The data for $T \gtrsim 4.2$ K are plotted in Fig. 3 as M/M_s against $T/\sqrt{B}$. Except for T = 4.2 K, the experimental results show almost perfect universal behavior. In the figure we included several theoretical predictions, which were calculated using the reported values for $CsNiF_3$, $J/k = 23.6$ K, $A/k = 9$ K, and the value $g = 2.08$ deduced from our experimental value of M_s. The dotted curve represents the numerical calculations [8] on the sG model. The decrease of M resulting from spin waves is reflected by the dashed curves, corresponding to T = 4.2 and 20 K, respectively. This prediction shows small deviations from universal behavior. The decrease of M resulting from solitons is represented by the dashed-dotted curve. The drawn curve reflects the combined effect of magnons and solitons, cal-

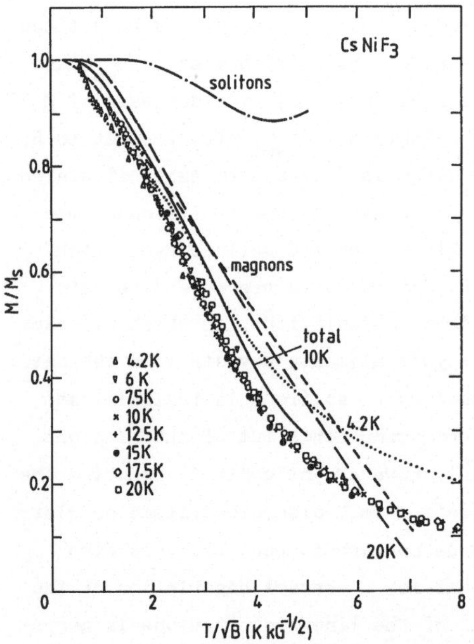

Fig. 3
Magnetization M of $CsNiF_3$ plotted in reduced form together with the decrease of M resulting from magnons and that resulting from kink solitons. The dotted curve reflects numerical results for the sG model [8]

culated for T = 10 K. It is obvious that the latter prediction gives a nice qualitative description of the data; in fact, a complete agreement can been obtained if in the calculations the reported value of J is reduced by 10%.

References

1. K. Kopinga, A.M.C. Tinus, and W.J.M. de Jonge, Phys. Rev. B25, 4685 (1982).
2. H.J. Mikeska, J. Phys. C11, L29 (1978); C13, 2913 (1980); E. Magyari and H. Thomas, J. Phys. C15, L333 (1982).
3. A.M.C. Tinus, K. Kopinga and W.J.M. de Jonge, Phys. Rev. B32, 3154 (1985).
4. A.C. Phaff, C.H.W. Swüste, W.J.M. de Jonge, R. Hoogerbeets, and A.J. van Duyneveldt, J. Phys. C17, 2583 (1984).
5. S.W. Lovesey, Theory of Neutron Scattering from Condensed Matter, (Clarendon, Oxford, 1984)
6. T. Goto, Phys. Rev. B28, 6347 (1983).
7. K. Sasaki and T. Tsuzuki, Solid State Commun. 41, 521 (1982).
8. T. Schneider and E. Stoll, Phys. Rev. B22, 5317 (1980).

Doublings of Soliton Modes and Soliton Magnetic Resonance (SMR)

*J.P. Boucher**

Centre d'Etudes Nucléaires de Grenoble, Département de Recherche Fondamentale, Service de Physique/Groupe Dynamique de Spin et Propriétés Electroniques, 85 X, F-38041 Grenoble Cedex, France

1. INTRODUCTION

In the field of non-linear excitations, antiferromagnetic (AF) Ising-like chains are expected to reveal new and basic features concerning the physics in one-dimension. In particular, in the case of classical spin chains, HALDANE has established that additional internal modes, should be observable in the soliton excitations [1]. Another important question was also raised by HALDANE on the nature of these internal modes, which should depend on whether the spin value is integer or half-integer. More recently, AFFLECK has related these predictions to the "dyon" concept which is used in particle physics [2]. It was also suggested that the application of an external magnetic field in the spin direction might offer a "useful way of disentangling the dyon levels". In the present work, we consider explicitly the case of antiferromagnetic Ising quantum spin (AFIQS) chains in an external magnetic field H. A derivation of the dynamical structure factors $S^{\alpha}(q,\omega)$ (α = x,y,z) associated with the solitons is proposed for both H parallel and H perpendicular to the spin direction. It will be shown that the effect of an external field results in a doubling of the soliton modes. Experimental evidence of this new feature is given for the compound $CsCoCl_3$ where the uniform (q = 0) soliton mode is observed by electron spin resonance (ESR) measurements. Due to the high sensitivity of this technique, such a soliton magnetic resonance (SMR) provides completely new information on the properties of solitons in real systems.

2. THEORY

The starting point is the following Hamiltonian :

$$\mathcal{H} = \sum_n 2J\,[S^z_n S^z_{n+1} + 2\epsilon J(S^x_n S^x_{n+1} + S^y_n S^y_{y+1})] - H_{\parallel} S^z_n - H_{\perp} S^x_n \tag{1}$$

with $|S| = \frac{1}{2}$ and ϵ, $H_{\parallel}/J$, $H_{\perp}/J \ll 1$. The first term where J is the exchange coupling describes the Ising energy. It is responsible for the antiferromagnetic ground state with the spins aligned along the z direction (which here coincides with the chain axis). In (1), $H_{\parallel} = g_{\parallel}\mu_B H$ and $H_{\perp} = g_{\perp}\mu_B H$ define the Zeeman energies when the field H is applied parallel and perpendicular to the chains, respectively. Following VILLAIN, who first calculated the soliton fluctuations in zero field [3], the present derivation is performed in the subspace of states $|p\rangle$ (p half-integer) which contains one soliton between the sites p − ½ and p + ½ in the chains. In this subspace, the matrix elements of the spin operators $\langle p|S^{\alpha}_n|p\rangle$ are easily calculated (cf.ref.[4]). One can also define new soliton creation-annihilation operators c_p and c^+_p as $S^{\alpha}_q = N^{-\frac{1}{2}}\sum e^{-iqn}\,\langle p|S^{\alpha}_n|p'\rangle\, c^+_p c_{p'}$, where the S^{α}_q

also member of Equipe de Recherche CNRS 216

are the Fourier components of the spin operators S_n^α. With these new operators, Hamiltonian (1) can be rewritten as

$$\mathcal{H} = \sum (J + 2\epsilon J \cos 2k + H_\perp \cos k)\, c_k^+ c_k + (i\, H_\parallel/2)\, c_k^+ c_{k-\pi} \tag{2}$$

with $c_k = N^{-1/2} \sum e^{-ikp}\, C_p$. The corresponding eigenvalues are obtained by diagonalization in the $|k\rangle$, $|k - \pi\rangle$ subspace. This diagonalization has been given first by SHIBA and ADACHI who found [5] :

$$\omega_k^\pm = J + 2\epsilon J \cos 2k \pm \sqrt{H_\perp^2 \cos^2 k + H_\parallel^2 / 4} \tag{3}$$

with $-\pi \leqslant k \leqslant +\pi$. As shown in Figs. 1a and b for $H_\parallel$ ($H_\perp = 0$) and $H_\perp$ ($H_\parallel = 0$) respectively, the application of an external field yields a splitting of the soliton energy into two branches. For $H_\parallel$, the splitting which is independent of the wave vector k corresponds to a simple Zeeman effect. For $H_\perp$, the effect of the field is to change both the soliton energy (ω_k) and its velocity ($d\omega_k/dk$).

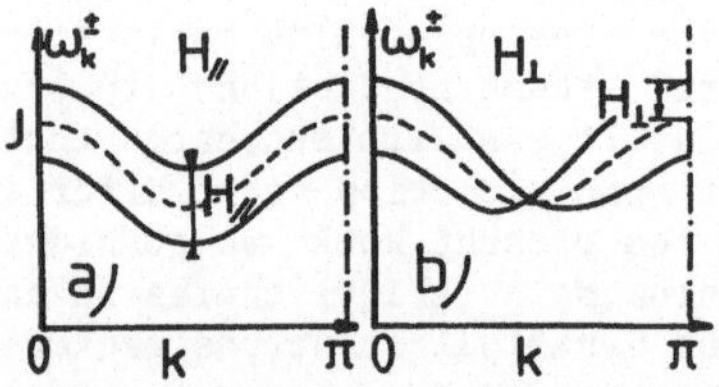

Fig. 1 - Soliton energy branches in a field

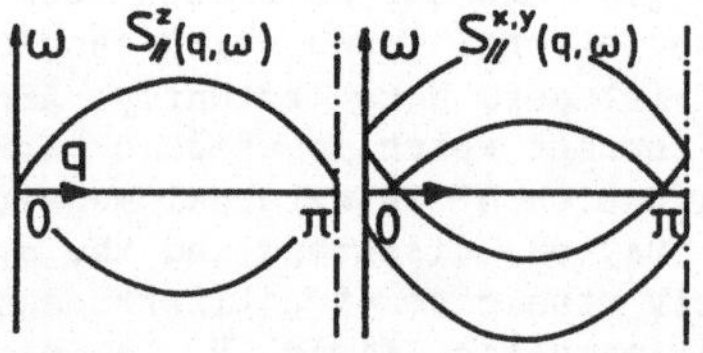

Fig. 2 - Singularity lines for $S_\parallel^\alpha(q,\omega)$

Within the present one-soliton model, exact analytical expressions can be derived for the dynamical structure factors [4]

$$S^\alpha(q,\omega) \sim \int dt\, e^{-i\omega t} \langle S_q^\alpha(t) S_{-q}^\alpha \rangle \; .$$

However, with the approximations H, $\epsilon J \ll T$, one can write for $H_\parallel$

$$S_\parallel^z(q,\omega) \sim \frac{n_s(1 - n_s)}{4\cos^2 q/2}\, N^{-1} \sum P_\parallel^k\, \delta[\omega - \Omega_q \sin(2k - q)] \; , \tag{4a}$$

$$S_\parallel^x(q,\omega) = S_\parallel^y(q,\omega) \sim n_s(1 - n_s)\, N^{-1} \sum P_\parallel^k \cos^2(k + q/2)$$
$$\{\delta[\omega - \Omega_q \sin(2k - q) + H_\parallel] + \delta[\omega - \Omega_q \sin(2k - q) - H_\parallel]\} \; , \tag{4b}$$

where $n_s \sim \exp(-J/T)$ is the soliton density, $\Omega_q = 4\epsilon J \sin q$,
$P_\parallel^k = \exp[-2\epsilon J \cos(2k/T)]/Z_\parallel$, $Z_\parallel = \cosh(H_\parallel/2T)\, I_0(2\epsilon J/T)$ ($I_0(x)$ is the zero-order modified Bessel function). One observes that, while the fluctuations along the chains are practically not affected by the external field - in particular there is no frequency shift in (4a) - the transverse components (4b) undergo a doubling of the soliton modes which are split by $\pm H_\parallel$. As in zero field, the integration over k gives rise to characteristic square-root singularities in the frequency spectrum, the positions of which are shown in Fig.2. Within the same approximations, one can write for $H_\perp$

$$S_\perp^x(q,\omega) \sim n_s(1-n_s) N^{-1} \sum P_\perp^k \cos^2(k+\frac{q}{2})\, \delta\left[\omega - \Omega_q \sin(2k-q) - 2H_\perp \sin(k-\frac{q}{2}) \sin\frac{q}{2}\right] \; , \tag{5a}$$

$$S^y_\perp(q,\omega) \sim n_s(1-n_s)N^{-1}\sum P^k_\perp \cos^2(k+\frac{q}{2})\,\delta\left[\omega - \Omega_q \sin(2k-q) - 2H_\perp \cos(k-\frac{q}{2})\cos\frac{q}{2}\right], \qquad (5b)$$

$$S^z_\perp(q,\omega)\ \frac{n_s(1-n_s)}{4\cos^2 q/2}\, N^{-1}\sum P^k_\perp\ \delta\left[\omega - \Omega_q \sin(2k-q) - 2H_\perp \cos(k-\frac{q}{2})\cos\frac{q}{2}\right], \qquad (5c)$$

with $P^k_\perp = \exp[-(2\epsilon J\cos 2k + H_\perp \cos k)/T]/Z_\perp$ and $Z_\perp = N^{-1}\sum P^k_\perp$. For these functions, the number of square-root singularity lines depends on the value of $H_\perp$ with respect to $8\epsilon J$. For $H_\perp < 8\epsilon J$, 4 singularity lines can be observed (Fig.3a) : for $S^z_\perp(q,\omega)$ and $S^y_\perp(q,\omega)$ gaps with frequencies $\omega = \pm 2H_\perp$ are open at $q = 0$ for two lines while the other two lines are restricted to the interval between $q = 2\sin^{-1}(H_\perp/8\epsilon J)$ and $q = \pi$. With $H_\perp$ increasing, these lines are reduced to a smaller and smaller zone near $q = \pi$, where the soliton mode condenses for $H_\perp = 8\epsilon J$. For a larger field ($H_\perp \geqslant 8\epsilon J$), only the first singularity lines remain present (Fig.3b). Similar behaviour is observed for the function $S^x_\perp(q,\omega)$ on changing q to $\pi - q$. For this function the gaps occur at $q = \pi$ and the condensation point at $q = 0$.

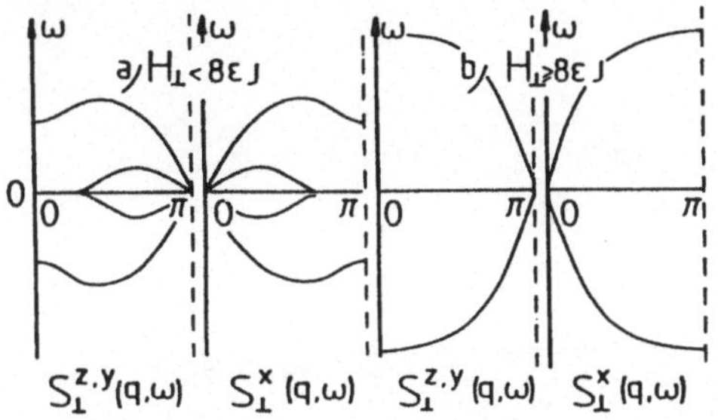

Fig. 3 - Singularity lines for $S^\alpha_\perp(q,\omega)$

Interactions between solitons, between solitons and magnons and with impurities are expected to round off the square-root singularities predicted by the one-soliton model [6-8]. To account for such rounding effects we introduce a "soliton damping" in the expressions for the $S^\alpha(q,\omega)$ by the following substitution $\delta[\omega - \omega(q,k,H)] \longrightarrow \Delta / \{\Delta^2 + [\omega - \omega(q,k,H)]^2\}$. For simplicity, Δ is assumed to be independent of k, H and ω. In the following, this description is referred to as the constant Lorentzian approximation (CLA). Examples of such "realistic" soliton modes are shown in Figs. 4 and 5 for $H_\parallel$ and $H_\perp$, respectively. Clearly, if the damping is not too strong, the doubling of the soliton mode described above should be accessible experimentally. In principle, neutron scattering measurements (INS) should permit the observation of the $S^\alpha(q,\omega)$ in the full Brillouin zone. However, as a direct check of the doubling effect, we can observe the uniform ($q = 0$) soliton mode by the ESR technique. For $H_\parallel$, only the modes $S^x_\parallel(q = 0,\omega)$ and $S^y_\parallel(q = 0,\omega)$ are shifted by $\pm H_\parallel$ (Fig.4) (Zeeman effect). On the other hand, for $H_\perp$, the modes $S^z_\perp(q = 0,\omega)$ and $S^y_\perp(q = 0,\omega)$ now extend continuously from zero to the characteristic values $\pm 2H_\perp$ where we expect to observe a maximum.

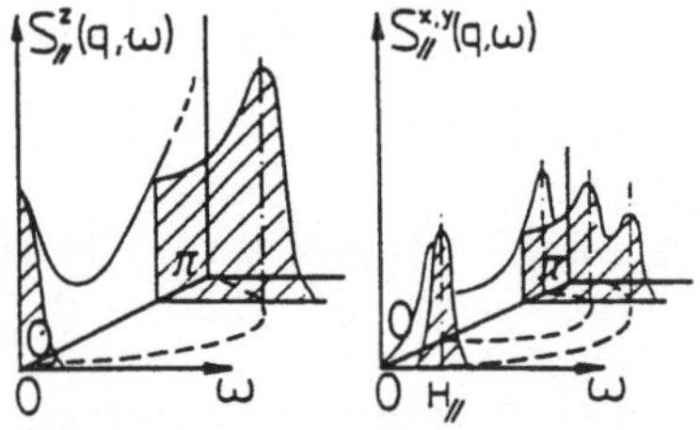

Fig. 4 - Soliton modes for $H_{/\!/}$

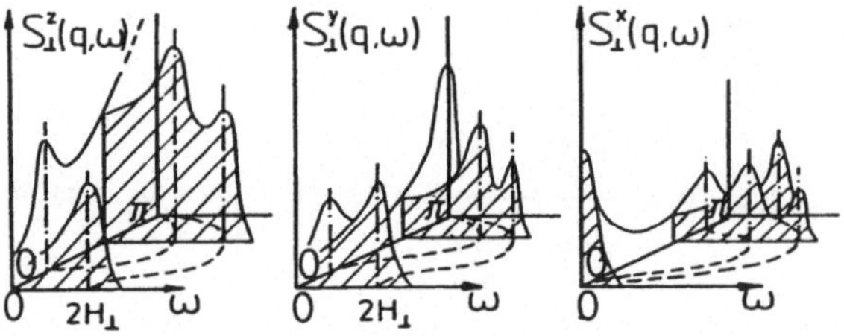

Fig. 5 - Soliton modes for $H_{\perp}$

3. SOLITON MAGNETIC RESONANCE

The SMR measurements presented below have been performed on the compound $CsCoCl_3$, which is known to be a good realization of AFIQS chains with J = 75 K and $\epsilon \simeq 0.12$. Soliton fluctuations have been observed in this compound by INS measurements above the magnetic ordering temperature which occurs at $T_{N1} \simeq 21$ K [6,9]. ESR signals were previously observed by Adachi at the frequency of 9 GHz but below T_{N1}, where the one-dimensional soliton concept becomes questionable [10]. Moreover, the lineshape analysis, which appears to be a crucial test for identifying the soliton modes, could not be performed as the damping effect was not considered.

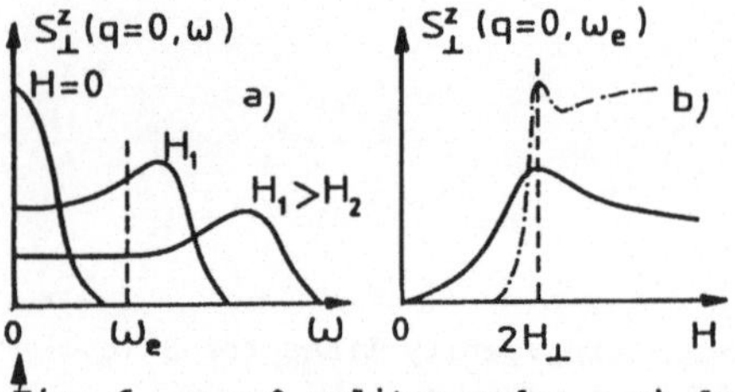

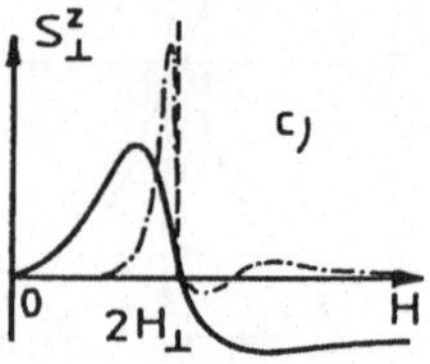

Fig. 7

Fig. 6 - q = 0 soliton modes : a) for different fields H, b) versus H for heavy (——) and light (–·–) damping

Fig. 7 - SMR signals for heavy (——) and light (–·–) damping

In Fig.6a different examples of $S^z_{\perp}(q = 0,\omega)$ are given for different values of the field. ESR measurements are usually performed at a fixed frequency ω_e (ω_e = 35 GHz in the present work) by varying the field. The result is shown in Fig.6b for heavy and light dampings. In order to increase the sensitivity, the field modulation procedure was used. Consequently, the signal S^{α} which is observed is given by the derivative of the absorption mode with respect to H, i.e. $S^{\alpha} = dS^{\alpha}(q = 0,\omega_e)/dH$. The ESR lines corresponding to Fig.6b are shown in Fig.7. The SMR signals are seen to display very characteristic asymmetric lineshapes.

In our ESR experiments, the external magnetic field was applied perpendicular ($H_{\perp}$) or parallel ($H_{/\!/}$) to the chain direction (referred to as the z axis). For $H_{\perp}$, the fluctuations were probed along the two axes $\alpha = y$ and $\alpha = z$, perpendicular to the field. For $H_{/\!/}$, the fluctuations were probed in a direction perpendicular ($\alpha = x,y$) to the chains. The experimental frequency was $\omega_e \simeq 35$ GHz. A few measurements were also performed at 9 GHz. For the first time, SMR signals have been observed *above* T_{N1}, in a temperature range ($21 \lesssim T \lesssim 50$ K) where the 1D soliton concept is meaningful. These signals could be followed and studied near the transition $T_{N1} \simeq 21$ K and well below down to a few Kelvins. At all temperatures,

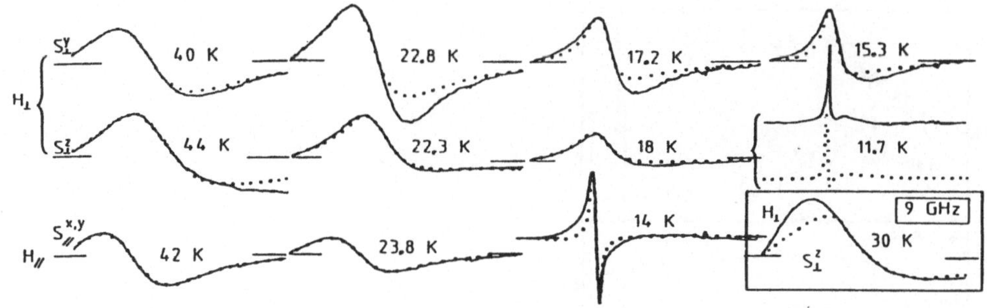

Fig. 8 - SMR signals in $CsCoCl_3$ at 35 GHz in the field range 0-16 kOe. Inset : the same at 9 GHz in 0-8 kOe

these ESR modes were observed with a particularly good signal-to-noise ratio as shown in Fig.8. On the same figure, one observes that the present CLA description (the dotted lines) for the uniform soliton modes reproduces quite well the peculiar experimental lineshapes. At each temperature, the adjustment between theory and experiment was made with three parameters I, Δ and $g_{/\!/}$ and $g_\perp$ for scaling the intensity, the damping and the resonant frequency of the SMR signals, respectively. For each type of fluctuation, $S_\perp^z$, $S_\perp^y$ or $S_{/\!/}^{x,y}$, I was taken equal to 1 at one temperature chosen around 30 K where one expects the soliton model for 1D systems to be valid. The other values were determined with respect to this point. In Fig.9a a plateau is observed between 25 and 42 K. It would define the temperature range where the 1D soliton description can be applied. The decrease of I observed below 25 K could be associated with the partial magnetic ordering which occurs at $T_{N1} \simeq 21$ K. On the other hand, at temperature above ~ 42 K, the present soliton gas model, which assumes a low soliton density ($n_s \ll 1$) becomes questionable ($n_s \simeq 0.17$ at 42 K) and a decrease of the soliton mode intensity can be expected [7,11].

The values for $g_{/\!/}$ and $g_\perp$ (Fig.9b) are strongly different, in agreement with the Ising character of the spin system. The slight difference for $g_\perp$ obtained from $S_\perp^z$ and $S_\perp^y$ can be attributed to the excessive simplicity of the CLA. The same remark holds for the values of Δ obtained from $S_\perp^z$ and $S_\perp^y$ (Fig.9c). However, the three determinations of Δ display a similar and an appreciable temperature dependence. It is also interesting to consider the quantities $\Delta H_{/\!/} = \Delta/g_{/\!/}\mu_B$ and $\Delta H_\perp = \Delta/g_\perp\mu_B$. In Fig.9d, they are seen to have very similar values.

At the transition and below T_{N1}, the ESR signals could be easily followed. Although a drastic narrowing -probably due to the change in the soliton dynamics- is observed, the lineshapes remain characteristic of soliton modes. This result confirms the persistence of soliton excitations below T_{N1}, in this compound.

For the data obtained at 9 GHz, the situation is much more confused. Above T_{N1}, very broad signals were observed for $S_\perp^y$ and $S_{/\!/}^{x,y}$, so broad that no analysis could really be done. Surprizingly, a much narrower signal was observed for $S_\perp^z$. An example is shown in the inset of Fig.8. It exhibits the same lineshape feature characteristic of a soliton mode. The determination of $g_\perp$ which, essentially, defines the point where the SMR signals cross the base line, remains reliable. The corresponding values are shown in Fig.9b. Above T_{N1}, they differ from the values obtained at 35 GHz by a factor of the order of 2 while they collapse below. This remarkable result shows that, at low frequency, there is an important renormali-

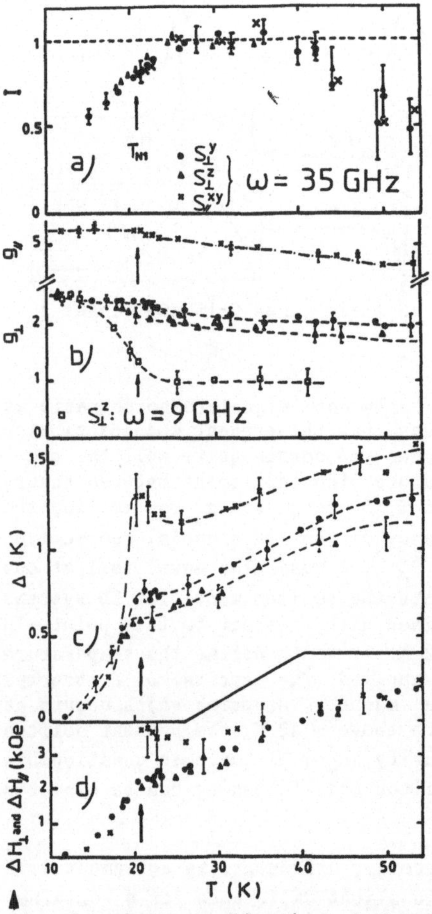

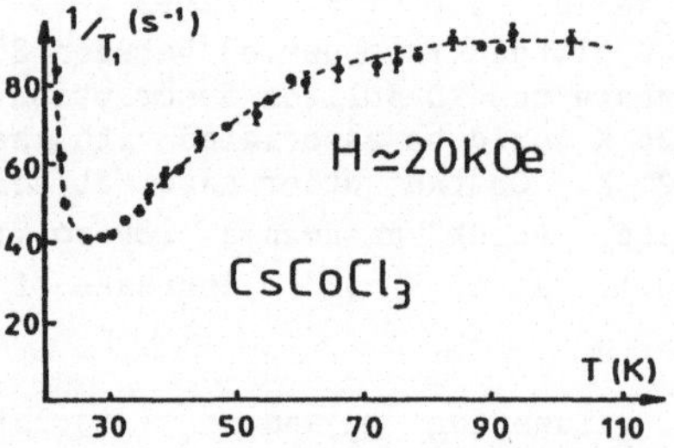

Fig.10

Fig. 9 - a) Renormalisation of the SMR intensity ; b) Experimental values of $g_{/\!/}$ and $g_\perp$; c) Soliton dampings ; d) Quantities $\Delta H_{/\!/}$ and $\Delta H_\perp$ in $CsCoCl_3$

Fig.10 - Nuclear relaxation rate $1/T_1$ of Cs Nuclei in $CsCoCl_3$

sation of the resonant frequency. In fact, this feature appears also to be present at 35 GHz but it is much less pronounced.

Above T_{N1}, the main questions raised by the present SMR measurements concern the soliton damping Δ and the frequency renormalisation, which may have the same dynamical origin. The result of Fig.9d suggests that Δ is isotropic and induced by magnetic coupling. According to SASAKI and MAKI, it cannot be attributed to the collisions between solitons, which broaden the soliton modes only at $q = \pi$ [8]. We suggest that the damping at $q = 0$ is induced by the fluctuations of the solitons in the surrounding chains. For this purpose it is interesting to consider the temperature dependence of the nuclear relaxation rate T_1^{-1} of the Cs nuclei which are located in the neighbourhood of the chains (Fig.10). The same behaviour is observed for Δ and T_1^{-1}. We believe that it is due to the fact that they probe similar soliton fluctuations.

When T_{N1} is approached, one observes an increase of Δ (and of T_1^{-1}). This effect can be attributed to the interchain correlations and associated with critical dynamics. Finally, the observation of solitons in ordered phases (below T_{N1}) raises a basic question. We may ask if it is a general result or if it is specific to this compound which presents a strong interchain frustration.

4. CONCLUSION

The present description provides a comprehensive picture of the effect of an external magnetic field on the soliton excitations in Ising-like quantum spin chains. New specific features have been established for both $H_{\parallel}$ and $H_{\perp}$. For $H_{\parallel}$, an additional Zeeman splitting is found. For $H_{\perp}$, one is led to a simultaneous change of both the energy and the velocity of the solitons. The dynamical structure factors have been calculated explicitly [4]. Our description, the CLA, is used as a first approach. It must now be completed. However, it is shown that this description provides a good test for identifying soliton modes by ESR and that series of new results can be obtained with this technique : they concern the soliton behaviour in real systems in a 1D paramagnetic phase, near a magnetic transition and in ordered phases.

The present analysis has been derived in the case of quantum spin chains. We may wonder if they can be extrapolated to classical spin chains. The Zeeman splitting obtained for $H_{\parallel}$ is the analog of the dyon levels described by AFFLECK [2]. It remains to find the classical analog of the soliton doubling for $H_{\perp}$. Finally, the present work which concerns a spin ½ system is a first contribution to the crucial question concerning the different behaviours expected for antiferromagnetic chains with integer and half-integer spins [1].

I would like to express my thanks to G. Rius and F. Devreux for their essential contributions to the experimental and theoretical parts of this work.

1. F.M.D. Haldane: Phys. Rev. Lett. 50, 1153 (1983)
2. I. Affleck: Phys. Rev. Lett. 57, 1048 (1986)
3. J. Villain: Physica B 79, 1 (1975)
4. F. Devreux and J.P. Boucher: to be published
5. H. Shiba and K. Adachi: J. Phys. Soc. Japan 50, 3278 (1981)
6. J.P. Boucher, L.P. Regnault, J. Rossat-Mignod, Y. Henry, J. Bouillot and W.G. Stirling: Phys. Rev. B 31, 3015 (1985)
7. J.P. Boucher, L.P. Regnault, R. Pynn, J. Bouillot and J.P. Renard: Europhys. Lett. 1, 415 (1986)
8. K. Sasaki and K. Maki: Phys. Rev. B35, 263 (1987)
9. H. Yoshizawa, K. Hirakawa, S.K. Satija and G. Shirane: Phys. Rev. B23, 2298 (1981)
10. K. Adachi: J. Phys. Soc. Japan 50, 3904 (1981)
11. K. Kakurai, M. Steiner, R. Pynn, B. Dorner, J. Magn. Magn. Mater. 54-57, 835 (1986)

Relaxation Phenomena in Non-linear Systems; A Mössbauer Effect Spectroscopy and Susceptibility Study

M. ElMassalami, H.H.A. Smit, H.J.M. de Groot, R.C. Thiel, and L.J. de Jongh*

Kamerlingh Onnes Laboatorium, Rijksuniversiteit Leiden, Nieuwsteeg 18, NL-2311 SB Leiden, The Netherlands

1. INTRODUCTION

In the last few years our group has been engaged in a systematic study of the dynamics of kinks (solitons) in magnetic chain systems by means of Mössbauer Effect Spectroscopy (MES). In this paper we report on the progress made since the San Miniato Conference /1/, as well as on very recent frequency-dependent susceptibility studies /2/, which have greatly contributed to our understanding of the problems involved.

We first recall that with the Mössbauer effect one probes the autocorrelation function $<S^z(o)S^z(t)>$ of the electronic spins on the magnetic chains by means of the hyperfine interaction between the spin of the Mössbauer nucleus and the electron spin in the same atom (in our case the Fe atom). Since the fluctuations in the electron spin cause line broadening and even a collapse of the magnetic hyperfine splitting in the MES spectra, one has direct access to the spin dynamics by studying the associated relaxation phenomena in the spectra. Evidently, one is limited here to the frequency window appropriate to MES, which for ^{57}Fe amounts to 10^6 - 10^9 Hz. In this respect, we remark that these frequencies are quite low compared to the typical range available in neutron scattering ($\gtrsim 10^{10}$ Hz). Thus the MES provides a welcome complementary technique for investigating the low-frequency dynamics. As will be explained below, for the ferromagnetic chain systems, we have been able to extend this frequency range even much further downward by means of frequency-dependent ac-susceptibility, which involves frequencies of 10 - 10^5 Hz.

Table I Some magnetic properties of the investigated magnetic chain compounds. S_{eff} denotes the effective spin value at low temperatures, T_c is the transition temperature for the pure systems (x = 0), and $2E_k$ is the creation energy of a kink pair (or soliton-anti-soliton).

	Impurity M	S_{eff}	T_c(x=0) (K)	J or J_z (K)	D_x (K)	D_z (K)	$2E_k$ (K)
$Fe(N_2H_5)_2(SO_4)_2$	-	2	6.0	-2.0	0.8	5.6	34
$FeC_2O_4 \cdot 2H_2O$	-	½	11.7	-82.6	Ising		165
$Fe(BiPy)Cl_3$	-	5/2	?	- 3.0	0.08	0.19	21
$RbFeCl_3 \cdot 2H_2O$	-	½	12.0	-39	Ising		78
$Fe_{1-x}M_xCl_2Py_2$	Cd,Co,Mn Cu,Mg,Ni	½	6.6	+25	Ising		50
$Co_{1-x}M_xCl_2Py_2$	Fe,Cd	½	3.17	+11.7	Ising		23
$Ni_{1-x}M_xCl_2Py_2$	Fe,Cd,Mn	½	6.8	+21.4	Ising		43

* Present address: Physics Department, P.O. Box 321, University of Khartoum, Khartoum, Sudan.

In Table I we summarize the properties of the investigated magnetic chain systems. As may be seen (sign of J), both antiferromagnetic and ferromagnetic chain systems have been studied. For $MCl_2(Py)_2$ (M = Fe,Co,Ni), $FeC_2O_4 \cdot 2H_2O$, and $RbFeCl_3 \cdot 2H_2O$ the relevant populated electronic levels of the magnetic ions are a low-lying doublet /3,2/, and the magnetic properties can be described in terms of an effective spin $S' = \frac{1}{2}$ formalism with strong Ising-type exchange anisotropy. In $Fe(N_2H_5)_2(SO_4)_2$, however, specific heat studies show that the low-temperature state should be described by a spin S = 2 and a strong axial (Ising-type) single-ion anisotropy term /4/. Lastly, in $Fe(bipyridine)Cl_3$ we have an Fe^{3+} magnetic chain, for which the Ising-type anisotropy is very much smaller, since we are dealing with an S-state ($3d^5$) ion /5/. In Fig.1a,b the ferro- and antiferromagnetic classical sine-Gordon solitons are depicted, which have been shown /6/ to be associated with an interaction hamiltonian of the form

$$\mathcal{H}_{cl} = \sum_i [-2J\vec{S}_i \cdot \vec{S}_{i+1} - D_x(S_i^x)^2 + D_z(S_i^z)^2] ,$$

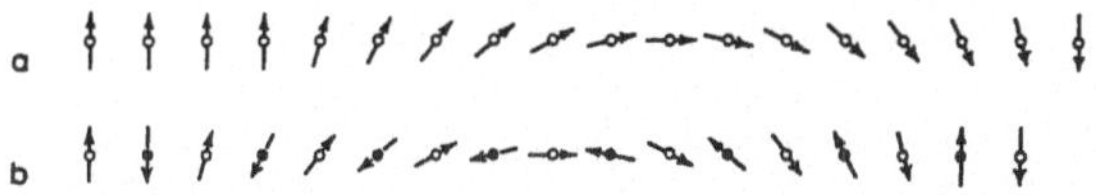

Fig. 1 Schematic representation of the ferro- and antiferromagnetic classical sine-Gordon solitons.

if the continuum limit is taken. Here the $\vec{S}_i$ are classical vectors, the term in $D_z(S_i^z)^2$ is a planar anisotropy term forcing the spins into the XY plane, whereas the term $D_x(S_i^x)^2$ represents the Ising anisotropy favouring the X-axis within this plane.

When the Ising anisotropy D_x becomes very large compared to the exchange J, the wall-width becomes very small and the continuum approximation necessary for the sine-Gordon description evidently breaks down. Furthermore, in the case of effective spin $S = \frac{1}{2}$, one is in the extreme quantum limit and the use of classical spin vectors also becomes dubious. For most of the investigated compounds it will therefore probably be more appropriate to use the Ising-type hamiltonian

$$\mathcal{H}^I = - \sum_i [2J_z S_i^z S_{i+1}^z + 2J_x S_i^x S_{i+1}^x + 2J_y S_i^y S_{i+1}^y] - \sum_i g\mu_B B_z S_i^z , \qquad (J_x, J_y << J_z),$$

where we have added an applied field term for later use. As in the classical sine-Gordon model, an orthorhombic exchange anisotropy is needed to enable a dynamics in the ferromagnetic Ising-type chain /7/, whereas for the antiferromagnetic case, a uniaxial anisotropy ($D_z = 0$ or $J_x = J_y$) suffices.

2. MES EXPERIMENTS

In general the excess linewidth $\Delta\Gamma$ of the MES due to electron spin fluctuations is given by

$$\Delta\Gamma = \int_0^\infty dt \, < S^z(0) S^z(t) > \, e^{i\omega_N t} ,$$

where ω_N is the nuclear Larmor frequency. In the case of a stochastic up and down flipping of the electronic spin, with flip rate Γ_ω as appropriate to first approximation for a gas of free kinks, the autocorrelation function becomes

$$< S^z(0) S^z(t) > \propto S^2 e^{-\Gamma_\omega t}$$

and the expression for $\Delta\Gamma$ reduces to the usual Lorentzian form $\Delta\Gamma \propto \Gamma_\omega/(\Gamma_\omega^2 + \omega_N^2)$. As discussed earlier /3/, instead of extracting $\Delta\Gamma$ from individual lines in the spectra, a much more elegant way is to actually fit the Mössbauer spectrum at each temperature using the Blume-Tjon method /8,9/ for calculating the spectra in the presence of relaxation, with the rate Γ as the only adjustable parameter. In this way

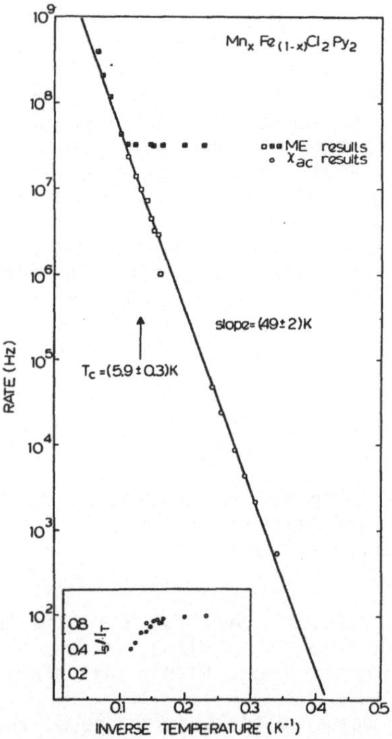

Fig. 2 Relaxation rate versus inverse temperature for $Mn_{0.01}Fe_{0.99}Cl_2Py_2$. Full and open squares represent the relaxation rate of the relatively fast and slowly relaxing subspectra, respectively. For temperatures at which only one subspectrum could be distinguished half-filled squares are drawn. Relaxation rates below 10^5 Hz are derived from the ac-susceptibility measurements. The intensity of the relatively slowly relaxing subspectrum is shown in the insert.

one obtains Γ_ω as a function of temperature, and since $\Gamma_\omega \propto n_k v_k$, where n_k and v_k are respectively the density and the average velocity of the kinks, it follows that by MES we may investigate the dynamics of the kinks, e.g. as a function of temperature, applied field, etc. As explained earlier /8/, the MES experiment is not able to see the structure of the wall, since the typical wall velocities are too high ($\simeq 10^{11}$ sites/second) compared to the MES frequency window. Therefore, even in the case of a finite wall-width, a MES experiment will experience the passage of a kink as an instantaneous flipping of the electron spin, and thus of the hyperfine field component proportional to it.

In comparing the experimental dependence of Γ_ω on temperature with predictions of the model of a free gas of noninteracting kinks, we were confronted with two major discrepancies, which have puzzled us for quite some time. Firstly, in the free-kink-gas approximation, the kink density is given by $n_s \propto \exp(-E_k/k_BT)$, where E_k is the excitation energy of a single kink ($E_k = 4S^2(D_xJ)^{\frac{1}{2}}$ and $E_k \simeq |J_z|$ for the hamiltonians $\mathcal{H}_{C\ell}$ and $\mathcal{H}_I$, respectively). Since $\Gamma_\omega \propto n_s$ one would expect to derive an activation energy E_k from the $\ln \Gamma_\omega$ versus $1/T$ plots. However, in <u>all</u> the experimental systems studied we found an activation energy corresponding to $\overline{2\,E_k}$ (see e.g. Fig.2). The value of E_k could be derived independently from determinations of the exchange and anisotropy parameters by means of magnetic specific heat, susceptibility and far-infra-red absorption data. The experimental susceptibilities of our ferromagnetic Ising-type chains, are described at low temperatures by $\chi \propto n_k^{-1} \propto \exp(J/k_BT)$, as is predicted by simple entropy considerations and is observed by elastic neutron scattering in analogous antiferromagnetic chain systems. One should conclude from this that the relaxation rate Γ_ω in the MES experiment is not proportional to the kink density n_k, but is merely sensitive to the kink-pair excitations.

Secondly, upon closer inspection we find that the Mössbauer spectra in the relaxation regime actually consist of two different, coexisting subspectra (Fig.3). Most prominent is the already discussed magnetically split subspectrum, whose line broadening can be ascribed to relatively slow fluctuations of the spins, with the asso-

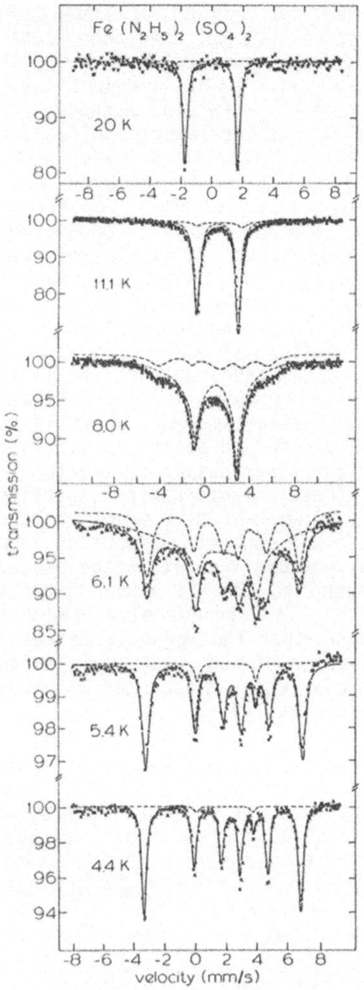

Fig. 3 Representative Mössbauer spectra of $Fe(N_2H_5)_2(SO_4)_2$ versus temperature. The two subspectra (see text) resulting from the Blume and Tjon model are shown separately.

ciated activation energy 2 |J|. In addition there is a fast relaxing weaker subspectrum, which has the form of a paramagnetic doublet indicating that the fluctuations are very rapid with respect to ω_N. The relaxation rate of the doublet amounts to 5×10^7 Hz, and is independent of temperature (see e.g. Fig.2). Furthermore, since the fastest relaxation channel is most effective, the coexistence of both the fast and slow fluctuation processes tells us that only a limited fraction of the spins can be involved in these fast fluctuations, their percentage decreasing with temperature.

In Fig.3, our MES data for $Fe(N_2H_5)_2(SO_4)_2$ show the just mentioned subspectra very clearly. Here the presence of the doublet contribution is unmistakable since the magnetic hyperfine splitting in $Fe(N_2H_5)_2(SO_4)_2$ is about 5 times larger than for e.g. $FeCl_2(Py_2)$ /3/. Reanalysing the older data in retrospect, we find that also in the earlier investigated materials the doublet contribution can be distinguished.

3. AC-SUSCEPTIBILITY

As mentioned at the beginning, the ferromagnetic chain systems offer the additional possibility of studying the relaxation phenomena by means of frequency-dependent ac-susceptibility, χ_{ac} /2/. To this end, a large number of samples of $X{:}MCl_2(py)_2$, where M = Fe, Co, Ni and the dopant X = ^{57}Fe, Cd, Co, Ni, Cu, Mn, have been studied. As already emphasized in earlier work /7,10-13/ on n-magnon bound states (alias spin-clusters), the energy levels of the 2J-multiplet are field dependent through the associated Zeeman energies. This implies that a small perturbing magnetic field should be able to modulate these energy levels and thus the relaxation effects can also be probed by the χ_{ac} measurement, in a way quite similar to the vast work on paramagnetic relaxation processes /14/. Pursuing the analogy still further, one may argue that the kink distribution can be used to define a temperature for the kink system (T_k) just in the same spirit as the Boltzmann distribution in a paramagnetic spin system is used to define a spin temperature. Furthermore, this kink distribution (and hence T_k) may also be disturbed adiabatically, i.e. without a change in the kink density. Then the kink-kink relaxation (KKR) is simply the relaxation process involved in re-establishing a new kink distribution, in analogy with the well-known spin-spin relaxation. Also, the kink-lattice relaxation (KLR) is then the relaxation process which is associated with the energy exchange between the lattice and the kink system, analogous to the spin-lattice relaxation process. Basically, what it amounts to is that the thermodynamics of the Ising-type spin chain is completely determined /10/ by: (i) the total number of overturned spins (n_s); (ii) the total number of clusters (n_c) and (iii) the statistical distribution of the total n_c spin clusters over the m-sized spin clusters, $n|m$. These thermodynamical quantities presumably will fluctuate around their equilibrium values. In this view, the above mentioned fast fluctuations of the kinks, which yield the doublet subspectrum in MES data, would correspond to the adiabatic KKR process, whereas the activation process, governing the MES split subspectrum, would correspond to the KLR process and the relaxation rate would then reflect the activation gap. As argued below, there are two energy gaps, namely $2\,|J|$ or $|J|$, depending on whether the magnetic chains are cyclic or segmented. It follows then there are two types of KLR processes, and consequently two exponential terms ($\exp(-2J/k_BT)$ and $\exp(-J/k_BT)$, associated with pair-kink-lattice and single-kink-lattice relaxation, respectively).

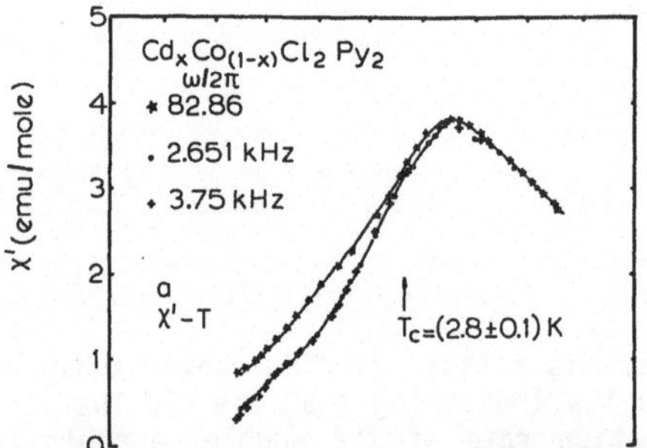

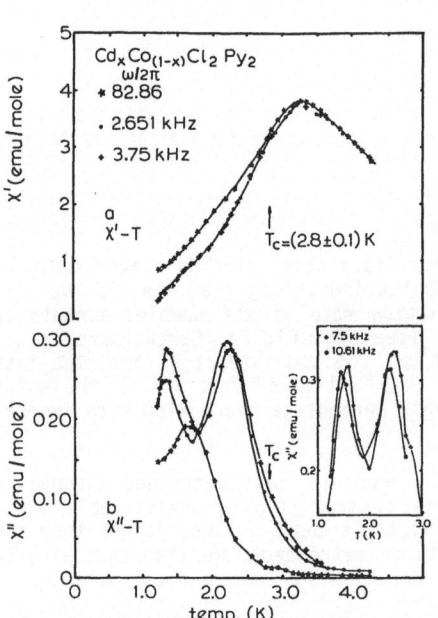

Fig. 4 Representative curves for the real (χ') and imaginary (χ") parts of the ac-susceptibility as a function of temperature for $Cd_{0.01}Co_{0.99}Cl_2Py_2$.

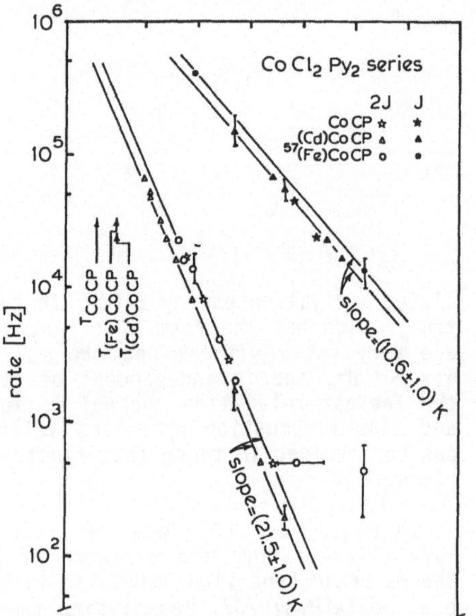

Fig. 5 Relaxation rates vs inverse temperatures for the series $Co_{1-x}M_xCl_2Py_2$, as obtained from ac-susceptibility data.

The latter two processes are clearly observed in our studies of the frequency dependent χ_{ac}, in the doped samples. Figure 4 shows a temperature scan of the real susceptibility χ' and the imaginary susceptibility χ'', keeping the frequency of the oscillating magnetic field fixed. The frequency dependence of the maxima of the peaks as determined from the curve of χ'' - T are plotted in Fig.5. Another way to derive the same relaxation rates is to fix the temperature and scan the frequency. A typical example is shown in Fig.6, where the solid line is a theoretical fit according to the well-known Debye relations. From this frequency scan, the usual Argand diagrams can be extracted, shown in the same figure. For more details on the experimental procedure, on the derivation of the Debye curves, or for more results see Ref./2/.

By performing such experiments at different temperatures, we obtain the kink-lattice relaxation rates as a function of temperature (it is not possible to scan the KKR process with the available susceptometer, 10 Hz < $\omega/2\pi$ < 100 kHz). The results are included in Fig.2, and it can be seen that they form an excellent continuation towards lower frequencies of the MES data.

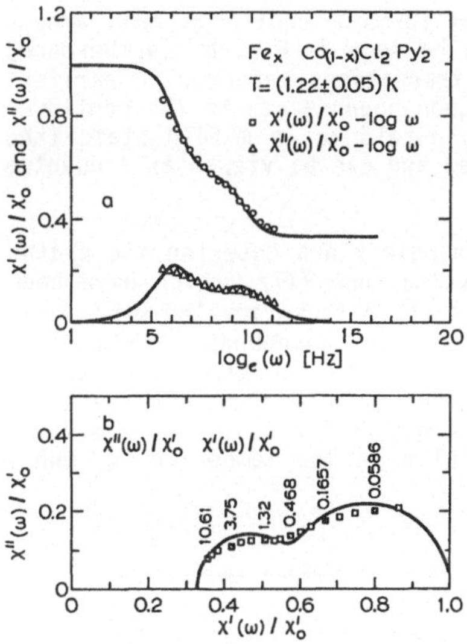

Fig. 6 Debye curves (a) and the Argand curves (b) for $Fe_{0.03}Co_{0.97}Cl_2Py_2$ at T = 1.22 K. The indicated angular frequencies in (b) are in kHz.

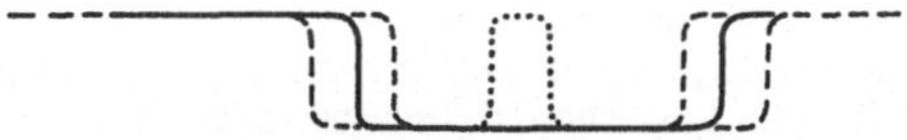

Fig. 7 Sketch of the proposed "kink-lattice" model. Dashed curves illustrate the kink confinement, and the dotted curve a kink pair excitation.

4. DISCUSSION

The proposed model for the interpretation of the MES and the ac-susceptibility results supposes the existence of two distinct types of relaxation processes that are associated with the experimentally observed two distinct types of excitations, namely: (1) a relaxation process associated with the rapid fluctuation of the kinks over regions of finite extent (referred to as Kink-Kink Relaxation KKR), and (2) the relaxation process associated with the excitation or de-excitation of kink pairs (or single kinks) that occur in the chain segments in between the just mentioned regions (referred to as Kink-Lattice Relaxation KLR). The two types of fluctuations and their region of influence are illustrated in Fig.7. The kink distribution along the chain might be visualized as a "kink-lattice" pattern, illustrating the confinement of the fluctuating walls. However, neither the MES experiments nor the ac-susceptibility can say anything about the distribution of the kinks, which would yield additional information about the driving mechanism for this observed confinement of the kinks. In this respect, one can think of a regular lattice if repulsive interactions between the kinks due to quantum effects are responsible /15/. On the other hand an irregular "kink-lattice" arises if the damping finds its origin in scattering off lattice defects and/or interchain coupling /16/. It should be mentioned that in the Ising limit, this "kink-lattice" pattern should reduce to the statistical kink distribution which can be calculated from combinatorial techniques /10/.

It is very clear that such a model is far from the free-gas model of non-interacting kinks, since the former predicts two types of magnetic fluctuations, while the latter would involve only one type of magnetic fluctuation due to the dynamics of the moving kinks /3,18/. Moreover, the above model takes into account the experimentally observed excitation of kinks in the form of kink-antikink pairs /7,10,11/ rather than admitting them only as individual kinks (to our knowledge the pair excitation has not been considered in the statistical mechanical theory of sine-Gordon solitons).

As an illustration of the existence of the pair excitations, consider the cyclic $S = \frac{1}{2}$ Ising-type chains for which it is well known that the pair-kink excitations are the elementary excitations /7,10-13/. They have been called n-magnon bound states, n-fold spin-clusters or volume excitations where n denotes the number of overturned spins between kink and anti-kink (c.f. Fig.8a). As is well known the presence of the transverse exchange terms (J_{xx} and J_{yy}) as a perturbation on the Ising-chain introduces the dispersion character and forms a continuum band around the basic excitation energy $2\,|J|$ /7,11-13/, as can be seen in Fig.8b. Furthermore, Fig.8b illustrates also the two distinct types of transitions referred to earlier: (i) transitions that involve the creation of a n-magnon bound state in the continuum band and (ii) transitions within this band from an n-fold to an m-fold state (the motion of the kink pair as a whole is also possible, and can be viewed as a quantum analogue of solition motions).

For a finite chain segment, on the other hand, single kinks entering the chains from the ends provide another set of elementary excitations /17/, which have been called surface excitations (for the antiferromagnetic chain these are the kinks considered by Villain /18/). It follows that for pure chain compounds with a small number of lattice defects, one would only expect to observe the kink-pair excitations, whereas for compounds with e.g. nonmagnetic impurities that cut the chains in finite segments, one would expect the additional presence of the single-kink excitations. The latter should become apparent in particular at low temperatures, where

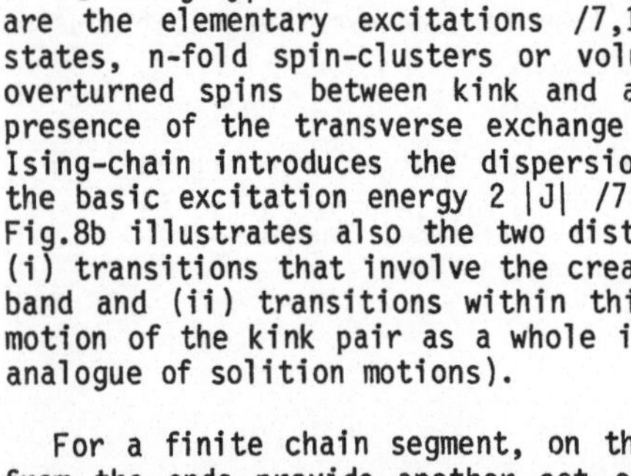

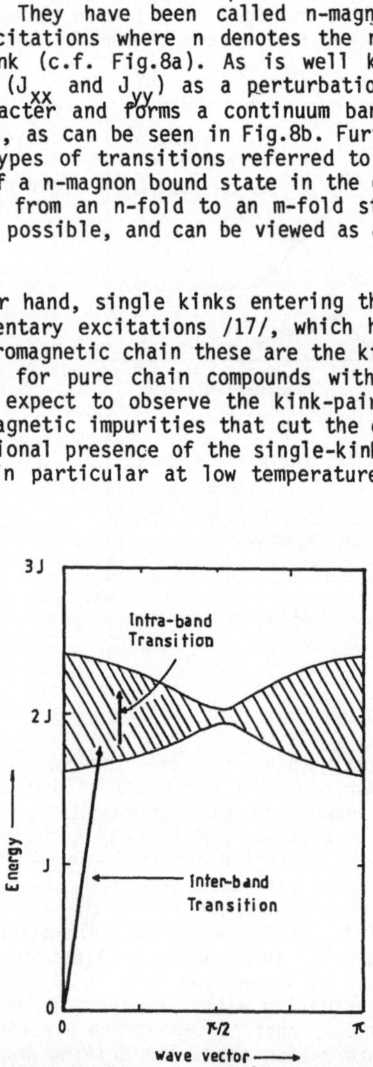

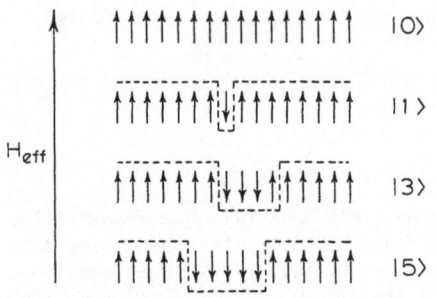

Fig. 8a Examples of the elementary excitations in a ferromagnetic cyclic chain. |o, denotes the ground state; |n denotes the n-fold spin cluster enclosed by a kink anti-kink pair.

b) Transitions that contribute to the kink-lattice relaxation (interband: slowly relaxing subspectrum in MES) and kink-kink relaxation (intraband: fast relaxing doublet in MES). Assuming cyclic boundary conditions, this band is centred around 2J. For finite chains there is an additional mode around J.

the relaxation due to surface excitations ($\exp(-J/k_BT)$) may dominate the relaxation due to volume activation ($\exp(-2J/k_BT)$) /17,19/.

In conclusion, the above model is shown to be able to explain all the experimentally observed relaxational behaviour of the studied magnetic chain systems: the identification of three relaxation processes and the description of their temperature dependency. Moreover, it has been shown that the combination of MES and χ_{ac}, though basically quite different measuring techniques, enables a study of kink-dynamics over a frequency range of nearly ten decades.

ACKNOWLEDGEMENTS: Part of this work belongs to the research program of the "Stichting voor Fundamenteel Onderzoek der Materie" (Foundation for Fundamental Research on Matter) and was made possible by financial support from the "Nederlandse Organisatie voor Zuiver-Wetenschappelijk Onderzoek" (Netherlands Organization for the Advancement of Pure Research). The investigations are sponsored by the Leiden Materials Science Group ("Werkgroep Fundamenteel Materialen Onderzoek"). We wish to thank Prof. J. Reedijk (Chem. Dept., Leiden University) for providing the samples studied in this work and Dr. A.J. van Duyneveldt and Dr. J.C Verstelle for making available the differential susceptibility apparatus

REFERENCES

1. H.J.M. de Groot, R.C. Thiel and L.J. de Jongh: in Magnetic Excitations and Fluctuations ed. by S.W. Lovesey, U. Balucani, F. Borsa, and V. Tognetti, Springer Ser. in Solid State Science, vol.54 (Springer, Berlin 1984).
2. M. ElMassalami, H.J.M. de Groot, H.H.A. Smit, R.C. Thiel, L.J. de Jongh: to be published; M. ElMassalami, thesis, University of Leiden (1987).
3. H.J.M. de Groot, L.J. de Jongh, R.C. Thiel and J. Reedijk: Phys. Rev. B30 (1984) 4041; H.H.A. Smit, H.J.M. de Groot, M. ElMassalami, R.C. Thiel and L.J. de Jongh: to be published.; H.H.A. Smit, H.J.M. de Groot, R.C. Thiel, L.J. de Jongh, M.F. Thomas and C.E. Johnson: Solid State Commun. 53 (1985) 573.
4. F.W. Klaaijsen, thesis, University of Leiden, 1974.
5. H.H.A. Smit, T.A.M. Haemers, R.C. Thiel and L.J. de Jongh: to be published.
6. H.J. Mikeska: J. Phys C11 (1978) L29; J. Phys. C13 (1980) 2313; J. Appl. Phys. 52 (1981) 1950.
7. (a) J.B. Torrance Jr. and M. Tinkham: Phys. Rev. 187 (1969) 5587; ibid 187 (1969) 595; (b) D.F. Nicoli and M. Tinkham: Phys. Rev. B9 (1974) 3120.
8. H.J.M. de Groot, W.J. Huiskamp, L.J. de Jongh, H.H.A. Smit and R.C. Thiel: in Trends in Mössbauer Spectroscopy ed. by P. Gütlich and G.M. Kalvius (Mainz University Press, 1983) 158; F. Borsa, Phys. Lett. 80A (1980) 309.
9. M. Blume and J.A. Tjon, Phys. Rev. 165 (1968) 446; M. Blume Phys. Rev. 174 (1968) 351.
10. M. Date and M. Motokawa: Phys. Rev. Lett. 16 (1966) 1111; J. Phys. Soc. Jpn. 24 (1968) 41.
11. W.J.M. de Jonge, C.H.W. Swuste and K. Kopinga: in Physics in 1-d ed. J. Bernasconi and T. Schneider, Berlin, Springer-Verlag 1981 (Springer Series in Solid State Sciences) p.169 and references therein; L.A. Bosch, G.J.P.M. Lauwers, K. Kopinga and W.J. de Jonge: J. Magn. Magn. Mater. 54-57 (1986) 1161.
12. H. Yoshizawa, K. Hirakawa, S. Satija and G. Shirane: Phys. Rev. B23 (1981) 2298.
13. N. Ishimura and H. Shiba: Prog. Theor. Phys. 63 (1980) 743.
14. J.C. Verstelle and D.A. Curtis: Handbuch der Physik (1968); (b) A. Morrish: The Physical Principle of Magnetism, John Wiley & Sons, Inc. (1965) p.144; (c) C.P. Slichter: Principles of Magnetic Resonance, Harper & Row (1963) p.115; (d) R.L. Carlin and A.J. van Duyneveldt: Magnetic Properties of Transition Metal Compounds; (e) C.J. Gorter in Paramagnetic Relaxation Elsevier Publishing Company Inc., Amsterdam (1974); f) H.B.G. Casimir and F.K. du Pré, Physica 5 (1938) 507.
15. V.E. Korepin: Commun. Math. Phys. 76 (1980) 165.
16. K. Maki: Phys. Rev. B24 (1981) 335.
17. M. Tinkham: Phys. Rev. B188 (1964) 967.
18. J. Villain: Physica 79B (1975) 1; in Ordering of strongly fluctuating condensed matter systems ed. Riste, Plenum Press, N.Y. (1980) p.81.
19. J. de Neef: J. Phys. Soc. Jpn. 37 (1974) 71.

A Raman Scattering Study of the Spin-Phonon Interaction in Rutile Structure Antiferromagnets

D.J. Lockwood[1] *and M.G. Cottam*[2]

[1]Division of Physics, National Reserach Council, Ottawa K1A 0R6, Canada

[2]Department of Physics, University of Essex, Colchester C04 3SQ, United Kingdom

1. Introduction

Light scattering studies of magnetic compounds provide spectral information on magnetic excitations such as the magnons and on lattice vibrations. In particular cases, the phonon and magnetic excitations show evidence of direct symmetry-allowed coupling, but more often the spin-phonon interaction is observed less obviously in the Raman spectrum of the phonons. The phonon intensity, frequency and linewidth may be modified by the exchange interaction between magnetic ions. A summary of the light scattering work on spin-phonon coupling in magnetic solids is given in COTTAM and LOCKWOOD [1]. Theories for the spin dependence of the zero wavevector optic phonon frequencies, and the corresponding Raman intensities, have been proposed by several authors, e.g. [2-4]. Much of the work is either phenomenological or has been applied specifically to ferromagnets of the EuO and $CdCr_2Se_4$ structures.

Rutile structure antiferromagnets such as MnF_2 (T_N = 68 K, S = 5/2) and FeF_2 (T_N = 78 K, S = 2) have been widely studied by many techniques, because of the simplicity of their magnetic ordering, and their magnetic excitations are by now well known. However, little is known of the spin-phonon interaction in these compounds from light scattering studies. There have been brief reports on the experimental results for some phonons in FeF_2 [5,6] and MnF_2 [7] and these have been interpreted in terms of simple phenomenological models developed for other compounds. Here we present preliminary results of a detailed experimental and theoretical investigation of the temperature dependences of the four Raman-active phonons in MnF_2 and FeF_2. Spin-phonon coupling effects are observed for all of these phonons, and we infer that the magnon scattering may also be influenced by this coupling [8].

2. Experimental Results for MnF_2 and FeF_2

The phonon Raman spectrum of MnF_2 was measured as a function of temperature from 4 to 300 K using the method described earlier [9]. The data were analysed to obtain the line parameters of peak frequency, integrated intensity and linewidth (full width half maximum) for each of the A_{1g}, B_{1g}, B_{2g} and E_g phonons. The results presented here for the phonons in FeF_2 were taken from the earlier publications [5,6].

Results obtained for the temperature dependences of the A_{1g} and E_g phonon frequencies are given in Fig. 1. Noticeable departures of the frequencies from their expected low-temperature behaviours are observable even for temperatures $T > T_N$. Note also that the relative shift in frequency of the A_{1g} mode in MnF_2 is opposite in sign to that of the FeF_2 A_{1g} mode. The results for the B_{2g} mode in MnF_2 resemble the A_{1g} mode behaviour in that crystal. The B_{1g} mode frequencies in MnF_2 [7] and FeF_2 [6] show an almost linear decrease in frequency with decreasing temperature for $T > T_N$ and then flatten out for $T < T_N$. This anomalous mode softening results from the lattice thermal contraction [6]. Intensity versus temperature results for the A_{1g} and E_g phonon intensities are given in Fig. 2.

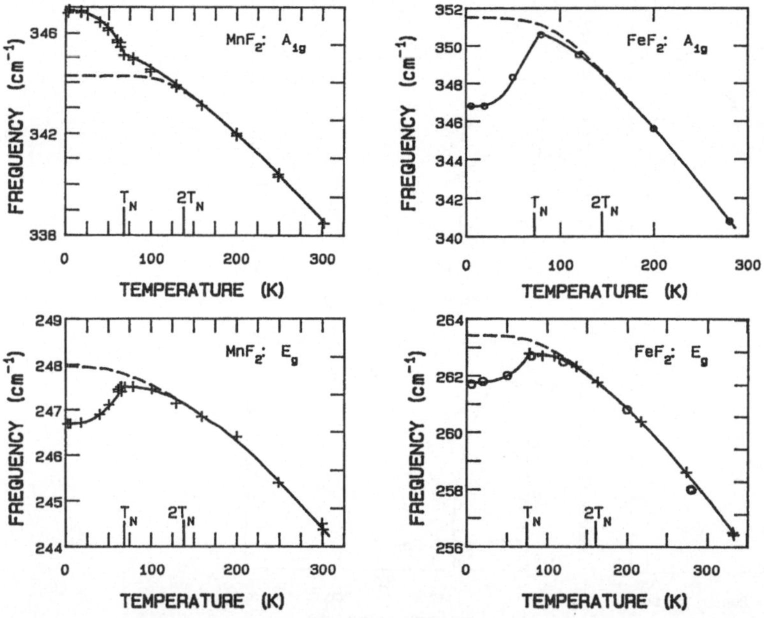

Figure 1. Frequency versus temperature data for the A_{1g} and E_g phonons in MnF_2 and FeF_2 (the data given by open circles are from [5]). The full curves are guides to the eye, while the broken curves represent the behaviour in the absence of spin-phonon coupling

Again there are noticeable renormalization effects for temperatures $T \lesssim T_N$, although in this case the renormalization is always upwards in intensity. The intensities of the B_{1g} and B_{2g} phonons in MnF_2 are much less sensitive to the magnetic ordering, whereas the B_{1g} phonon in FeF_2 shows a pronounced effect [6]. The phonon linewidths for the four modes in MnF_2 increase smoothly with increasing temperature and show no noticeable effects of magnetic exchange interactions.

3. Theory for Spin-Dependent Effects

The calculation of the spin dependence of the phonon Raman frequencies and integrated intensities can be formulated by analogy with previous work [2,3] on ferromagnetic systems. In rutile-structure antiferromagnets the exchange dependent term in the Hamiltonian can be represented as

$$\sum_{i,j} J_{ij}(\underline{r}^{(n)})\underline{S}_i \cdot \underline{S}_j$$

where i and j denote magnetic sites on opposite sublattices and J_{ij} is the dominant intersublattice exchange interaction. Because of superexchange effects, J_{ij} also depends on the position coordinates $\underline{r}^{(n)}$ (with n = 1,2,3,4) of the non-magnetic F^- ions in the unit cell. The displacement eigenvectors describing the four Raman-active phonon modes can be taken from KATIYAR [10]. Corresponding to the Brillouin zone centre (or Γ point) these particular eigenvectors involve only the coordinates $\underline{r}^{(n)}$ of the non-magnetic ions, and the magnetic ions are stationary. This leads to a modulation of J_{ij} by the Raman-active phonons, and by using a Taylor series expansion and expressing the coordinates $\underline{r}^{(n)}$ in terms of phonon variables, the appropriate spin-phonon coupling for rutile antiferromagnets can be derived. The result for the phonon frequency ω_p can eventually be written as

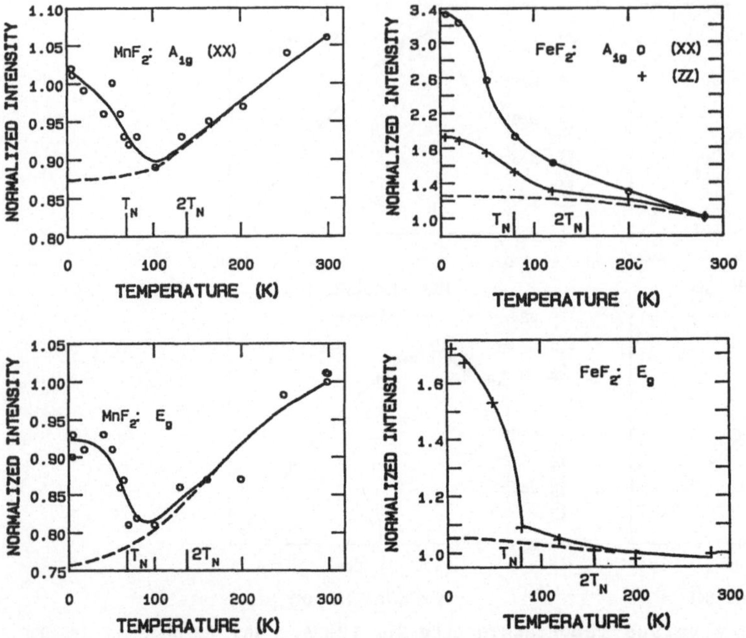

Figure 2. Integrated intensity versus temperature data for the A_{1g} and E_g phonons in MnF_2 and FeF_2. The curve notation is as given in Fig. 1. For the A_{1g} phonon, results are given for two polarizations (XX) and (ZZ), where X and Z are parallel to the crystal a and c axes, respectively. The FeF_2 data [5] have been corrected for the Bose population factor

$$\omega_p = \omega_p^0 + \lambda\langle \underline{S}_i \cdot \underline{S}_j \rangle$$

where ω_p^0 is the phonon frequency in the absence of spin-phonon coupling and $\langle \underline{S}_i \cdot \underline{S}_j \rangle$ denotes a thermal average for adjacent opposite-sublattice spins. The above result has an analogous form to that for ferromagnets [2], but in the present case we can relate the coefficient λ to parameters of the microscopic theory.

Calculations of the integrated intensity for Raman scattering from the phonons can likewise be carried out by expanding the polarizability tensor in terms of spin operators and the displacements of the F^- ions. The final result for the Stokes intensity I can once again be written in an analogous form to previous theory for ferromagnets [3,4] as

$$I = (n_p+1)[\,|A+B\langle \underline{S}_i \cdot \underline{S}_j \rangle|^2 + C^2\langle S^z \rangle^2]$$

where n_p is the Bose factor for the phonons, and A, B and C are constants (i.e. independent of spin averages).

4. Comparison and Discussion

The short-range order parameter $\phi(T) = |\langle \underline{S}_i \cdot \underline{S}_j \rangle|/S^2$ that enters into the theory of Sect. 3 must be evaluated for a detailed comparison of theory and experiment. In the above reduced form $\phi(T)$ will vary from 1, at $T = 0$ in the mean-field approximation, to 0, as $T \to \infty$. As a first estimate for $\phi(T)$, we have applied mean field theory for $T < T_N$, which gives $\phi(T) = (\langle S^z \rangle/S)^2$ with $\langle S^z \rangle$ calculated from a

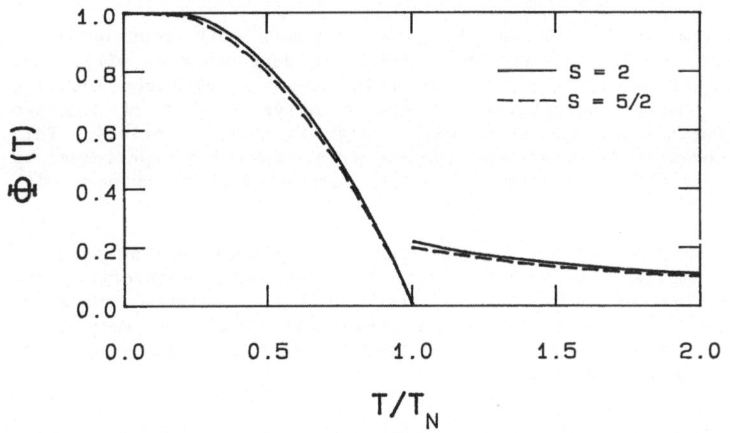

Figure 3. Numerical estimates for the short-range order parameter $\phi(T)$ for spin $S = 5/2$ and 2 in simple rutile-structure antiferromagnets

Brillouin function. Also we have used a modified two-spin cluster approach (e.g. see [11]) to obtain a simple estimate of the short-range order at $T > T_N$. The results of the numerical calculations for the two regimes $T < T_N$ and $T > T_N$ are given in Fig. 3, where it can be seen that in the reduced units the two spin cases give remarkably similar results. As expected, the mean field $\phi(T)$ goes to zero at T_N and so does not join continuously with the higher-temperature cluster result. However, from this knowledge of $\phi(T)$ at $T = 0$ and for $T > T_N$ we have deduced some spin-phonon coupling coefficients from the experimental results. As shown in Sect. 3, the frequency shift of the phonon from its expected behaviour due to the magnetic exchange interaction is $\Delta\omega_p(T) = \lambda\langle \underline{S}_i \cdot \underline{S}_j \rangle = -\lambda S^2 \phi(T)$. By estimating $\Delta\omega_p(0)$ and $\Delta\omega_p(2T_N)$ from the frequency versus temperature results and knowing $\phi(0)$ and $\phi(2T_N)$ from the numerical calculations we have obtained approximate values for the coupling coefficients λ, as given in Table 1.

Table 1. Spin-phonon coupling coefficients (in cm^{-1}) in rutile-structure antiferromagnets deduced from the frequency versus temperature behaviour

Compound	Phonon mode			
	A_{1g}	B_{1g}	B_{2g}	E_g
MnF_2	-0.4	-0.3	-0.3	0.2
FeF_2	1.3	-0.4	-	0.5

The values of λ are small compared with the phonon energies, but are not negligible, especially when compared with the magnon energies. The differences in magnitude of the values for λ for each phonon in MnF_2 and FeF_2, and between phonons of different symmetries, are not great. The main difference between MnF_2 and FeF_2 lies in the sign of λ for the A_{1g} phonon. In a fuller account of this work we shall interpret these features in terms of the parameters of the microscopic theory and the phonon eigenvectors. More accurate values of λ and of the detailed shape of the $\Delta\omega_p$ versus temperature curve will come from better estimates of the short-range order parameter, especially in the vicinity of T_N for $T < T_N$. A Green function analysis of $\phi(T)$ is being carried out for the whole temperature range, and further comparisons of theory and experiment will be presented elsewhere.

The intensity results of Fig. 2 are in general accord with what would be expected from the theory of Sect. 3 and the temperature dependence of $\phi(T)$ shown in Fig. 3. However, it is not feasible at this stage to evaluate coupling coefficients for the phonon intensities, as the relative weights of the two spin-dependent contributions are not known well enough in these materials. This is an area where further work is necessary. We are also conducting experiments on other rutile structure antiferromagnets to provide further data to compare with theory.

The fact that the Raman scattering from phonons is influenced by the magnetic exchange interactions implies, as a corollary, that the Raman scattering from magnons is likewise influenced by the spin-phonon coupling. Estimates using [8] suggest that the effect on the magnon Raman frequencies might be very small compared with exchange contributions, but it is possible that the influence on the magnon linewidth may be more pronounced.

References

1. M.G. Cottam and D.J. Lockwood: Light Scattering in Magnetic Solids (Wiley, New York 1986) p. 220
2. W. Baltensperger and J.S. Helman: Helvetica Physica Acta 41, 668 (1968)
3. N. Suzuki and H. Kamimura: J. Phys. Soc. Japan 35, 985 (1973)
4. E.F. Steigmeier and G. Harbeke: Phys. Kondens. Materie 12, 1 (1970)
5. J.L. Sauvajol, R. Almairac, C. Benoit and A.M. Bon: In Lattice Dynamics, ed. by M. Balkanski (Flammarion, Paris 1978) p. 199
6. D.J. Lockwood, R.S. Katiyar and V.C.Y. So: Phys. Rev. B28, 1983 (1983)
7. D.J. Lockwood: In Proc. IXth Int. Conf. Raman Spectroscopy (Chem. Soc. Japan, Tokyo 1984) p. 810
8. M.G. Cottam: J. Phys. C7, 2901 (1974)
9. D.J. Lockwood and M.G. Cottam: Phys. Rev. B35, 1973 (1987)
10. R.S. Katiyar: J. Phys. C3, 1087 and 1693 (1970)
11. J.S. Smart: Effective Field Theories of Magnetism (Saunders, Philadelphia 1966)

Raman Scattering Studies of Optical Magnons and of High-Energy Magnetic Excitons

P. Moch[1], *A.T. Abdalian*[1], *and B. Briat*[2]

[1]Laboratoire P.M.T.M., CNRS, Université Paris-Nord, avenue J.B. Clément, F-93430 Villetaneuse, France

[2]Laboratoire d'Optique Physique, ESPCI, 10 rue Vauquelin, Parix Cedex 05, France

1. Introduction

Since the pioneering work of FLEURY et al. /1/, Raman spectroscopy has revealed itself as a powerful tool for the study of magnetic excitations in magnetic insulators. However, there have been only a few studies concerning ferromagnets for the following reasons : first, such compounds are less common than antiferromagnets ; second, two-magnon scattering, which allows one to evaluate the exchange interactions, shows a negligibly small intensity in ferromagnets, in contrast to the case of antiferromagnets ; third, the magnon frequency at zero wave vector, which is essentially related to anisotropy, is generally too small to allow one-magnon Raman spectroscopy. On the other hand, it is well known /2/ that not only magnons but, more generally, magnetic excitons connected with transitions from the ground state to an excited single-ion level when the magnetic exchange is taken into account, propagate in such crystals and can be optically detected : up to now the extension of Raman studies to magnetic excitons other than magnons has been mainly restricted to measurements involving excitons related to an orbitally degenerate single ion crystal-field ground state /3,4/. Since inter-term transitions within a d^n configuration are generally allowed for Raman scattering while, in centro-symmetric crystals, they are forbidden for absorption or fluorescence, at least in an electric dipole process, Raman spectroscopy is expected to be a powerful method to derive information about the d^n excitonic manifold /5/ provided that a large enough Raman shift spectral range is accessible : such is the case when using an ultraviolet argon laser excitation, which allows one to measure Raman shifts up to 18000 cm^{-1}, a significant fraction of the $3d^n$ configuration.

In the following we present examples of Raman studies concerning the two above-mentioned fields : the peculiar canted structure of the planar ferromagnet Rb_2CrCl_4 gives rise to two distinct magnon branches and the corresponding one-magnon Raman spectra can be observed, allowing one to derive exchange and anisotropy constants. Various Ni and Co alkali-fluorides show high-energy exciton Raman lines, connected with the $^3A_{2g}\rightarrow{}^3T_{2g}$, $^3A_{2g}\rightarrow{}^3T^a_{1g}$ transitions for the $3d^8$ configuration of Ni^{2+} and with the $^4T^a_{1g}\rightarrow{}^4T_{2g}$ transition for the $3d^7$ configuration of Co^{2+}: we have selected some typical information resulting from the data.

2. Magnon scattering in Rb_2CrCl_4

The orthorhombic D_{2h}^{18} nuclear structure /6/ derives from a tetragonal one through a Jahn-Teller distortion of the $CrCl_6$ octahedra, two neighbouring units showing an elongation along the $\vec{a}$ and $\vec{b}$ tetragonal axes. The phonon Raman spectrum has been obtained and interpreted but will not be discussed here /7/. The appropriate spin Hamiltonian /8/, is :

$$H=-\bar{J}\sum_{ij:nn}\vec{S}_{i1}.\vec{S}_{j2}-\bar{P}\left[\sum_i S_{ia1}^2+\sum_j S_{jb2}^2\right]+\bar{D}\left[\sum_i S_{ic1}^2+\sum_j S_{jc2}^2\right]-g\mu\,\vec{B}.\left[\sum_i\vec{S}_{i1}+\sum_j\vec{S}_{j2}\right]\;;$$

i and j stand for the ith (j^{th}) ion on sublattices 1 and 2, respectively. $\bar{J}$ is the nearest neighbour (nn) exchange interaction between spins in the plane perpendicular to the tetragonal axis $\vec{c}$; $\bar{P}$ and $\bar{D}$ are the in-plane and the out-of-plane anisotropy parameters. The Zeeman term can be refined using an anisotropic Landé factor instead of g. The equilibrium spin orientation and the spin-wave dispersion curves can be derived from H. For $\vec{B}=\vec{0}$, the spins of the two sublattices lie within the plane perpendicular to $\vec{c}$ and are slightly canted from a strictly parallel orientation : one finds two spin-wave branches, an acoustic one and an optic one, with respective energies E_a and E_o at zero wave vector :

$$E_a=\left[2(P+2D)P^2/J\right]^{1/2}\;,\qquad \left(P=\bar{P}(S-1/2)/S\;;\;D=\bar{D}(S-1/2)/S\;;\;J=2\bar{J}\right)$$

$$E_o=16J+(P+2D)\;.$$

Due to the small values of P and D compared to J, E_o is close to 16J, the energy found at the zone boundary for a non-canted ferromagnet. The presence of two sublattices induces a folding of the magnon branch and, hence, an optic gap near 16J. We have observed this optic mode and its variation versus magnetic field for different directions of $\vec{B}$, in the Raman spectrum. For high enough fields the acoustic magnon line is also obtained (using an iodine cell to reabsorb the elastically scattered light). Figure 1 illustrates the results for the optic mode, which behaves in the way predicted by the theory.

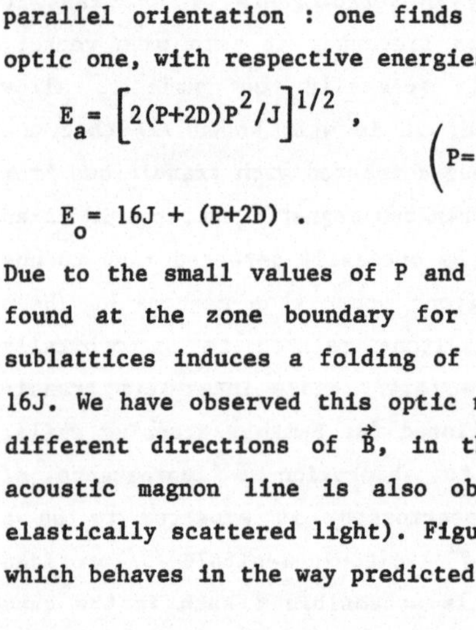

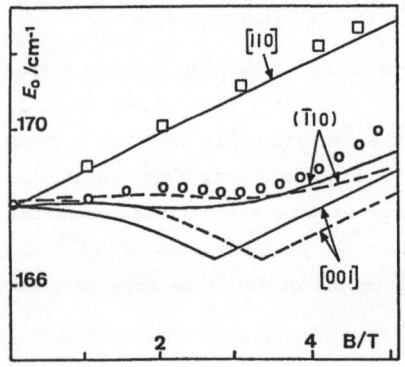

Figure 1 : Field dependence of E_o for various directions of $\vec{B}$: along [001]; 10° off this axis in the ($\bar{1}$10) plane ; along [110] (or [100]). J=10.42 cm^{-1}, g=2 ; two sets of parameters are used for P and D : P=1.65 cm^{-1}, D=0.22 cm^{-1} (straight lines) ; P=1.45 cm^{-1}, D=0 (dotted lines). Experimental results : □ ($\vec{B}$ along [110]) ; o ($\vec{B}$ along [001]).

From our data we derived an improved value of J(J=10.42 ± 0.02 cm^{-1}) and a reasonable estimate of (P+2D) (1.4 cm^{-1} at 8K). Moreover, we find a slightly anisotropic Landé factor with, presumably, $g_{\perp} > g_{//}$. Finally, using the general exciton theory /2/, it can be proved that, at $\vec{k}=\vec{0}$, the acoustic and optic magnons behave as γ_2^+ and γ_1^+ respectively, where the labelling of the representations refers to the appropriate C_{2h} unitary subgroup of the magnetic group of the crystal (for $\vec{B}=\vec{0}$). The selection rules for the polarised Raman spectra are then easily derived and agree with our experimental data, the optic mode, for instance, appearing in diagonal polarisation only.

3. High-energy excitons in Ni and Co alkali-fluorides

Among the crystals that we have studied, we have chosen to briefly present the cubic $KNiF_3$ and $KCoF_3$ perovskites and the tetragonal K_2NiF_4 structure, which show cubic crystal-fields close to each other /9-11/ (≃7000 - 8000 cm^{-1}) ; in this crystal-field the ground states are the orbital singlet $^3A_{2g}$ and the orbital triplet $^4T^a_{1g}$, for the $3d^8$ Ni^{2+} ion and for the $3d^7$ Co^{2+} ion respectively. The studied excitons involve the first three excited states $^3T_{2g}$ (≃7000 cm^{-1}), $^3T^a_{1g}$ (≃12000 cm^{-1}) and $^1E^a_g$ (≃15000 cm^{-1}) of Ni^{2+} and the first excited state $^4T_{2g}$ (≃7000 cm^{-1}) of Co^{2+}. As shown in Fig.2, at least for the spin-allowed transitions (ΔS=0), the Raman spectra at low temperature (≃10K) consist of one or several sharp exciton lines with, on their high-energy wing, a more or less intense broad vibronic band which will not be discussed here. The frequencies of the sharp lines and their selection rules for polarised spectra are given in Table 1. Theoretically /2/, the excitons have to be built from the single-ion sublevels split, according to the site symmetry, by the combined perturbing effects of the spin-orbit coupling, of the low symmetry crystal-field (when it exists), and of the diagonal part (molecular field) of the magnetic interaction. In fact, for high-energy excitons /8/, the calculations as well as the experimental results show that the

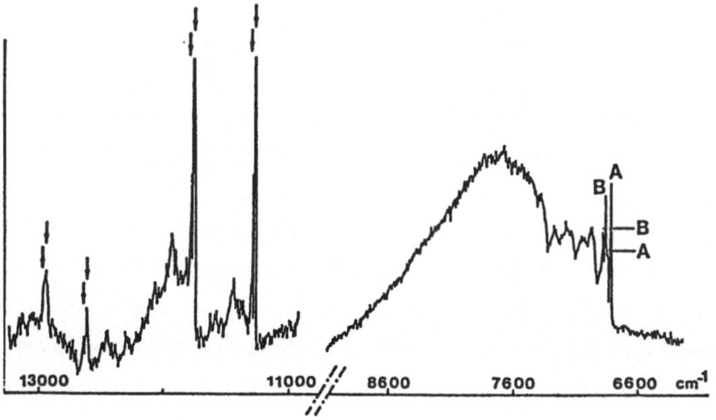

Figure 2 : High-energy exciton Raman spectrum of $KNiF_3$: overview showing the zero-phonon lines and the vibronic bands (T ≃ 10K ; unpolarised spectrum). Due to the simultaneous use of the 3511Å and 3514 Å exciting wavelengths each sharp line appears as a doublet

*Table 1 : Assignments of the observed zero-phonon excitonic Raman lines in $KNiF_3$, K_2NiF_4, $KCoF_3$. When the frequencies are not underlined the assignment is tentative. * : possibly phonon-assisted. The theoretically predicted selection rules are given when they could be used for the analysis of the data. z is the direction of the sublattice magnetisation (in the perovskite case it can be along any cubic axis)*

Co^{2+} $(3d^7)$

O_h	SO	C_{4h}	Selection rules	ν (cm^{-1}) $KCoF_3$
$^4T_{2g}$	Γ_6	γ_6	xz,yz	6650
		γ_5	xx,yy,zz,xy	6668
	Γ_8^a	γ_5	xx,yy,zz,xy	6675
		γ_7	xx,yy,xy	
		γ_8	xz,yz	6688
		γ_6	xz,yz	
	Γ_8^b	$\gamma_5,\gamma_6,\gamma_7,\gamma_8$		
	Γ_7	γ_7,γ_8		

Ni^{2+} $(3d^8)$

O_h	SO	C_{4h}	Selection rules	ν (cm^{-1}) $KNiF_3$	ν (cm^{-1}) K_2NiF_4
$^3T_{2g}$	Γ_3	γ_1	xz,yz	6803	6898
		γ_2	xz,yz	6842	
	Γ_4	$\gamma_3,\gamma_1,\gamma_4$		*6943-7006	*7114
	$\Gamma_5;\Gamma_2$	$\gamma_3,\gamma_2,\gamma_4;\gamma_2$		*7073-7116-7346	
$^3T_{1g}^a$	Γ_1	γ_1	xz,yz	11260	11574
	Γ_4	γ_2	xz,yz	11747-11751	11956
		γ_3,γ_4	xx,yy,zz,xy		12100
	Γ_3	γ_1,γ_2	xz,yz	12610	
	Γ_5	γ_3,γ_4	xx,yy,zz,xy	12936	
		γ_2	xz,yz		

single-ion approach is sufficient and that the off-diagonal interaction, which for low-energy excitons induces dispersion versus wave vector and in some cases Davydov splitting, has negligible effects. The antiferromagnetic ordering reduces the site symmetry of the magnetic ions from O_h to C_{4h} in $KNiF_3$ or $KCoF_3$ and from D_{4h} to (approximately) C_{4h} in K_2NiF_4. In the excited orbital multiplets the spin-orbit coupling is the leading perturbation : it is then convenient to obtain the spin-orbit components prior to the calculation of their further perturbations due to the tetragonal crystal-field and to the molecular field. A magnetic field $\vec{B}$ applied along the sublattice magnetisation generally induces a splitting which can be calculated, at least approximately, for each exciton, but $\vec{B}$ has negligible effects when it is perpendicular to the sublattice magnetisation. In the perovskite crystals the multidomain structure is affected by a magnetic field applied along a cubic axis and, as a result of this, for fields exceeding about 1T the sublattice magnetisation lies perpendicular to $\vec{B}$: consequently we found no shift of the Raman lines in these crystals but we were able to follow the displacements of the domain walls in the [0-2] T interval through the intensity variations of the polarised spectra (in $KNiF_3$). Conversely, in K_2NiF_4, with $\vec{B}$ applied along the easy $\vec{c}$ axis, we observed Zeeman splittings (Fig.3) helpful for identifying the excitons. Except

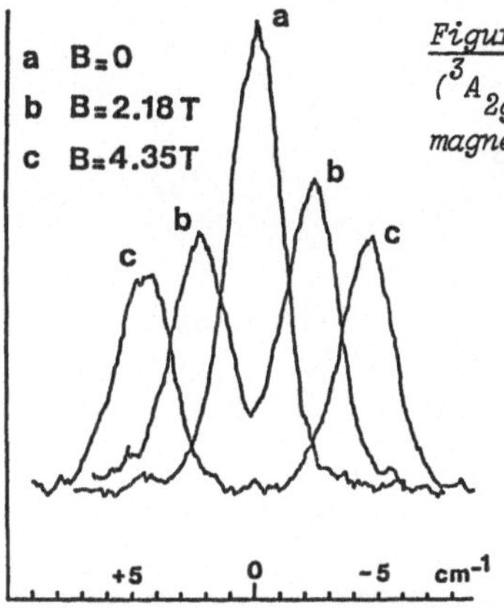

Figure 3 : Splitting of the 11574 cm^{-1} ($^3A_{2g} \rightarrow {}^3T^a_{1g}$) line in K_2NiF_4 under an applied magnetic field.

for the $^3A_{2g} \rightarrow {}^3T_{2g}$ transitions in Ni compounds /9,11-13/, which are magnetic dipole allowed for absorption, the excitonic frequencies reported in Table 1 are obtained for the first time. However, in some cases, it was possible to compare our results to absorption data involving phonon-assisted transitions /9,10,11,14/. A detailed analysis including Raman and absorption measurements, which will not be developed here /15,16/, allowed us to identify most of the lines (Table 1) and to draw conclusions about the exchange interaction approximated by an exchange field H'_E in each excited state. Concerning this last point, the most convincing determination is related to $^3T_{2g}$ in $KNiF_3$ where we derive $H'_E = 80cm^{-1}$ from the energy difference between the two observed prominent sharp Raman lines, a value about five times smaller than in the ground state. In this crystal we also found arguments for assuming that H'_E is smaller than 10 cm^{-1} in $^3T^a_{1g}$. Unfortunately, in K_2NiF_4 the large tetragonal crystal-field prevents the evaluation of H'_E. Finally, in $KCoF_3$, for H'_E in $^4T_{2g}$ we found a plausible value of 20 cm^{-1}, to compare to 150 cm^{-1} in the $^4T^a_{1g}$ ground state.

1. P.A. Fleury, S.P.S. Porto, L.E. Cheesman and H.J. Guggenheim: Phys. Rev. Letters 17, 84 (1966)
2. R. Loudon: Adv.Phys. 17, 243 (1968)
3. D.J. Lockwood, I.W. Johnstone, H.J.Labbe and B. Briat: J.Phys. C16, 6451 (1983)
4. J.P. Gosso, P. Moch, M. Quilichini, J.Y. Gesland and J. Nouet: J. Physique 40, 1069 (1979)
5. R.M. MacFarlane: Solid State Commun. 15, 535 (1974)
6. E. Janke, M.T. Hutchings, P. Day and P.J. Walker: J. Phys. C16, 5959 (1983)
7. A.T. Abdalian, B. Briat, C. Dugautier and P. Moch:to be published in J. Phys. C
8. M.T. Hutchings, J. Als Nielsen, P.A. Lindgård and P.J. Walker: J. Phys C14 5327 (1981)
9. J. Ferguson, H.J. Guggenheim and D.L. Wood: J. Chem. Phys. 40, 822 (1964)
10. J. Ferguson, D.L. Wood and K. Knox: J. Chem. Phys 39, 881 (1963)
11. J.S. Tiwari, A. Mehra and K.G. Srivastava: Japan. J. Appl. Phys. 7, 506 (1968); 7, 1227 (1968)
12. T. Bandai: J. Phys. Soc. Japan 50, 1538 (1981)
13. W. Kleemann and J. Pommier: Phys. Stat. Sol.(b) 66, 747 (1974)
14. J. Ferguson, E.R. Krausz and H.J. Guggenheim: Molec. Phys. 29, 1021 (1975)
15. A.T. Abdalian and P. Moch: J. Phys. C19, 7307 (1986)
16. A.T. Abdalian and P. Moch: to be published

Spin-Dynamics Studies of Excitations in *xy*-Chains

D.P. Landau[1], *R.W. Gerling*[1,2], *and M.S.S. Challa*[1,3]

[1]University of Georgia, Athens, GA 30602, USA
[2]Permanent address: Institut für Theoretische Physik der Universität Erlangen-Nürnberg, D-8520 Erlangen, Fed. Rep. of Germany
[3]Present address: Department of Physics and Astronomy, Michigan State University, East Lansing, MI 48824, USA

1. Introduction

Over the past decade, there has been increasing interest in the study of solitons in magnetic chains. Motivation for this interest has come from the beautiful neutron scattering studies of $CsNiF_3$ by Steiner and Kjems [1] and from the theoretical work by Mikeska and co-workers [2] who mapped the anisotropic Heisenberg chain onto the sine-Gordon equation. Extensive investigations using a variety of methods followed, but the situation was complicated by differences between physical systems and theoretical models, quantum corrections, and finite size effects [3]. We have been using computer simulation methods to study excitations in classical XY-chains in a symmetry-breaking field

$$H = J \sum_{(i,j)} (S_i^x S_j^x + S_i^y S_j^y) + h \sum_i S_i^x, \tag{1}$$

where the $\underline{S}_i$ are three component classical vectors of length $|\underline{S}_i| = 1$. From Monte Carlo simulations, we have observed equilibrium distributions of solitons but could not detect contributions to the specific heat [4]. (Monte Carlo studies of the anisotropic Heisenberg chain [5] showed specific heat peaks but the field dependence differed markedly from theoretical predictions.) In this paper we shall present the first results obtained using a new vectorized spin-dynamics method on the Cyber 205.

2. Method

An important sampling Monte Carlo technique was used to generate equilibrium configurations at a given temperature T and field h. A chain, typically of length L = 20,000, with a periodic boundary was divided into two interpenetrating next-nearest-neighbor sublattices and spin updates were performed on all spins in each sublattice using vectorized code. (This is the 1-dimensional analog of "checkerboard" algorithms in higher dimensions.) The first 3000 MCS/site were discarded to insure that the system had reached equilibrium and then 10 spin configurations, each separated by 200 MCS/site were retained as starting configurations for the spin dynamics calculation. The equation of motion for the ith spin is

$$\dot{\underline{S}}_i = \underline{S}_i \times [J(S_{i+1}^x + S_{i-1}^x)\hat{e}_x + J(S_{i+1}^y + S_{i-1}^y)\hat{e}_y + h\hat{e}_x], \tag{2}$$

where $\hat{e}_x$, $\hat{e}_y$, $\hat{e}_z$ are unit vectors in the x, y, and z directions respectively. The coupled equations of motion for each spin were then integrated using a 4th oder predictor-corrector method and time interval $\Delta = 0.01/J$. Since this method is not self-starting the first few time intervals were obtained using a 4th order Runge-Kutta method. Each starting configuration was followed out to time $t_{max} = 100/J$; time-displaced space-displaced correlation functions were determined and then averaged together over the 10 chains. Fourier transforms of these correlation functions over space and time were calculated to yield the scattering factor $S(q,\omega)$. To eliminate spurious effects due to the time and space cutoffs, a Gaussian "resolution function" was included in the Fourier transforms giving rise to an "instrument" contribution to the linewidth.

3. Results

The scattering function with polarization along the chain axis $S_z(q,\omega)$ shows very clear spin-wave peaks. For $kT = 0.1J$, $h = 0.1J$ the linewidth for $q = 0$ is essentially the same as the instrumental width and becomes larger by about a factor of 2 as q increases. The linewidth inceases by a factor of 4 or more as T is increased to 0.5J, and for still higher temperature the spin-wave peaks are even more washed out. The most interesting results are expected to be those for $S_y(q,\omega)$ which are predicted to show central peaks due to solitons without contributions from two-magnons [2]. In Fig. 1 we show results

Fig. 1.

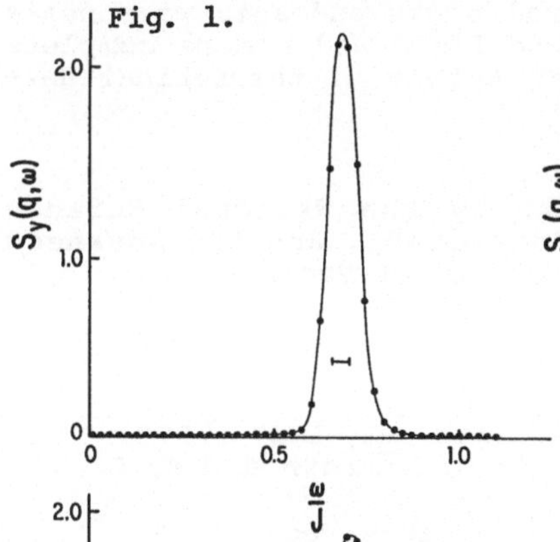

Fig. 2.

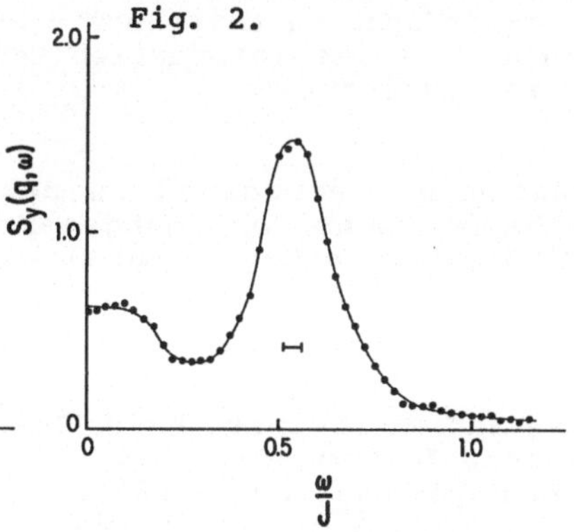

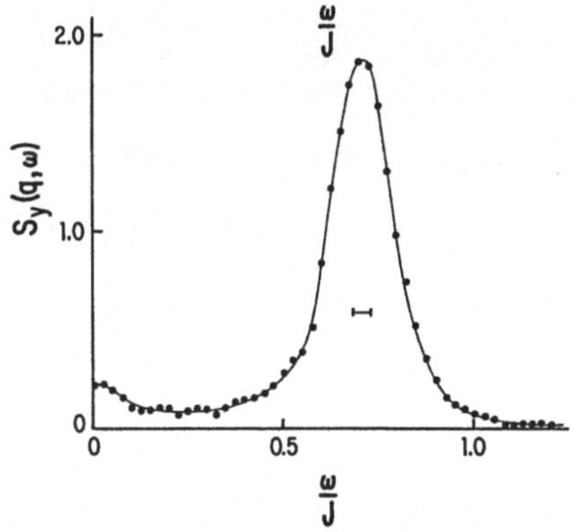

Fig. 3.

Fig. 1. $S_y(q,\omega)$ as determined for $kT = 0.1J$, $h = 0.1J$. The smooth curve is a line to guide the eye. The "instrumental" width is shown schematically by the bar.

Fig. 2. $S_y(q,\omega)$ as determined for $kT = 0.4J$, $h = 0.1J$. The smooth curve is a line to guide the eye. The "instrumental" width is shown schematically by the bar.

Fig. 3. $S_y(q,\omega)$ as determined for $kT = 0.4J$, $h = 0.2J$. The smooth curve is a line to guide the eye. The "instrumental" width is shown schematically by the bar.

for $kT = 0.1J$, $h = 0.1J$ for $qa = \pi/8$. Only a single spin-wave peak at $\omega/J = 0.685$ is visible, and it has a linewidth which is is about twice the instrumental width. For $kT = 0.4J$ and $h = 0.1J$, the data look quite different. As shown in Fig. 2, a very pronounced central peak is present in addition to the spin-wave peak. The spin-wave peak is shifted down in frequency to $\omega/J = 0.525$, its linewidth is just over double that observed for $kT = 0.1J$, and the integrated intensity of the peak is almost 50% higher for $kT = 0.4J$. The most interesting feature of Fig. 2 is the rather broad central peak. Unlike the spin-wave peak, which is quite smooth, the central peak still possesses remnants of the oscillations produced by the finite time cutoff of the time integration. When the field is increased to $h = 0.2J$, the central peak in $S_x(q,\omega)$ for $qa = \pi/8$ becomes dramatically smaller while the spin-wave peak shifts to larger ω and its integrated intensity increases by 30% (see Fig. 3).

4. Conclusions

Using a new ultrafast vectorized program on the Cyber 205, we have seen clear central peaks in $S_y(q,\omega)$ for XY-chains with a magnetic field in the x-direction. Both the central peaks and spin-wave peaks show a clear dependence on temperature and field and a more complete study will provide ample information for a test of theoretical descriptions of the excitations.

This research was supported in part by the National Science Foundation, the Deutsche Forschungsgemeinschaft, and the Advanced Computational Methods Center at the University of Georgia.

References

1. J. K. Kjems and M. Steiner, Phys. Rev. Lett. 41, 1137 (1978)
2. H. J. Mikeska, J. Phys. C11, L29 (1978); E. Allroth and H. J. Mikeska, Z. Phys. B43, 209 (1981)
3. See e.g., Magnetic Excitations and Fluctuations, eds. S. W. Lovesey, U. Balucani, F. Borsa, and V. Tognetti (Springer-Verlag, Berlin, 1984)
4. R. W. Gerling and D. P. Landau, J. Mag. Magn. Mater. 45, 267 (1984) and to be published
5. O. G. Mouritsen, H. Jensen, and H. C. Fogedby, Phys. Rev. B30, 498 (1984)

Numerical Transfer-Kernel Results for the Classical *xy*-Chain in a Magnetic Field

R.W. Gerling, T. Delica, and H. Leschke*

Institut für Theoretische Physik I, Universität Erlangen-Nürnberg, D-8520 Erlangen, Fed. Rep. of Germany

1. Introduction

In the last decade the interest in one-dimensional magnetic systems has been great. The reason for this is two-fold: (i) there exist beautiful experimental results [1-6] for quasi one-dimensional systems and (ii) these systems are theoretically much easier to treat than the corresponding two- or three-dimensional models.

We study in the present paper the thermodynamic properties of the classical xy-chain. This model was extensively studied by means of computer simulations [7]. Here we use the (principally exact) transfer kernel method [8]: the partition function and correlation functions are calculated from the eigenvalues and eigenfunctions of an integral kernel which is constructed from the Hamilton function. For zero field JOYCE [9] was able to calculate the eigenvalues and eigenfunctions analytically. For nonzero field the transfer kernel is approximated by a transfer matrix which is then diagonalised numerically.

The results of this method are compared and contrasted with results of Monte Carlo simulations for the very same system.

2. Method

The Hamilton function of the xy-model reads

$$H = - J \sum_{i=1}^{N} (S_i^x S_{i+1}^x + S_i^y S_{i+1}^y) - h \sum_{i=1}^{N} S_i^x, \tag{1}$$

where the $\underline{S}_i$ are classical three-dimensional vectors of unit length. J is the exchange coupling constant and $h=g\mu_B H$ is a magnetic field which for simplicity is measured in energy-units. With the symmetric function

$$h(\underline{S},\underline{S}') := J(S_i^x S_{i+1}^x + S_i^y S_{i+1}^y) - h/2(S_i^x + S_{i+1}^x) \tag{2}$$

the Hamilton function (1) can be rewritten as a sum over symmetric pair-interaction functions $h(\underline{S},\underline{S}')$. Assuming a cyclic boundary $\underline{S}_{N+1}:=\underline{S}_1$ and using the transfer kernel

$$K(\underline{S},\underline{S}') := \exp(-\beta h(\underline{S},\underline{S}') > 0 \tag{3}$$

we can write the partition function for the N spin system as

*also: Center for Simulational Physics
University of Georgia, Athens, GA 30602, USA

$$Z_N := \int \Big(\prod_{i=1}^{N} d\underline{S}_i \Big) \prod_{i=1}^{N} K(\underline{S}_i, \underline{S}_{i+1}) = \int d\underline{S}\, K^{(N)}(\underline{S},\underline{S}) =: \mathrm{Tr}(K^{(N)}), \tag{4}$$

where we have introduced the N-th iterate

$$K^{(N)}(\underline{S},\underline{S}') := \int d\underline{S}_2 \dots d\underline{S}_N K(\underline{S},\underline{S}_2) \dots K(\underline{S}_N,\underline{S}') \tag{5}$$

of the transfer kernel. Using the spectral resolution

$$K^{(N)}(\underline{S},\underline{S}') = \sum_{n=0}^{\infty} \lambda_n^N \psi_n(\underline{S}) \psi_n^*(\underline{S}') \tag{6}$$

where the λ_n and the ψ_n are the eigenvalues and eigenfunctions respectively, we get for the partition function

$$Z_N = \mathrm{Tr}(K^{(N)}) = \sum_{n=0}^{\infty} < \psi_n | K^{(N)} \psi_n > = \sum_{n=0}^{\infty} \lambda_n^N . \tag{7}$$

In this sum the main contribution for large N stems from the largest eigenvalue λ_0, which is according to a theorem of Jentzsch [10] positive and not degenerate. Therefore we get in the thermodynamic limit for the free energy of the system

$$F = \lim_{N\to\infty} -(N\beta)^{-1} \ln(Z_N) = -\beta^{-1} \ln(\lambda_0). \tag{8}$$

Within the same framework we can express expectation values of functions of the spin components, e.g. the correlation functions $<S_i^k S_{i+1}^k>$ in terms of the eigenfunctions and the eigenvalues of the integral kernel [8].

To approximate an integral equation of the type

$$\int d\underline{S}' K(\underline{S}|\underline{S}') \psi(\underline{S}') = \lambda \psi(\underline{S}) \tag{9}$$

by a matrix eigenvalue equation we have to approximate the integral by an appropriate integration formula

$$\int d\underline{S}\, f(\underline{S}) = \sum_{i=1}^{L} w_i f(\underline{S}_i) \tag{10}$$

where the $\underline{S}_i$ are the points and the w_i are the weight factors of the integration formula.

Replacing the eigenfunction on both sides of the integral equation (9) by vectors $\underline{\psi}$ whose i-th component is defined by $\psi_i = (w_i)^{\frac{1}{2}} \psi(\underline{S}_i)$, i=1,...,L, we get the matrix eigenvalue equation

$$\sum_{i=1}^{L} M_{i,j} \psi_i = \lambda \psi_j \tag{11}$$

with the symmetric matrix

$$M_{i,j} := (w_i)^{\frac{1}{2}} K(\underline{S}_i | \underline{S}_j) (w_j)^{\frac{1}{2}}. \tag{12}$$

The magnetic field in the easy plane breaks the rotational symmetry in the xy-plane so that the one-variable integration formula (10) cannot be used. We have to use a two-variable integration formula. In addition we have to perform an integration on the three-dimensional unit sphere. A good integration formula for this problem is the so-called McLaren formula [11]: it uses 72 points on the unit sphere. The points are the corners of rotated tetrahedron, octahedron and icosahedron. The great advantage of the McLaren formula is the fact that the points are equally distributed over the unit sphere.

3. Results

To test the quality of our algorithm we have calculated the zero-field results (which are known exactly from JOYCE [9]) for the specific heat and the entropy using the McLaren formula. The results for zero magnetic field are shown in Fig. 1. Above about $k_BT = 0.08|J|$ the results agree perfectly. Below this temperature there is quite a significant difference. In the specific heat we see a clear spurious maximum and in the entropy there exists a plateau. Both effects can be understood as an artefact of the numerical approximation. By using a discretisation of the integral we introduce discrete but classical spins.

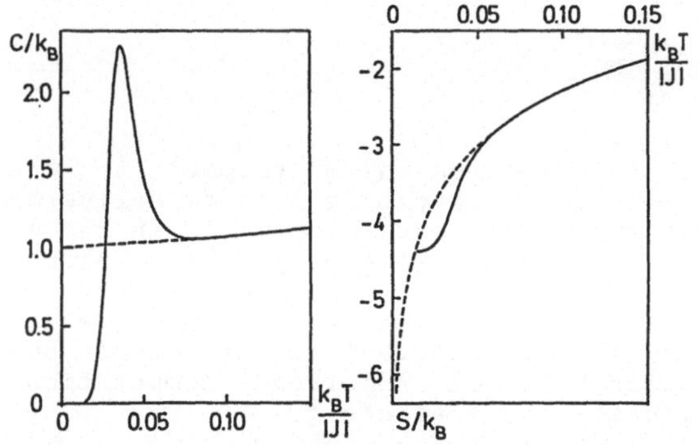

Figure 1: Comparison of the exact results (dashed lines) for the specific heat and the entropy for zero field with the corresponding results (solid lines) using the McLaren formula to discretise the integrals in the integral equation

This effect shows up in all quantities which are related to fluctuations of the system i.e. second derivatives of the free energy. With increasing magnetic field the effect is shifted to higher temperatures. For small temperatures the system occupies the groundstate (or better the discrete state with the lowest energy) and it does not fluctuate around it, because the thermal energy is not large enough to allow the spins to reach the next of the discrete points on the unit sphere. If the thermal fluctuations become large enough, the system starts to occupy these states like an avalanche. From this one can understand the field dependence of the effect: the field aligns the spins parallel and therefore suppresses the fluctuations.

However, this method can be easily handled and it gives results quickly. This fact is of importance for the calculation of correlation functions, where all eigenfunctions are involved.

For low temperatures and high fields we used a more sophisticated method [12,13] to get useful results. By an expansion of the origi-

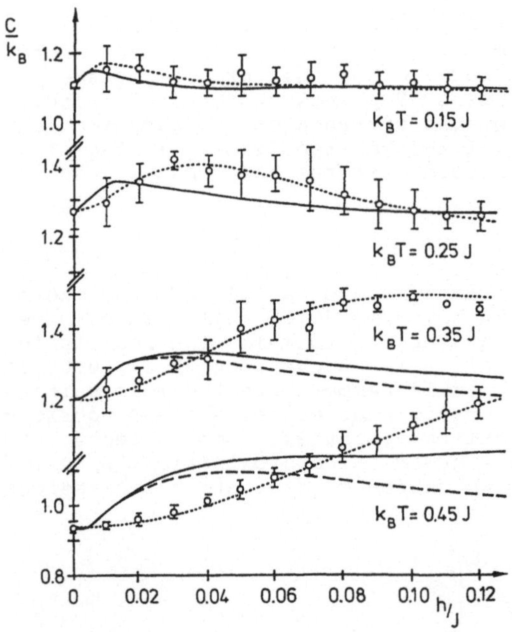

Figure 2: The specific heat of a ferromagnetic coupled chain as a function of magnetic field for different temperatures as indicated. The datapoints are from a Monte Carlo simulation [7]. The dotted line is the result from the transfer kernel method. The dashed line is the prediction of the soliton contribution from the sine-Gordon theory. The solid line uses the expression for the specific heat from the sine-Gordon theory but the density is taken from a fit of Monte-Carlo results for the soliton density.

nal integral equation in the (θ,ϕ)-angles with respect to the functions $(2\pi)^{-\frac{1}{2}}\exp(im\phi)$ ($m= 0, \pm1, \pm2, \ldots$) one is able to derive an equivalent but infinite system of coupled integral equations in the ϕ-angles only. In practice it is sufficient to take only a small number of expansion coefficients into account. The remaining system of integral equations can be approximated by a matrix eigenvalue equation using standard Gauss formulas. In contrast to the method described first this method includes a lot of numerical computation. Therefore we used it only to calculate the free energy.

Figure 2 shows results for the specific heat for different temperatures for the ferromagnetic coupled chain. The results of the Monte Carlo calculations of GERLING and LANDAU [7] agree perfectly with the result from the transfer-kernel calculation (dotted line). The dashed line is the analytic result from the sine-Gordon model. From earlier work we know that in the Monte Carlo simulation the density of solitons is higher than predicted by the sine-Gordon theory because out-of-plane fluctuations are treated properly. The solid line was gained by using the result of a fit of the Monte Carlo soliton density in the sine-Gordon expression for the specific heat. For the two lowest temperatures no difference can be seen. For higher temperatures it increases the specific heat, but it does not shift the maximum of the specific heat to much higher fields. For temperatures lower then $k_BT=0.15J$ the sine-Gordon theory seems to produce the correct specific heat.

4. Conclusion

The transfer-kernel method is a powerful tool to calculate the thermodynamic properties of magnetic systems. It has clear advantages over the Monte Carlo method. The Monte Carlo simulation is nevertheless the only method which gives direct access to a single spin configuration; and from looking directly at spin configurations, one can learn quite a lot about the physics of the system.

We could demonstrate that the sine-Gordon theory describes for low temperatures the qualitative features of the one-dimensional xy-system. To get a quantitative description one has definitively to include the discreteness of the system, the out-of-plane fluctuations and the interaction with the spin-waves.

This research was supported in part by the Deutsche Forschungsgemeinschaft. We thank D.P. Landau for stimulating discussions and fruitful advice.

5. References

1. M. Steiner, J. Villain and C.G. Windsor, Adv. in Physics 25, 87 (1976); M. Steiner, K. Kakurai and J.K. Kjems, Z. Phys. B53, 117 (1983)
2. A.P. Ramirez and W.P. Wolf, Phys. Rev. Lett. 49, 227 (1982)
3. J.P. Boucher, L.P. Regnault, J. Rossat-Mignot, J.P. Renard, J. Bouillot and W.G. Stirling, J. Appl. Phys. 52, 1956 (1982); J.P. Boucher, L.P. Regnault, A. Pires, J. Rossat-Mignot, Y. Henry, J. Bouillot, W.G. Stirling and J.P. Renard, In Magnetic Excitations and Fluctuations, ed. by S.W. Lovesey, U. Balucani, F. Borsa and V. Tognetti (Springer, Berlin, Heidelberg 1984)
4. F. Borsa, Phys. Rev. B25, 3430 (1982); F. Borsa, M.G. Pini, A. Rettori and V. Tognetti, Phys. Rev. B28, 5173 (1983)
5. K. Kopinga, A.M.C. Tinus, and W.J.M. de Jonge, Phys. Rev. B29, 2868 (1984), A.M.C. Tinus, W.J.M. de Jonge and K. Kopinga, Phys. Rev. B32, 3154 (1985)
6. T. Goto and Y. Yamaguchi, J. Phys. Soc. Jap. 50, 2133 (1981)
7. R.W. Gerling, Habilitations thesis, Universität Erlangen-Nürnberg (1986); R.W. Gerling and D.P. Landau, to be published
8. T. Delica, Diplom thesis, Universität Erlangen-Nürnberg (1986); T. Delica, R.W. Gerling and H. Leschke, Phys. Srip. 35, 57 (1987)
9. G.S. Joyce, Phys. Rev. Lett. 19, 581 (1967)
10. K. Jörgens, Lineare Integraloperatoren (Teubner, Stuttgart 1970)
11. A.D. McLaren, Math. Comp. 17, 361 (1963); A.H. Stroud, Approximate Calculation of Multiple Integrals, (Prentice-Hall, New Jersey 1971)
12. M. Blume, P.Heller and N.A. Lurie, Phys. Rev. B11, 4483 (1975)
13. F. Boersma, W.J.M. de Jonge and K. Kopinga, Phys. Rev. B23, 186 (1981)

Monte Carlo Simulation of the Quantum Sine-Gordon Chain

S. Wouters and H. De Raedt

Physics Department, University of Antwerp,
Universiteitsplein 1, B-2610 Wilrijk, Belgium

I. INTRODUCTION

One-dimensional (1D) lattice-dynamical models exhibiting soliton-like features have been studied by a variety of methods. In particular the sine-Gordon (sG) and ϕ^4 models have been the subject of much theoretical research [1 − 15]. The former is also relevant to experiments on magnetic chain materials [16 − 24]. Most theoretical studies have been carried out within the framework of classical statistical mechanics. As these 1D models contain up to nearest-neighbor interactions only, the transfer matrix method has been employed to compute numerically exact results [2]. Partially motivated by significant discrepancies between experiment and classical theory [25], the question to what extent quantum effects are important has been addressed. The Bethe-Ansatz solution of the *continuum* sG chain, [4,10] semi-classical approximation schemes [5,7,14] as well as variational path-integral [9,13] calculations all indicate that there remains a quantitative disagreement between theory and experiment which may be due to the fact that the sG model neglects the out-of-plane motion of the spins. Recent Quantum Monte Carlo simulations and transfer-matrix calculations suggest that the S=1/2 model, assumed to be applicable to CHAB [$(C_6H_11NH_3)CuBr_3$], cannot adequately reproduce experimental facts [26].

In view of the vast amount of knowledge on these 1D lattice-dynamical models it is of interest to compare results of Quantum Monte Carlo (QMC) simulations with predictions of theoretical methods. In this paper we confine ourselves to a discussion of the lattice sG model defined by the Hamiltonian [2]

$$H = Da \sum_{l=1}^{L} \left[\frac{p_l^2}{2D^2} + \omega_0^2 (1 - \cos y_l \,) + \frac{c_0^2}{2a^2} (y_{l+1} - y_l \,)^2 \right] \quad , \tag{1}$$

where y_l is the displacement at site l, p_l its conjugate momentum, a is the lattice constant, c_0 and ω_0 are the characteristic velocity and frequency respectively, and D sets the energy scale. In our numerical work we will set $D = 1$, $a = 1$, $\hbar = 1$, measure the temperature in units of the classical soliton rest-energy $E_S = 8D\omega_0 c_0$, i.e. $t \equiv k_B T / E_S$, and express the strength of anharmonicity and quanticity in terms of the parameters [8,9] $R = c_0/\omega_0$ and $Q = \omega_0 / E_S$.

Of particular interest is the effect of the non-linear term in (1) on the thermodynamic properties. Kink and breather solitons give rise to extra contribution in for instance the specific heat. Experiments on 1D magnetic materials have demonstrated the existence of such contributions [19,21,22,23]. From theoretical point of view a proper treatment of these effects requires a non-perturbative technique. One of the few non-perturbative methods for calculating the thermodynamic functions of quantum system (1) is based on Monte Carlo simulation.

II. QUANTUM MONTE CARLO METHOD

Our aim is to construct a QMC method that *directly* yields the relevant anharmonic contributions to the thermodynamic properties. We have emphazised the term "directly" because it is essential in the following sense. According to the standard procedure for setting up a QMC scheme [27] one would start by discretizing the Feynman path-integral expression [28] for the partition function Z and perform the integration over the multi-dimensional configuration space by Monte Carlo sampling. However for the problem at hand, this naive approach is bound to fail. To obtain the anharmonic contribution two, possibly large, numbers have to be substracted and as these numbers are subject to statistical uncertainty numerical precision is lost, rendering this approach hopelessly inefficient.

The seriousness of the problem is nicely illustrated by considering a free particle. After discretizing the Feynman path-integral the expression for the energy reads

$$E_m = \frac{m}{2\beta} - \frac{m}{2\beta^2}\sum_{j=1}^{m}\langle\langle(x_j - x_{j+1})^2\rangle\rangle \quad , \tag{2}$$

where m stands for the number of "imaginary-time" steps used to discretize the path-integral, and $\langle\langle\ldots\rangle\rangle$ denotes the average taken with respect to a multi-dimensional Gaussian distribution. In this particular case we know the exact answer $E_m = E = 1/2\beta$. However computing E_m via Monte Carlo implies that, according to (2) we will have to subtract two numbers of $\mathcal{O}(m)$ to obtain a result of $\mathcal{O}(1)$. Except when $m = 1$ a significant amount of precision is lost in this manner. Although we have illustrated this fundamental problem in QMC by using the free particle as an example, the conclusion is valid whenever a simple discretization of the Feynman path-integral is employed.

To circumvent this difficulty, we decompose Hamiltonian (1) into harmonic and anharmonic part and exploit the fact that the exact propagator of a single harmonic oscillator is known [28]. It is then possible to show that the partition function can be written as

$$Z = Z_a \lim_{m\to\infty} Z_m \quad , \tag{3a}$$

where Z_a is the *exact* partition function of the harmonic system,

$$Z_m \equiv \int \prod_{j=1}^{m}\prod_{l=1}^{L} dx_{l,j}\, \exp\left[-x_{l,j}^2 - \tau\omega_0^2(1-\cos y_{l,j})\right] \tag{3b}$$

contains all anharmonic effects, and $y_{l,j} = y_{l,j}(x_{1,1},\ldots,x_{L,m})$ is related to the $x_{l,j}$ via a 2D Fourier transform. In practice this transform is carried out by means of the Fast Fourier Transform algorithm in order to reduce the number of operations per Monte Carlo step from $\mathcal{O}(N^2)$ to $\mathcal{O}(N\log N)$. The major advantage of this formalism is that harmonic and anharmonic contribution have been separated analytically, i.e. without loss of accuracy. Note that we have refrained from writing $1-\cos y = 1-\cos y - y^2/2 + y^2/2$ and absorb the last term in the harmonic part, as is usually done [4, 8, 9, 13]. The reason is that by doing so, a basic symmetry of (1) is broken, namely its invariance for translations over 2π. Although this symmetry is restored in the limit $m \to \infty$, for all practical purposes m is finite (and as small as possible) and breaking this symmetry results in very large statistical fluctuations in the Monte Carlo data.

III. SIMULATION RESULTS

The anharmonic contribution to the specific heat δC is obtained from the knowledge of

$$C_m \equiv \beta^2 \frac{\partial^2}{\partial \beta^2} \ln \langle\langle \exp[-\tau\omega_0^2 \sum_{l=1}^{L}\sum_{j=1}^{m}(1-\cos y_{l,j})]\rangle\rangle \quad . \tag{4}$$

The quantity in brackets is computed by generating sets of normally distributed random numbers. The final result is then corrected for the harmonic contribution hidden in $\cos y$. In all our calculations $m = 2$ was found to be sufficient to guarantee convergence with respect to m. In QMC simulations of other models a much larger value of m is usually required. There are two reasons why this is not the case for the sG model. By construction our approach yields exact results if $\omega_0 = 0$, independent of the temperature. This has to be compared to the conventional QMC approach where m has to be large in order to recover the exact low-temperature results of the harmonic system [27]. Moreover for the range of parameters relevant to experiment the ω_0 is small compared to c_0/a.

Some preliminary results for $R\delta C$ are shown in Fig.1. The data points represent averages of at least 20000 samples. It has been verified that the length of the chain $L = 256$ is sufficient to eliminate boundary effects.

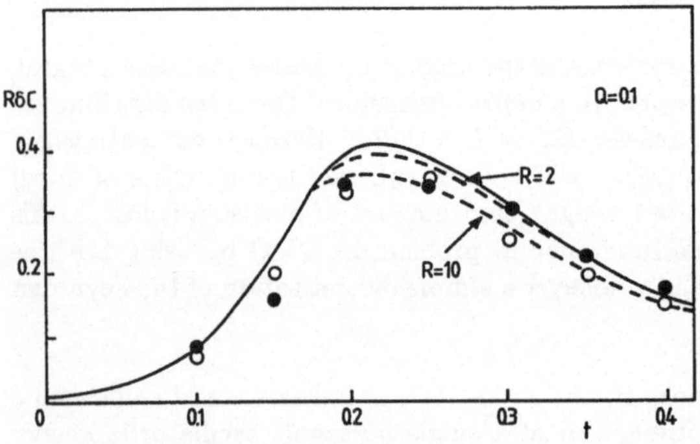

Fig.1. Non-linear contribution to the specific heat of the sine-Gordon chain. Solid dots: QMC data for $R = 2.9$. Open circles: QMC data for $R = 10$. Dashed lines: variational path-integral calculation. [9,13] Full line: exact transfer-matrix results for the classical sG chain [2]

The overall agreement with the variational path-integral treatment [9,13] is satisfactory. Comparison with a semi-classical theory [7] learns that there is a disagreement with respect to the shift of the peak position, relative to the classical result. The reduction of the specific heat found by another semi-classical treatment [14] is substantially larger than in our case. As shown in Fig.2, our results for $R = 2.9$ are also in good agreement with the Bethe-Ansatz solution of the sG model in the continuum limit [4], provided we set

$$\frac{\pi}{\mu} = \frac{2}{\pi}\cos^{-1}\left(-\frac{1}{2R}\right) \quad , \tag{5a}$$

and express the temperature in units of the quantum-renormalized soliton energy E_s^* given by [3]

$$\frac{E_s}{E_s^*} = \frac{n}{n-1}(8R)^{1/(n-1)} \quad , \quad \text{where [4]} \tag{5b}$$

$$\frac{\mu}{\pi} = \frac{n-1}{n} \quad , \tag{5c}$$

yielding for $R = 2.9$, $\mu/\pi \approx 9/10$ and $E_s/E_s^* \approx 1.58$.

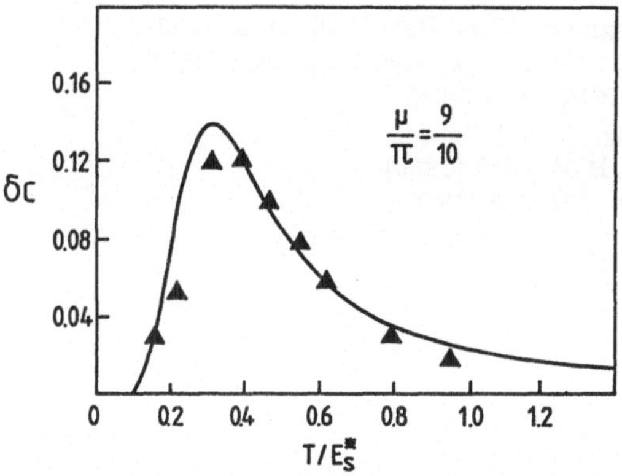

Fig.2. Non-linear contribution to the specific heat of the sine-Gordon chain. Solid symbols: QMC data for $R = 2.9$. Full line: Bethe-Ansatz results for the continuum sG chain [4]

This indicates that although the width of the classical kink ($R = 2.9$) is of the order of the lattice spacing, the continuum approximation to the lattice sG model is quite good.

Acknowledgement

This work is partially supported by the "Supercomputer Project" of the Belgian National Science Foundation (N.F.W.O.) and NATO grant no. RG.86/0026. One of us (H.D.R) thanks the N.F.W.O. for financial support.

REFERENCES

1. H.J. Mikeska: J. Phys. C 11, L29 (1978).
2. T. Schneider, and E. Stoll: Phys. Rev. B 22, 5317 (1980).
3. K. Maki: Phys. Rev. B 24, 3991 (1981).
4. M. Fowler, and X. Zotos: Phys. Rev. B 25, 5806 (1982).
5. H.J. Mikeska: Phys. Rev. B 26, 5213 (1982).
6. M. Imada, K. Hida, and M. Ishikawa: J. Phys. C 16, 35 (1983).
7. T. Tsuzuki: Prog. Theor. Phys. 70, 975 (1983).
8. M. Moraldi, A. Rettori, and M.G. Pini: Phys. Rev. A 31, 1971 (1985).
9. R. Giachetti, and V. Tognetti: Phys. Rev. Lett. 55, 912 (1985).
10. M.D. Johnson, and N.F. Wright: Phys. Rev. B 32, 5798 (1985).
11. J. Timonen, M. Stirland, D.J. Pilling, Yi Cheng, and R.K. Bullough: Phys. Rev. Lett. 56, 2233 (1986).
12. T. Schneider: in Solitons, edited by S.E. Trullinger, V.E. Zakharov, and V.L. Pokrovsky, (North-Holland, Amsterdam, 1986).
13. R. Giachetti, and V. Tognetti: Phys. Rev. B 33, 7647 (1986).
14. H.C. Fogedby, K.Osano, and H.J. Jensen: Phys. Rev. B 34, 3462 (1986).
15. G. Kamieniarz, and C. Vanderzande, (preprint, 1987)
16. J.K. Kjems, and M. Steiner: Phys. Rev. Lett. 41, 1137 (1978).
17. J.P. Boucher, L.P. Regnault, J. Rossat-Mignod, J.P. Renard, J. Bouillot, and W.G. Stirling: Solid State Commun. 33, 171 (1980).
18. J.P. Boucher, and J.P. Renard: Phys. Rev. Lett. 45, 486 (1980).
19. A.P. Ramirez, and W.P. Wolf: Phys. Rev. Lett. 49, 227 (1982).
20. M. Steiner, K. Kakurai, and J.K. Kjems: Z. Phys. B 53, 117 (1983).
21. F. Borsa: Phys. Rev. B 25, 3430 (1982).

22. F. Borsa, M.G. Pini, A. Rettori, and V. Tognetti: Phys. Rev. B 28, 5173 (1983).
23. K. Kopinga, A.M.C. Tinus, and W.J.M. de Jonge: Phys. Rev. B 29, 2868 (1984).
24. A.P. Ramirez, and W.P. Wolf: Phys. Rev. B 32, 1639 (1985).
25. M.G. Pini, and A. Rettori: Phys. Rev. B 29, 5246 (1984).
26. G.M. Wysin, and A.R. Bishop: Phys. Rev. B 34, 3377 (1986).
27. H. De Raedt, and A. Lagendijk: Phys. Rep. 127, 233 (1985).
28. R.P. Feynman, and A.R. Hibbs: in Quantum Mechanics and Path Integrals, (McGraw-Hill, New York 1965).

Index of Contributors